# Compact Plasma and Focused Ion Beams

# Compact Plasma and Focused Ion Beams

## Sudeep Bhattacharjee

CRC Press
Taylor & Francis Group
Boca Raton London New York

CRC Press is an imprint of the
Taylor & Francis Group, an **informa** business

CRC Press
Taylor & Francis Group
6000 Broken Sound Parkway NW, Suite 300
Boca Raton, FL 33487-2742

First issued in paperback 2017

© 2014 by Taylor & Francis Group, LLC
CRC Press is an imprint of Taylor & Francis Group, an Informa business

No claim to original U.S. Government works

ISBN-13: 978-1-4665-5788-8 (hbk)
ISBN-13: 978-1-138-03367-2 (pbk)

**Visit the Taylor & Francis Web site at**
**http://www.taylorandfrancis.com**

**and the CRC Press Web site at**
**http://www.crcpress.com**

*This book is dedicated to a few people whom I owe the most: My father, Bibhuti Bhusan Bhattacharjee, for his inspiration, work ethic, and sincerity; my mother, Sipra Bhattacharjee, for love and compassion; and my beloved wife, Jayeeta Bhattacharjee, for teaching me the way to tread in life.*

# Contents

# Foreword

I take great pleasure in writing the foreword to this book by Professor Sudeep Bhattacharjee. He visited our laboratory at the Institute of Physical and Chemical Research (RIKEN), Japan, through a Japanese Government, Ministry of Education, Science and Culture (*Monbusho*) Scholarship in 1996 to carry out doctoral work.

In his book, Professor Bhattacharjee has shown how compact and over-dense plasmas in the microwave regime can be generated and sustained using multicusp magnetic fields. Detailed investigations of the physics of wave interactions with such bounded plasmas have been carried out and elaborated in the book, including adequate theoretical treatments and modeling. Important physical properties of the plasma sustained in different types of wave guide geometries have been described while in operation in continuous and pulsed wave modes. The book reports several new findings in this class of plasmas where the wave propagation belongs to the $k \perp B$ mode, as different from classical research in large mirror machines where primarily $k \parallel B$ mode has been investigated. As an emerging and timely application, the latter half of the book is devoted to generation of multielement focused ion beams (FIB) from these compact plasmas, to widen the scope of research and applications of current day FIB systems.

The reader can be assured that Professor Bhattacharjee is highly competent to write this book, for he has contributed original research papers on the subjects presented in the book. I hope the reader will enjoy it as much as I have. The book will also be useful to graduate students or researchers wanting to learn experimental plasma physics in general and or entering the research area of wave-assisted plasmas.

**Hiroshi Amemiya**
*RIKEN Expert Corporation (Rec-Rd)*
*Wako, Saitama, Japan*

I am very pleased to see the publication of this valuable treatise on compact plasma production and its applications by Dr. Sudeep Bhattacharjee of IIT (Indian Institute of Technology), Kanpur. Sudeep, a young, dynamic, and highly innovative plasma physicist, whom I have had the pleasure of knowing since his graduate student days, has done outstanding work in this field over the past several years. The book provides a consolidated and unified description of this subject based on the author's own work and experience. The book will be useful not only to researchers in this area but will also serve as an excellent introduction to those contemplating to apply these ideas to various current and new applications. My congratulations to the author on this excellent contribution.

**Abhijit Sen**
*Institute for Plasma Research, India*

A new generation of FIB systems are being developed which will change the way we regard surface engineering. As well as using the beams for surface analysis, they can be used, either with active beam control or using a 5-axis stage, for micron level subtractive and additive manufacturing. This book addresses one of the popular methods used to create dense plasmas— microwave sources. It is addressed to those wishing to have a more profound understanding of the physics of microwave plasma generation in small sources and the details of the extraction of kilovolt ion beams from such a source. I can recommend this book to all who wish to enter this new area of technological development.

**Rod Boswell**
*Australian National University, Canberra*

It is my great pleasure to write a brief foreword for this book by Professor Sudeep Bhattacharjee who was with us at the Cyclotron Laboratory, RIKEN from 1999 to 2001. This book by Professor Bhattacharjee is a compilation of research works conducted by him and his collaborators over the past two decades on compact wave-assisted plasmas in the microwave regime and application of the basic research. The book is rich in the knowledge on physics of wave interaction with plasmas sustained in compact geometries in multicusp magnetic fields and, furthermore, suggests a novel application of ion beams obtained from these plasmas. I am confident that the readers will benefit much from this book. The author is congratulated on completing this outstanding piece of work.

**Takahide Nakagawa**
*Nishina Center for Accelerator Based Science, RIKEN, Japan*

Recent research has brought the application of microwaves from the classical fields of heating, communication, and generation of plasma discharges to the generation of compact plasmas that can be used for applications such as FIB and small plasma thrusters. Such microwave plasmas are created within dimensions well below wave-propagation waveguide cutoff and wavelength and present a new regime of plasma research challenges. The book provides a timely account of this emerging and exciting field of microwave plasmas, and it is written by an author who has worked in the frontier of the field for nearly two decades.

The text is clear and concise. No print is wasted on lengthy expansions of basic textbook introduction to plasma physics. Instead, the focus turns immediately to the most relevant topics necessary to understand this particular type of plasma production. From the fundamentals of the cutoff problem for wave propagation in waveguides and plasma diagnostics, the author goes on to explain in detail the plasma production by microwaves in a compact geometry and narrow tubes. Further on, a thorough discussion on wave interaction with bounded plasmas provides a deeper understanding of the physics. The last part of the book is devoted to an up-to-date account of their recent research on pulsed microwaves and the application of compact microwave plasmas for multi-element FIB. This important book is a must for graduate students and researchers in experimental plasma physics who want to get up to speed with the latest research in compact microwave plasmas!

**Ashild Fredriksen**
*University of Tromso, Norway*

# *Preface*

This book is a result of the research work carried out since 1996 by the author and his students and collaborators on the generation of compact wave-assisted plasmas in the microwave regime, the physics of wave interaction with bounded plasmas, and utilizing these narrow cross-sectional plasmas for novel applications in nanoscience and technology.

The way compact plasmas are defined in this book presents the case where the size of the plasma is much smaller than the free space and guided wavelength of the waves and in most cases smaller than the cutoff dimension of the plasma chamber. The cutoff dimension depends upon the geometrical cross-section of the waveguide and the wave frequency. Compact plasmas find wide applications in many areas such as high-current ion beams, thrusters for space propulsion, intense sources for downstream processing, inner surface processing of narrow dimensional tubes, plasma spectroscopy, laser guiding through intense collimated plasma channels, and many other applications that require a high-density, narrow cross-sectional plasma.

An emerging area where the basic research on compact plasmas has been applied is the generation of multi-element-focused ion beams for new research in nanoscience and technology, which is described in the latter half of the book (Chapters 19 through 23). This book describes in detail the experimental aspects of wave-assisted plasma generation, sustenance, and the physics of wave interaction with compact and bounded plasmas confined in multicusp magnetic fields.

The book starts with an introduction to plasmas, where some basic definitions and properties of plasmas are described, so that the book may serve as a resource to even nonspecialists in the field. Thereafter, the cutoff problem is described in Chapter 2. Conventionally, microwave plasmas are generated in large-diameter waveguide-type vessels, where waves can freely propagate to generate and sustain the plasma. It is well known that electromagnetic waves cannot propagate through a waveguide that has a dimension below cutoff. However, to generate compact plasmas, there is a need to reduce the size of the plasma chamber and at the same time prevent the reflection of the waves due to geometrical cutoff of the waveguide, which is required for the very sustenance of the plasmas—two contradictory requirements that need to be met and addressed, hence the challenge.

It is shown that microwave plasmas can be sustained in such a waveguide-type plasma chamber having a dimension below cutoff based upon the research carried out by the author and collaborators. The mechanisms for plasma production and sustenance are described in detail. Multicusp magnetic fields are employed to confine these plasmas. Several types of measurements of the plasma are carried out. The plasma and wave diagnostics techniques are described in Chapter 3. Measurements indicate that the generated plasmas have densities that are higher than the cutoff density for the launched electromagnetic waves. Therefore, the research is faced with another problem, as the density cutoff limitation for wave propagation must be overcome to support sustenance of high-density, narrow cross-sectional plasmas. With regard to the density cutoff limitation, conventionally it is known that microwaves cannot propagate through an overdense plasma (with plasma frequency ($\omega_p$) > the wave frequency ($\omega$)); however, the research carried out shows that this may be possible to be overcome to a large extent. Chapter 4 is devoted to the mechanism of plasma initiation inside the compact geometry for waves launched in the $k \perp B$ mode, where $k$ is the wave vector and $B$ is the static magnetic field. It may be noted that most of the earlier research has been carried out in the $k \parallel B$ mode, for example, in mirror machines. However, there is not much work that has been done in the $k \perp B$ mode, which is the subject of the research presented in this book. The initiation of the plasma at the waveguide entrance by the evanescent wave fields is discussed next. Here the statistical process of plasma initiation is described and provides a new outlook as compared to the conventional definition of microwave breakdown. The discharge initiation time and threshold fields are derived from the crossover of number of electrons undergoing elastic and inelastic collisions when diffusing out of the chamber under the action of the wave electric and magnetostatic fields.

The properties of the generated microwave plasma in narrow cross-sectional tubes of different shapes such as circular, rectangular, and square cross-sections, having some cross-sectional areas, are investigated and the results described in Chapters 5 through 7. Chapters 8 through 11 discuss the application of pulsed electromagnetic waves for plasma production. Pulsed power allows one to operate at higher pressures and provides greater flexibility in terms of varying the amplitude, pulse duration, repetition frequency, and the duty cycle that cannot be realized in plasmas produced by continuous mode microwaves. It has been seen that such pulses can be employed for production of intense charged particle beams.

Chapters 12 through 15 discuss wave interaction with compact and bounded plasmas. The wave penetration and absorption mechanisms in such plasmas are investigated in detail. Conventional understanding that waves cannot propagate through overdense plasmas does not hold in bounded plasmas, where the length scales of magnetic field nonuniformity and radial plasma inhomogeneity are much smaller than the free space ($\lambda_o$) and guided wavelengths ($\lambda_g$). Contrary to predictions of plane–wave dispersion theory and the Clemow–Mullaly–Allis (CMA) diagram, finite propagation occurs through the central plasma regions, where $\alpha_p^2 = \omega_p^2/\omega^2 \geq 1$ and $\beta_c^2 = \omega_{ce}^2/\omega^2 \ll 1 \ (\sim 10^{-4})$; $\omega_{ce}$ being the electron cyclotron frequency. Wave screening as predicted by the plane–wave model does not remain valid. The waves are found to satisfy a modified upper-hybrid resonance (UHR) relation and are damped at the local electron cyclotron resonance (ECR) location.

The next important aspect of the research on wave interaction is covered in Chapter 14, and is on birefringence of the waves penetrating through the overdense plasma. It is found that rotation of the polarization axis $\theta_c \sim 20°$ (with respect to vacuum) of the penetrating electromagnetic wave occurs through the bounded supercritical plasma. Birefringence of the radial and polar wave electric field components ($E_r$ and $E_\theta$) have been identified as the cause for the rotation, similar to a magneto-optic medium, however, with distinct differences owing to the presence of wave-induced resonances. The above results, including both the wave penetration and polarization, have been corroborated with numerical simulations obtained by solving Maxwell's equations by incorporating the plasma and magnetostatic field inhomogeneities within a conducting boundary, and show a reasonable agreement with the experimental results. This is elaborated in Chapters 13 and 14.

Chapters 16 through 18 are devoted to some new experiments using pulsed-mode microwaves. A quasisteady state has been found to exist in the interpulse plasma, which is a unique regime where the plasma is found to remain in a steady state even in the power off phase of the discharge. Furthermore, it is observed that the interaction of pulsed microwaves can lead to excitation of electron plasma waves that damp with time and lead to a rapid growth of the plasma beyond the waves by conveying the wave energy leading to rapid ionization of the gas. This aspect is quite interesting and is currently under further investigation. Changing the pulse duration leads to transition from the interpulse plasma to an afterglow observed for larger pulse widths ($t > 50\ \mu s$).

The remaining part of the book is devoted to focused ion beams (Chapters 19 through 23). Chapter 19 provides an introduction to the subject. Focused ion beams are a major requirement for research and applications in nanoscience and technology. Commercially available liquid metal ion source (LMIS)-based focused ion beam (FIB) systems provide primarily gallium ions with small ion currents (~10 pA) at the substrate, thereby requiring high operational times for volume milling. Moreover, contamination issues associated with the use of Ga is a major problem in such applications since the metallic Ga gets embedded into the substrates. Compact microwave plasmas have been utilized for the production of multi-element-focused ion beams (MEFIB) and are a new development in the field. Earlier plasma-based efforts involved use of gas field ionization sources, penning sources, and, more recently, rf plasmas. Using microwaves, one is able to preferentially excite the plasma electrons without significantly affecting the ions, because ions are sluggish and do not respond appreciably to the high-frequency wave fields. The MEFIB system can supply focused ion beams of a variety of elements (e.g., Ar, Kr, $H_2$, Ne, etc.) that will widen the scope of research and applications.

The complete design of the MEFIB system is discussed in Chapter 22. A double Einzel lens system extracts the ions from the plasma and focuses them on the substrate. A beam limiting aperture in between can limit the size of the beam. Two commercially available simulation codes, AXCEL-INP and SIMION, have been used to optimize the geometry and dimensions of the extraction electrodes. The focused ion beam current is measured with the help of a Faraday cup, and the beam profile is measured by wire scanner and knife edge methods. The spot sizes are further confirmed by micrography of craters formed on a metal surface due to the impinging ions. The developed MEFIB system is currently capable of supplying ~10 μA hydrogen ion current focused to <10 μm spot size. The image size current density is comparable to LMIS-based FIB, which focuses a lot less current (~10 pA) into a smaller spot (~10 nm). Several new methods are being employed to further reduce the beam spot at the substrates, such as by employing: (i) capillary guiding of ion beams; (ii) smaller source side apertures (50–60 μm) of plasma electrodes, currently 500 μm apertures are used; (iii) multipole magnetic fields before the substrates; and (iv) higher focusing fields (~24–30 kV), currently 18 kV is used. Implementation of these methods include some of the ongoing plans.

Chapter 24 provides a summary and presents future prospects. One of the key issues at the submicron level is measurement of the spatial beam profiles, for which new diagnostic techniques for measuring instantaneous beam (intensity) profile need to be considered.

# Acknowledgments

I am indebted to my present and past doctoral students who have shared with me the trials and tribulations of carrying out this largely experimental research work with a significant research and development component, including theory and computer simulations, as described in this book. Among them I would like to particularly thank Dr. Indranuj Dey, Dr. Jose V. Mathew, Mr. Debaprasad Sahu, Mr. Samit Paul, Mr. Abhishek Chowdhury, and Ms. Shail Pandey. I am grateful to the Indian Institute of Technology (IIT) Kanpur for financial support under the Centre for Development of Technical Education (CDTE) program, which helped me to write this book. My very special thanks go to Mr. Roshan for typing out the entire manuscript as per the specified format and preparing the figures to the specified standard. I would like to heartily thank my students Shail, Abhishek, and Samit for proof reading part of the chapters in the book and Debaprasad for helping in indexing the book. The support of my research by the Department of Science and Technology (DST); the Council of Scientific and Industrial Research (CSIR), India; the Asian Office of the Aerospace Research and Development (AOARD), USA; and IIT Kanpur during 2005–2013 is gratefully acknowledged. I am particularly thankful to my doctoral supervisor Dr. Hiroshi Amemiya at the Institute of Physical and Chemical Research (RIKEN), Japan, where the research was initiated in 1996 and continued through 1999. I would also like to thank Dr. Takahide Nakagawa and Dr. Y. Yano, also at RIKEN, at the Nishina Center for Accelerator-based Science. The encouragement and support of my colleagues at IIT Kanpur, Professors R. K. Thareja, Y. N. Mohapatra, R. C. Budhani, Avinash Singh, M. K. Harbola, H. C. Verma, and V. N. Kulkarni (deceased), helped me a great deal in a variety of ways while conducting my research and helped in the genesis of this book.

I would like to thank Professors Abhijit Sen, G. K. Mehta, Y. C. Saxena, D. Kanjilal, John Booske, John Scharer, Rod Boswell, Christine Charles, and Ashild Fredriksen. They played an important role in providing the motivation to continue our research.

I take pleasure in acknowledging many research discussions with Dr. John Wohlbier, Dr. Aarti Singh, Mr. Chad Marchewka, and Mr. John Welter during my stay at the University of Wisconsin–Madison (2001–2004), and with Dr. Trevor Lafleur at Australian National University, Canberra (2010).

My special thanks go to Ms. Aastha Sharma, commissioning editor, CRC Press, Taylor & Francis Group, for her pursuance and perseverance in having me take on this interesting assignment. I would also like to thank Mr. David Fausel, my project coordinator, of Taylor & Francis, for the several lessons in preparing the manuscript, for his patience while preparing the manuscript, and once, when submitted, for processing it with great alacrity.

My sincere thanks to Mr. Robert Sims, project editor for my book assigned by Taylor & Francis Group, for kindly bearing with me during the proof stage and for the several instructions on this matter and to Mr. Karthick Parthasarathy, project manager at Techset Composition, for editing and layout and for keeping me posted with the proceedings.

Finally, I owe a great deal to my parents, Mr. Bibhuti Bhusan Bhattacharjee and Mrs. Sipra Bhattacharjee, for their constant encouragement and inspiration; to my wife, Jayeeta, for her constant love and support; and to my children, Sayak and Syon, for standing by me, encouraging me, and giving up their share of my time for the sake of research.

**Sudeep Bhattacharjee**

# *Author*

**Dr. Sudeep Bhattacharjee** is currently an associate professor in the Department of Physics at the Indian Institute of Technology Kanpur. He obtained his PhD on a Japanese Government Ministry of Education, Science and Culture (*Monbusho*) scholarship from the Institute of Physical and Chemical Research (RIKEN), Japan in affiliation with Saitama University. Subsequently, he worked as a distinguished postdoctoral researcher (*Kisotokken*) at RIKEN, followed by another postdoctoral stint at the University of Wisconsin, Madison, WI. Dr. Bhattacharjee is an experimental physicist whose research interests include plasma physics, focused ion beams (FIB), and physics of nanoscale systems. His current research includes development of a compact microwave plasma-based multi-element FIB system, physics of nanoscale systems using FIB, and electromagnetic wave interaction with compact and bounded plasmas. Dr. Bhattacharjee has won international and national awards and fellowships that include the *Monbusho* scholarship (Japan) and a Norwegian Government Specialist Exchange scholarship for research (2006). He recently received the prestigious Buti Foundation National Research Award (2010) for research excellence in plasma science and technology in India and the Australian Government Endeavour Research Award (2010) for undertaking research work in Australia.

Dr. Bhattacharjee teaches undergraduate- and graduate-level physics courses at IIT Kanpur. He has received the Directors' Commendation for Excellence in Teaching several times based upon student evaluation.

# 1

## Introduction

## Plasma

### What Is Plasma?

A plasma is a system of interacting free electrons and ionized atoms or molecules, which exhibit collective behavior due to long-range Coulomb forces whose dependence upon the distance $r$ goes as $1/r^2$. However, not all media containing charged particles can be classified as plasmas. There are certain criteria that must be fulfilled to be called a plasma. Before we delve ourselves more into plasmas, let us look into the brief history of plasmas. The word *plasma* originates from the Greek language and means "something molded." It was applied for the first time by Tonks and Langmuir in 1929 to describe the inner region of a glowing gas produced by electric discharge in a tube, while the ionized gas as a whole remains electrically neutral.

### Plasma as a Fourth State of Matter

It is known that when a solid is heated it becomes a liquid, and then a liquid a gas, and then when the gas breaks down it becomes a plasma. The basic difference lies in the strength of the bonds that hold the constituent particles together. Binding forces are strongest in a solid, weak in a liquid, and almost absent in the gaseous state, although the magnitudes are relative. Whether a given substance is found in one of the states depends upon the random kinetic energy of the atoms or molecules, that is, on its temperature. If sufficient energy is provided, a molecular gas will gradually dissociate into an atomic gas as a result of collisions between those particles whose thermal kinetic energy greatly exceeds the molecular-binding energy. At sufficiently elevated temperatures, an increasing fraction of the atoms will possess enough kinetic energy to be overcome by collisions, the binding energy of the outer-most orbital electrons, and an ionized gas or a plasma results.

## Physics of Plasma Production

A plasma is most commonly produced by raising the temperature of a substance until a reasonably high-fractional ionization is obtained. Under thermodynamic equilibrium conditions, the degree of ionization and the electron temperature are closely related by Saha's equation:

$$\frac{n_i}{n_n} = 2.405 \times 10^{21} T^{3/2} \frac{1}{n_i} \exp\left(-\frac{U}{KT}\right), \tag{1.1}$$

where $U$ is the ionization energy, $K$ is the Boltzmann constant, $n_i$ is the number density of ions, and $n_n$ is the number density of neutrals. Since 1 eV = $KT$ for $T = 11{,}600$ K, we can write Saha's equation as

$$\frac{n_i}{n_n} = 3.00 \times 10^{27} T^{3/2} \frac{1}{n_i} \exp\left(-\frac{U}{T}\right), \tag{1.2}$$

where $T$ is in eV and $n_i$ and $n_n$ are in m$^{-3}$ and the total number density $n_t = n_i + n_n$. Thus, a considerable degree of ionization can be achieved for temperatures that are well below the ionization energy.

## Ways of Creating Plasmas in the Laboratory

There are many different ways of creating plasma in the laboratory, and depending upon the method, plasmas may have a variety of properties such as (a) high or low density; (b) high or low temperature; (c) steady or transient plasma (cw or pulsed plasma); or (d) stable or unstable plasma. Some of the most commonly used methods are (a) photoionization using lasers; (b) electric discharge, which can be dc plasma; or (c) wave-generated plasmas such as those generated by radio frequency or microwaves. We briefly discuss each method below:

a. *Photoionization*: Ionization occurs by absorption of incident photons whose energy is greater than or equal to the ionization potential of the absorbing atom. The excess energy of the photon is transformed into kinetic energy of the electron–ion pair, for example, the ionization energy of atomic oxygen is 13.6 eV, which can be supplied by radiation of $\lambda < 91$ nm in the far UV region.

b. *Electric field ionization*: An electric field is applied across the ionized gas, which accelerates the free electrons to energies sufficiently high to ionize other atoms by collisions. The applied electric field transfers energy much more efficiently to the light electrons than to the relatively heavy ions. The electron temperature is therefore higher than the ion temperature $(T_e \gg T_i)$, since the transfer of thermal energy from the electrons to the heavier particles is very slow.

If the ionizing energy is suddenly cut off, the ionization decreases gradually due to recombination, also known as electron ion recombination and diffusion, until an equilibrium value is reached consistent with the temperature of the medium. The diffusion and recombination are the major losses occurring in the plasma.

## Particle Interactions

The particle dynamics are governed by the internal fields due to the nature and motion of particles and by externally applied fields. The types of fields that are present in the plasma include:

a. Charged particles surrounded by an electric field interact with other charged particles according to the Coulomb force law, with its dependence

$$F_E \propto \frac{q_1 q_2}{r^2}. \tag{1.3}$$

b. A magnetic field is associated with a moving charged particle which produces force on other moving charges.

$$F_B \propto q(V \times B), \tag{1.4}$$

c. Electric polarization fields

$$P = \alpha E, \tag{1.5}$$

where $\alpha$ is the polarizability, lead to distortion of a neutral particle's electronic cloud during a close passage of a charged particle.

## Some Basic Plasma Properties

a. *Conductivity*: Due to the highly mobile electrons, plasmas are generally very good conductors of electricity as well as thermal conductors.

b. *Diffusivity*: The presence of density gradients, such as near boundaries, causes particles to diffuse from dense regions to regions of lower density. There is an interesting difference between what happens here versus in ordinary fields. Electrons tend to diffuse faster than ions, generating a polarization electric field as a result of charge separation. The field enhances the diffusion of ions and slows down that of the electrons in such a way that electrons and ions diffuse at approximately the same rate. This phenomenon is called ambipolar diffusion. When there is an external magnetic field, the diffusion of charged particles across the field lines is reduced. Strong magnetic fields are helpful in plasma confinement. Here, the difference between classical and Bohm's diffusion can be studied.

c. *Wave phenomena:* Plasmas can sustain a variety of waves: (i) Longitudinal electrostatic plasma waves; (ii) high-frequency transverse electromagnetic waves; and (iii) Alfven and magnetosonic waves in the low-frequency region. Some important features include the following:

   i. Each of the various possible modes of wave propagation can be characterized by a dispersion relation that relates the frequency $\omega$ of the wave to the wave number $k$.

   ii. Dissipative's processes such as collisions produce damping of the wave amplitude. This means that energy is transferred from the wave field to the plasma particles. There is a noncollisional mechanism of wave attenuation in a plasma known as Landau damping. This is the trapping of some plasma particles (the ones that are moving with velocities close to the wave phase velocity) in the energy potential well of the wave, so there is transfer of energy from the wave to the particles.

   iii. Cherenkov's radiation is emitted whenever charged particles pass through the plasma with a velocity $v$ exceeding the velocity of light in the medium $v > v_t = c/n$, where $n$ is the refractive index of the medium, $c$ is the velocity of light in the medium, and $v_t$ is the threshold velocity to excite Cherenkov's radiation.

   iv. It is also possible to have modes with growing amplitudes. As a result of instabilities, there is a transfer of energy from the plasma particles to the wave field.

d. *Emission of radiation:* Radiation from plasma can be used to infer plasma properties. There are two categories of radiation: (i) radiation from emitting atoms and molecules, and (ii) radiation from accelerated charges. The two are briefly discussed below.

   i. Radiation from emitting atoms and molecules: The recombination of ions and electrons form neutral particles. Radiation is emitted as those excited particles formed during recombination decay to the ground state. This is observed from the line spectra of plasmas.

ii. Radiation from accelerated charges: Decelerated-charged particles making some kind of collisional interaction lead to Bremsstrahlung's radiation. Cyclotron radiation occurs in magnetized plasmas from acceleration of charged particles circulating around $B$ field lines. Blackbody's radiation emitted from plasma in thermodynamic equilibrium are important in astrophysical plasmas in view of the large size needed for a plasma to radiate as a black body.

---

## Criterion for the Definition of a Plasma

a. *Macroscopic neutrality:* In the absence of external disturbances, a plasma is macroscopically neutral. The net resulting electric charge is zero. In the interior of the plasma, the microscopic space charge fields cancel each other out and no net space charge exists over a macroscopic region. Departures from electrical neutrality can occur only over distances in which a balance is obtained between the thermal particle energy and the electrical potential energy. This distance is known as the Debye length. The thermal particle energy tends to disturb charge particle neutrality and the electrical potential energy tends to restore electrical neutrality.

$$n_e = \sum_i n_i. \tag{1.6}$$

b. *Debye's shielding:* The Debye length is an important physical parameter for the description of a plasma. It is a measure of the distance over which the influence of the electric field is felt by the other charged particles inside the plasma. Beyond the Debye length, this influence is not felt. Charged particles arrange themselves to shield all electrostatic fields within a distance of the order of Debye's length. The distance was first measured by Debye for an electrolyte:

$$\lambda_D = \sqrt{\frac{\varepsilon_o \kappa T}{n_e e^2}} = 69.0 \sqrt{\frac{T}{n_e}} (m), \tag{1.7}$$

where $n_e$ is in $m^{-3}$, $T$ is in degrees Kelvin.

When a boundary is introduced into the plasma, the perturbations produced extend only up to a distance of the order of $\lambda_D$ from the surface. Electrons are attracted to the vicinity of an ion and shield its electrostatic field from the rest of the plasma. Similarly, ions are attracted to the vicinity of an electron and shield its electrostatic field from the rest of the plasma.

Therefore, charged particles arrange themselves to shield all electrostatic fields within a distance of the approximate Debye's length. The electrostatic potential of an isolated particle of charge $q$ is $\varphi = q/r$. Because of the shielding effect, the potential in the vicinity of the charged particle in a plasma is altered and therefore the potential of a charge at rest in a plasma is given by

$$\varphi = \frac{q}{r} e^{-r/\lambda_D}. \tag{1.8}$$

In other words, in the neighborhood of any surface inside the plasma, there is a layer of width of the order of $\lambda_D$, known as the plasma sheath, inside which the condition of macroscopic equilibrium may not be specified. The sheath thickness is usually a few times Debye's length. Beyond the plasma sheath region, there is the plasma region where macroscopic neutrality is maintained.

We can define a Debye's sphere, that is, a sphere of radius $\lambda_D$, by the number of electrons inside a sphere, given by

$$N_D = \frac{4}{3}\pi\lambda_D^3 n_e = \frac{4}{3}\pi \left( \frac{\varepsilon_0 \kappa T}{n_e^{1/3} e^2} \right)^{3/2}. \tag{1.9}$$

The Debye shielding effect is characteristic of all plasmas and ensures macroscopic neutrality. A necessary and obvious requirement for the existence of a plasma is that the physical dimensions of the system be large compared to $\lambda_D$ that is, $L \gg \lambda_D$, where $L$ is the characteristic dimension of the plasma.

Since the shielding effect is the result of the collective particle behavior inside the Debye sphere, it is necessary that the number of electrons inside a Debye sphere be very large for a good shielding effect, $n_e \lambda_D^3 \gg 1$. This means the average distance between electrons, $\sim n_e^{-1/3}$, must be very small compared to $\lambda_D$. The quantity $g = 1/n_e \lambda_D^3$ is known as the plasma parameter and the condition $g \ll 1$ is called the plasma approximation. $g$ is also a measure of the ratio of the mean interparticle potential energy to the mean plasma kinetic energy.

c. *Plasma frequency:* When a plasma is instantaneously disturbed from the equilibrium condition, the resulting internal space charge fields give rise to collective particle motions that tend to restore the original charge neutrality. These collective motions are characterized by a natural frequency of oscillation known as the plasma frequency.

The angular frequency of these collective electron oscillations is called the (electron) plasma frequency, given by

$$\omega_{pe} = \left( \frac{n_e e^2}{m_e \varepsilon_o} \right)^{1/2}.$$

(1.10)

Collisions between electrons and neutrals tend to dampen these collective oscillations and gradually diminish their amplitude. $v_{pe} > v_{en}$ is necessary if the oscillations are to be only slightly damped where $v_{en}$ is the electron neutral collision frequency. The above equation constitutes another (the fourth) criterion and can be alternatively written as $\omega_{pe} \tau_{en} > 1$: the fourth criterion. $\tau_{en} = 1/v_{en}$ is the average time an electron travels between collisions with neutrals.

## Different Types of Plasmas

a. In nature: During the last century, it was realized that most of the matter in the known universe exists as a plasma.

i. *The sun and its atmosphere:* The energy output is derived from thermonuclear fusion. The interior temperature exceeds $1.2 \times 10^7$ K. The solar atmosphere is divided into (i) photosphere, ~6000 K and a few hundred km thick; (ii) chromosphere, ~10,000 km thick and 105 K temperature; and (iii) corona—tenous hot plasma extending millions of km into space (106 K). There is a magnetic field of ~$10^{-4}$ T on the surface but at hot spots ~0.1 T. This is a magnetized plasma.

ii. *Solar wind:* A highly conducting blast of plasma is continuously emitted by the sun at very high speeds into interplanetary space, as a result of the supersonic expansion of the hot solar corona. This plasma mainly consists of electrons and protons, electron density: $5 \times 10^6$ m$^{-3}$; electron and ion temperature: $T_e \sim 5 \times 10^4$ K and $T_i \sim 10^4$ K, $B \sim 5^{-9}$ T and drift velocity $v_e \sim 3 \times 10^5$ m/s.

iii. *The magnetosphere and the Van Allen radiation belts:* As the highly conducting solar wind hits the earth's magnetic field, it compresses the field on the sunward side and flows around it at supersonic speeds. The inner region, from which the solar wind is excluded and contains the compressed earth's magnetic field, is called the magnetosphere. Inside the magnetosphere, we can find the Van Allen radiation belts. These are energetic charged particles (mainly electrons and protons) that are trapped into regions where they execute complicated trajectories that spiral around geomagnetic field lines and drift slowly around the earth. There is also a plasma sheet that extends for several million kms.

iv. *Ionosphere:* The large natural blanket of plasma in the atmosphere, which envelopes the earth from an altitude of ~60 km to several thousands of km, is called the ionosphere. The plasma is created by absorption of extreme UV and x-ray radiation by the atmospheric species. The earth's magnetic field exerts a great influence on the dynamic behavior of the ionospheric plasma. An interesting phenomenon that occurs in the polar regions is the aurora. This is electromagnetic radiation induced by energetic particles of solar and cosmic origin as they are accelerated and penetrate into the atmosphere along geomagnetic field lines.

v. *Plasmas beyond the solar system:* Beyond the solar system, we find a great variety of natural plasmas in (i) stars; (ii) interstellar space; (iii) galaxies; (iv) intergalactic space and far beyond; (v) interstellar shock waves from remote supernova explosions; (vi) rapid variation of x-ray fluxes from neutron stars; and (vii) pulsars, which are rapidly rotating neutron stars with plasma-emitting synchrotron radiation from the surface.

Plasma behavior in the universe involves the interaction between plasmas and magnetic fields. The Crab Nebula, for example, is a rich source of plasma phenomena because it contains a magnetic field.

b. Laboratory plasmas

i. *Low-pressure cold cathode discharge:* Electrodes sealed in an evacuated vessel can be used to initiate and maintain a low-pressure, steady-state (or pulsed) arc discharge in the vessel by applying a voltage. ($n \sim 10^{10}$ cm$^{-3}$, low density, $T_e < 500$ K, low temperature). These devices were used in the early plasma studies. The glow discharge devices such as small neon tubes, panel lights, and voltage regulator tubes fall under this category.

ii. *Thermionic arc discharge:* An arc discharge is characterized by a lower voltage than a glow discharge, and relies on thermionic emission of electrons from the electrodes supporting the arc. An arc between two electrodes can be initiated by ionization and glow discharge, as the voltage across the electrodes is increased. The breakdown voltage of the electrode gap is a function of the pressure and type of gas surrounding the electrodes. An arc in gases near atmospheric pressure is characterized by visible light emission, high-current density, and high temperature. An arc is distinguished from a glow discharge partly by the approximately equal effective temperatures of both electrons and positive ions; in a glow discharge, ions have much less thermal energy than the electrons. Electrical resistance along the continuous electric arc creates heat, which ionizes more gas molecules (where degree of ionization is determined by temperature), and, as per the sequence solid–liquid–gas–plasma, the gas is gradually turned into a thermal plasma. A thermal *plasma* is in thermal equilibrium, which is to say that the temperature is relatively homogeneous throughout the heavy particles (i.e., atoms, molecules, and ions) and electrons. This is so because when thermal plasmas are generated, electrical energy is given to electrons, which, due to their great mobility and large numbers, are able to disperse it rapidly and by elastic collision (without energy loss) to the heavy particles.

iii. *Alkali metal vapor plasmas (Q machines):* Class of plasma experiments where the plasma has a very low-kinetic temperature (few eV). Originally Q meant quiescent, but later on it was shown that the plasma was rich in several plasma phenomena, including several interesting instabilities. A number of basic plasma studies were carried out using Q machines. Contact ionization occurs when atoms of alkali metals (e.g., Cs) come in contact with a heated tungsten plate. The Cs atom impinges on the plate and comes off from the plate as an ion. If the tungsten plate is raised to a sufficiently high temperature (~2000°C) to be an emitter of electrons, an electron accompanies each Cs ion resulting in a plasma. Plasma density is typically ~$10^6$–$10^9 cm^{-3}$, temperature ~1000 K, and the plasma is 99% ionized. Many basic plasma theories such as drift waves, diffusion, resistivity, beam–plasma instabilities, and sheath effects have been studied with them.

iv. *Microwave and rf-produced plasmas:* A gas at low pressures will break down under the action of an applied steady electric field. It will also break down and form a plasma if the applied electric field is alternating, such as with a high frequency. The *rf* term is often loosely used to describe waves starting from radio frequency to microwaves where the commonly employed frequencies are 13.56 MHz and 2.45 GHz, respectively. There are several advantages in this process of gaseous breakdown. It provides a possibility of having an electrodeless discharge and helps to result in clean plasma, free from contaminants into the plasma coming from the electrodes. When using an *rf* coil, the alternating electric field can easily penetrate the dielectric, such as glass, and the like, in which the plasma is confined and accelerates the electrons in the vessel to energies above the ionization potential of the gas. Plasma densities above $10^{10} cm^{-3}$ are typical. Sometimes *rf* plasma are produced in a magnetic field. These are known as cyclotron resonance plasmas. When the exciting *rf* field has the same frequency as the cyclotron frequency for the electrons (or even ions), we obtain maximum power transfer from the waves to the plasma. Even with modest *rf* power from microwave sources, high-energy electrons and densities can be produced.

v. *Laser-produced plasmas:* The focused output of a high-power Q-switched laser can be used to irradiate solids and compressed gases to produce dense, isolated, high-temperature plasmas. A focused spot from a 5 MW peak power laser can be used to irradiate a metal target in vacuum to form small dense plasma in the irradiated region and metallic ions with energies of ~1 keV are ejected from that region. There are experiments where high-temperature plasma (~100 eV) can be produced by irradiating 10- to 20-millimeter diameter lithium hydride pellets, suspended in an evacuated chamber by electromagnetic fields with the focused output of a 20 MW Q-switched ruby laser.

## Description of Plasmas

The motion of the plasma particles is governed by the interaction between the plasma particles and the internal fields produced by the particles themselves, as well as the externally applied fields. As the particles move, they will likely generate local concentration of positive or negative charges that can give rise to internal electric fields. The motion can also generate electric currents and therefore magnetic fields. The particle dynamics are adequately described by the laws of classical mechanics.

Quantum mechanics usually do not play much of a role because the momentum of the plasma particles generally is high, and the density is low enough to keep their de Broglie wavelengths much smaller than the interparticle distance. Quantum effects become important only at very high densities and very low temperatures.

The interaction of charged particles with electromagnetic fields is governed by the Lorentz force,

$$\frac{d\vec{P}}{dt} = q\left(\vec{E} + \vec{v} \times \vec{B}\right). \tag{1.11}$$

Thus, if for a typical particle of charge $q$, mass $m$, and moving with a velocity $v$, in the presence of electric ($E$) and magnetic induction ($B$) fields,

$$\vec{p} = m\vec{v}. \tag{1.12}$$

In principle, we can at least describe the dynamics of a plasma by solving the equation of motion for each particle in the plasma under the combined influence of externally applied fields and the internal fields generated by all the other plasma particles. If the total number of particles is $N$, we have $N$ non-linear coupled differential equations of motion to solve simultaneously. A self-consistent formulation must be used since the particle trajectories (fields) are intrinsically coupled. The electromagnetic fields obey Maxwell's equations:

$$\vec{\nabla} \times \vec{E} = -\frac{\partial \vec{B}}{\partial t}, \tag{1.13}$$

$$\vec{\nabla} \times \vec{B} = \mu_o \left( \vec{J} + \varepsilon_o \frac{\partial \vec{E}}{\partial t} \right) \tag{1.14}$$

$$\vec{\nabla} \cdot \vec{E} = \frac{\rho}{\varepsilon_o}, \tag{1.15}$$

and

$$\vec{\nabla} \cdot \vec{B} = 0, \tag{1.16}$$

where $\rho$ = total charge density, $J$ = total electric current density, $\varepsilon_0$ = electric permittivity, and $\mu_0$ = magnetic permeability.

The plasma charge and current densities can be expressed as

$$\rho_p = \frac{1}{\delta V} \sum_i q_i, \tag{1.17}$$

and

$$\vec{J}_p = \frac{1}{\delta V} \sum_i q_i \vec{v}_i, \tag{1.18}$$

where the summation is over all the charged particles contained inside a suitably chosen small volume element $\delta V$. Since we are dealing with discrete distribution of charges/current densities, $\rho_p$ and $j_p$ should be accurately expressed in terms of Dirac Delta functions (the fields become singular at the particle positions). If the $\delta V$ chosen is big enough to contain a fairly large number of particles, then the above equations for $\rho_p$ and $j_p$ should give smooth functions of $\rho_p$ and $j_p$ suitable for analytical calculations. The self-consistent approach is conceivable in principle but cannot be carried out in practice without introducing some averaging scheme since there are large numbers of variables involved.

To explain and predict the macroscopic phenomena observed in nature and in the laboratory, it is not of interest to know the detailed individual motion of each particle, since the observable macroscopic properties of the plasma are due to average collective behavior of a large number of particles. With the availability of a large number of fast computers, it is possible to numerically follow the nonlinear motion of many particles. This method is known as plasma simulation via particles.

## Theoretical Approaches

a. *Particle orbit theory:* Here, we study the dynamics of the charged particles in the given fields. It is useful in predicting the behavior of low-density plasma in the presence of external fields, for example, Van Allen's radiation belts, solar corona, cosmic rays, accelerators, and cathode ray tubes (CRTs).

b. *Kinetic theory:* This uses a statistical approach where the distribution function of the particles is taken into account. One solves the appropriate kinetic equations that govern the evolution of the distribution function in phase space (e.g., Vlasov's equation). The internal electromagnetic fields consistent with the charge density and the current density distribution inside the plasma are smeared out, and close collisions are neglected.

c. *Fluid theory:* Here one talks about a single-fluid, two-fluid, or many-fluid theory. The collisions between plasma particles are very frequent. Each species maintains a local equilibrium distribution function. Each species can be treated as a fluid described by local density, velocity, and temperature.

## Microwave-Generated Plasmas

Microwave plasmas are widely studied both for basic plasma research [1–22] and for various applications [23–32]. Depending upon the type of application, including whether the waves are in the continuous or pulsed mode, microwave discharges can be efficiently created and maintained over a wide pressure range.

At low pressures, the electron cyclotron resonance (ECR) discharge is known to be useful for producing a high-density plasma and has been widely applied using microwaves in the continuous mode [1–31]. As the name suggests, microwave energy is coupled with the natural resonant frequency of the electrons $\omega_c\,(= eB/m)$ in the presence of a static magnetic field $B$, where $e$ and

$m$ are the charge and mass of an electron. Resonance occurs when the wave frequency $\omega$ equals the electron cyclotron frequency $\omega_c$, and there is efficient transfer of energy from the wave to the plasma electrons. In an actual discharge, this condition can be satisfied within the discharge by an adjustment of the static magnetic field to the resonance condition, and a component of the electric field should be perpendicular to the magnetic field. The magnetic field may be provided by employing permanent magnets or by current carrying electromagnetic coils. The electrons are accelerated in the ECR regions and in turn ionize and excite the neutral gas. At low pressure, this results in a nearly collisionless plasma that can be varied from a weakly to a highly ionized state by changing the discharge pressure and input microwave power.

Recently, another class of microwave plasma has become very popular, belonging to pulsed modulated discharges [33–37]. In pulsed discharges, microwave power is applied to the plasma within a certain pulse duration and the pulses are repeated with a certain pulse repetition frequency. When the waves are pulsed, additional control over the plasma can be achieved by varying control parameters such as the pulse width, pulse repetition frequency, and the duty cycle. These control parameters are absent in plasmas produced with continuous mode microwaves, where the external control parameters are usually limited to the pressure and the microwave power. Pulsed discharges are favorable for the creation of particles possessing internal energy such as metastables, radicals, and multicharged ions. They are also known to be favorable for the production of negative ions. They can be efficiently utilized for the generation of industrially useful pulsed UV sources and pulsed ion beams on a substrate. Besides, the plasma in the power off phase is electric field-free, with lower noise levels whereby gas-heating effects can be avoided.

## Compact Microwave Plasma Sources

Microwave-generated ECR plasmas, whether in continuous or pulsed mode, are generally produced inside a waveguide-type vessel. The magnetic field helps in both generation and confinement of the plasma. The field for the ECR action is provided either by current-carrying coils [1–6] or by permanent magnets [11–21], or even by a combination of both as in ion sources [29–32]. The general interest of earlier workers has been to obtain large-diameter plasmas primarily to address the processing needs of the industry. Therefore, the cross-section of the plasmas have usually been large (~10 s of cm), greater than the wavelength of the microwaves in free space. However, many applications demand high-density plasma with a narrow cross-section in emerging research areas such as focused ion beams, plasma thrusters, laser propagation through intense plasmas, and so on. This has led to research for new concepts in the development of compact microwave plasma sources, where the diameter of the waveguide could be much smaller than the free-space wavelength of the waves.

Plasma production in a narrow waveguide suffers from problems based on wave propagation through the waveguide, which is important for plasma production and maintenance. The first problem is regarding the waveguide geometrical cutoff [38]. Basically, the waveguide's cross-section should not be reduced beyond a lower limit, which depends upon the cross-section of the waveguide and the wavelength of the microwaves. The second problem is associated with the plasma density cutoff [39]. The density cutoff limitation depends upon the plasma frequency $\omega_p$, which in its turn is related to the plasma density. It implies that the waves would be reflected from the plasma in the waveguide if the wave frequency $\omega < \omega_p$.

Therefore, the production and maintenance of a high-density (with densities > cutoff density) microwave plasma in a narrow waveguide with a dimension smaller than the geometrical cutoff value is difficult and undoubtedly a challenging problem both from the physics and applications point of view.

## Research Objective and Methods

The objective of obtaining a high-density, narrow cross-sectional plasma can be achieved by reducing the cross-sectional size of the waveguide. This is because the microwave power density in the waveguide increases and thereby favors the formation of a high-density plasma. However, the limitation of the waveguide geometrical cutoff explains why this concept did not interest earlier researchers. Although theory predicts a lower limit in the cross-sectional size of the waveguide, we investigated a new method of producing a plasma inside the waveguide despite the transverse dimension being smaller than the cutoff value.

The procedure adopted was to use a peripheral arrangement of permanent magnets around the waveguide. The magnets were arranged in the form of a multicusp (neighboring poles having opposite polarity) so that a minimum $B$ field is produced in the center of the waveguide. The minimum $B$ field is useful for plasma confinement as well as for having a significant volume of the plasma almost unmagnetized. Moreover, in this way, a compact device can be constructed. The waves were launched axially into the waveguide such that the launched mode belonged to the $K \perp B$ mode, where $K$ is the wave vector.

If such a realization of a high-density narrow cross-sectional plasma is made, it will have many possible applications, such as compact, high-current, focused ion beams, multicharged ion sources where it is useful to have a collimated beam, or for inner surface processing of narrow cross-sectional pipes. One could also consider it for environmental applications such as discharge cleaning and decomposition of harmful gases released from tubular chimneys and narrow pipes of industries and waste incinerators. For all these applications, high-density plasma with a high-electron temperature would be favorable. Additionally, one can consider pulsing the discharge, which is another way to increase the discharge efficiency by increasing the reaction rate or by having greater controllability of the plasma [33–37]. Besides the application aspect, the research will have important physical implications because overcoming of the waveguide geometrical and plasma density cutoff in such bounded plasmas is a subject of interest. Here the length scales of magnetic field inhomogeneity and plasma density nonuniformity are much smaller than the free-space wavelength of the waves. The wave propagation and absorption would be a subject worth investigating both for the production and maintenance of overdense plasma in the narrow waveguide, including self-interactions of the wave with the generated plasma.

Based on the above motivation, this book presents research results on the subject. Experiments were first conducted using microwaves in the continuous mode (2.45 GHz). Waveguides of three different cross-sections, namely near-circular, rectangular, and square, were constructed and used in the experiments. The results have been compared and discussed. The plasma production and sustenance mechanisms have been found.

As a next approach, pulsed microwaves (2.45 and 3 GHz) were also considered. Initially, for an understanding of the general properties of the pulsed plasma, the experiment was performed in a waveguide with a dimension larger than the cutoff value. The discharge in three gases of different nature was studied. Thereafter, a circular waveguide with a dimension smaller than the cutoff value was used to study the plasma production possibilities and the plasma properties in the case of pulsed microwaves. The plasma production mechanisms in the below cutoff dimension waveguide were also investigated.

After the initial research, more advanced experiments were carried out in CW and pulsed-mode microwaves, which shed light on some fundamental aspects of wave plasma interaction. These advanced studies have been described in Chapter 16 for CW mode microwaves and in Chapter 19 for pulsed-mode microwaves. The last several chapters are devoted to the application of compact plasmas for generation of multi-element-focused ion beams, which is a novel development in the field.

## Note

Parts of this chapter has been adapted from Bittencourt, JA, *Fundamentals of Plasma Physics*, 3rd Edition, Springer Verlag, NY, 2004.

# References

1. O.A. Popov, S.Y. Shapoval, and M.D. Hoder Jr. 1992. 2.45 GHz microwave plasmas at magnetic fields below ECR, *Plasma Sources Sci. Technol.* **1**: 7–12.
2. S.R. Douglass, C. Eddy Jr., and B.V. Weber. 1996. Faraday rotation of microwave fields in an electron cyclotron resonance plasma, *IEEE Trans. Plasma Sci.* **24**: 16–17.
3. S. Samukawa. 1993. Wave propagation and plasma uniformity in an electron cyclotron resonance plasma, *J. Vac. Sci. Technol. A* **11**: 2572–2576.
4. O.A. Popov. 1990. Electron cyclotron resonance plasmas excited by rectangular and circular microwave modes, *J. Vac. Sci. Technol. A* **8**: 2909–2912.
5. R. Hidaka, T. Yamaguchi, N. Hirotsu, T. Ohshima, K. Koga, M. Tanaka, and Y. Kawai. 1993. 8 inch uniform electron cyclotron resonance plasma source using a circular TE01 mode microwave, *Jpn. J. Appl. Phys.* **32**: 174–178.
6. E. Camps, O. Olea, C.G. Tapia, and M. Villagran. 1995. Characteristics of a microwave electron cyclotron resonance plasma source, *Rev. Sci. Instrum.* **66**: 3219–3227.
7. K. Rypdal, A. Fredriksen, O.M. Olsen, and K.G. Hellblom. 1997. Microwave-plasma in a simple magnetized torus, *Phys. Plasmas* **4**: 1468–1480.
8. P. Mak and J. Asmussen. 1997. Experimental investigation of the matching and impressed electric field of a multipolar electron cyclotron resonance discharge, *J. Vac. Sci. Technol. A* **15**: 154–168.
9. P. Mak, G. King, T.A. Grotjohn, and J. Asmussen. 1992. Investigation of the influence of electromagnetic excitation on electron cyclotron resonance discharge properties, *J. Vac. Sci. Technol. A* **10**: 1281–1287.
10. K. Shibata, N. Yugami, and Y. Nishida. 1994. Sheet-shaped plasma produced by electron cyclotron resonance heating, *Rev. Sci. Instrum.* **65**: 2310–2315.
11. W.D. Getty and J.B. Geddes. 1994. Size-scalable, 2.45-GHz electron cyclotron resonance plasma source using permanent magnets and waveguide coupling, *J. Vac. Sci. Technol.* **12**: 408–414.
12. T. Lagarde, J. Pelletier, and Y. Arnal. 1987. Influence of the multipolar magnetic field configuration on the density of distributed electron cyclotron resonance plasmas, *Plasma Sources Sci. Technol.* **6**: 53–60.
13. L. Pomathiod, R. Debrie, Y. Arnal, and J. Pelletier. 1984. Microwave excitation of large volumes of plasma at electron cyclotron resonance in multipolar confinement, *Phys. Lett. A* **106**: 301–304.
14. Y. Arnal, J. Pelletier, C. Pomot, B. Petit, and A. Durandet. 1984. Plasma etching in magnetic multipole microwave discharge, *Appl. Phys. Lett.* **45**: 132–134.
15. A. Hatta, M. Kubo, Y. Yasaka, and R. Itatani. 1992. Performance of electron cyclotron resonance plasma produced by a new microwave launching system in a multicusp magnetic field with permanent magnets, *Jpn. J. Appl. Phys.* **31**: 1473–1479.
16. L.A. Berry and S.M. Gorbatkin. 1995. Permanent magnet electron cyclotron resonance plasma source with remote window, *J. Vac. Sci. Technol. A* **13**: 343–348.
17. M. Pichot, A. Durandet, J. Pelletier, Y. Arnal, and L. Vallier. 1998. Microwave multipolar plasmas excited by distributed electron cyclotron resonance: Concept and performance, *Rev. Sci. Instrum.* **59**: 1072–1075.
18. H. Amemiya and S. Ishii. 1989. Electron energy distribution in multicusp-type ECR plasma, *Jpn. J. Appl. Phys.* **28**: 2289–2297.

19. H. Amemiya, K. Shimizu, S. Kato, and Y. Sakamoto. 1988. Measurements of energy distributions in ECR plasma, *Jpn. J. Appl. Phys.* **27**: 927–930.
20. M. Maeda and H. Amemiya. 1994. Electron cyclotron resonance plasma in multicusp magnets with a checkered pattern, *Jpn. J. Appl. Phys.* **33**: 5032–5037.
21. M. Maeda and H. Amemiya. 1994. Electron cyclotron resonance plasma in multicusp magnets with axial magnetic plugging, *Rev. Sci. Instrum.* **65**: 3751–3755.
22. S. Satoru and N. Sato. 1991. Plasma structures in an electron cyclotron resonance plasma processing device, *J. Appl. Phys.* **70**: 4165–4171.
23. J. Asmussen. 1989. Electron cyclotron resonance microwave discharges for etching and thin-film deposition, *J. Vac. Sci. Technol. A* **7**: 883–893.
24. J.E. Stevens, Y.C. Huang, R.L. Jarecki, and J.L. Cecchi. 1992. Plasma uniformity and power deposition in electron cyclotron resonance etch tools, *J. Vac. Sci. Technol. A* **10**: 1270–1275.
25. A. Saproo and T.D. Mantei. 1995. Performance and modeling of a permanent magnet electron cyclotron resonance plasma source, *J. Vac. Sci. Technol. A* **13**: 883–886.
26. A. Ghanbari, M.S. Ameen, and R.S. Heinrich. 1992. Characterization of a large volume electron cyclotron resonance plasma for etching and deposition of material, *J. Vac. Sci. Technol. A* **10**: 1276–1280.
27. H. Amemiya, S. Ishii, and Y. Shigueoka. 1991. Multicusp type electron cyclotron resonance ion source for plasma processing, *Jpn. J. Appl. Phys.* **30**: 376–384.
28. T.D. Mantei and S. Dhole. 1991. Characterization of permanent magnet electron cyclotron resonance plasma source, *J. Vac. Sci. Technol. B* **9**: 26–28.
29. Z.Q. Xie and C.M. Lyneis. 1994. Plasma potentials and performance of the advanced electron cyclotron resonance ion source, *Rev. Sci. Instrum.* **65**: 2947–2952.
30. G.D. Alton and D.N. Smith. 1994. Design studies for an advance ECR ion source, *Rev. Sci. Instrum.* **65**: 775–787.
31. S. Fieder and H.P. Winter. 1994. Development of a compact 2.45 GHz ECR ion source, *Rev. Sci. Instrum.* **65**: 775–787.
32. R. Geller. 1990. ECRIS: The electron cyclotron resonance ion sources, *Annu. Rev. Nucl. Part. Sci.* **40**: 15–43.
33. A. Rousseau. L. Tomasini, G. Gousset, C. Boisse-Laporte, and P. Leprince. 1994. Pulsed microwave discharge: A very efficient H atom source, *J. Phys. D: Appl. Phys.* **27**: 2439–2441.
34. T. Mieno and S. Samukawa. 1995. Time variation of plasma properties in a pulse-time-modulated electron cyclotron resonance discharge of chlorine gas, *Jpn. J. Appl. Phys.* **34**: 1079–1082.
35. S. Samukawa and K. Terada. 1994. Pulse time modulated electron cyclotron resonance plasma etching for highly selective, highly anisotropic, and less charging polycrystalline silicon patterning, *J. Vac. Sci. Technol. B* **12**: 3300–3305.
36. R.L. Papenbreer and J. Engemann. 1992. Pulsed of a Mpdr-Type ECR-plasma source for high-power applications, *Rev. Sci. Instrum.* **63**: 2550–2552.
37. S. Samukawa, H. Ohtake, and T. Mieno. 1996. Pulse time modulated electron cyclotron resonance plasma discharge for highly selective, highly anisotropic, and charge free etching, *J. Vac. Sci. Technol. A* **14**: 3049–3058.
38. C.G. Montgomery, R.H. Dicke, and E.M. Purcell (editors). 1948. *Principles of Microwave Circuits* (McGraw-Hill Book Company, Inc., New York).
39. M.A. Heald and C.B. Wharton. 1965. *Plasma Diagnostics with Microwaves* (John Wiley & Sons Inc., New York, London, Sydney).

# 2

## Review of the Cutoff Problem

### A Brief History of Earlier Work

Since the early days of research in ECR plasmas in 1980s, heating and confinement in electron cyclotron resonance (ECR) plasmas were brought about by electromagnetic coils. Coils arranged on either side of the experimental device can produce an axial magnetic field. Depending upon the application, the axial field can be linear, diverging, or in the form of a magnetic mirror. A linear field [1–3] has been used in plasma experiments, while a diverging field [4,5] is known to be beneficial for particle directionality to a downstream plasma processing area or for controlling the energy of the ions flowing along the field lines [6]. Recently, there are several experiments on utilizing a diverging field for plasma thrusters [7–9]. The mirror field is routinely used in ion sources [10–13]. In these field configurations, the waves are mainly launched in the $K || B$ mode, often referred to as the whistler mode.

However, a current-carrying coil is bulky and occupies large space, besides the fact that the current models are expensive both in their construction and from the viewpoint of power supply requirements. This has led researchers to use permanent magnets. The availability of permanent magnets in small sizes makes them useful for the development of compact ECR plasma sources [14–17]. The magnets can be easily mounted on a waveguide to fulfill the field requirements. Permanent magnets provide the experimenter greater flexibility in their arrangements [18–28], which is certainly difficult with electromagnetic coils. Moreover, the short-range nature of the fields is favorable for industrial applications, where stray fields are undesirable at a downstream location of processing substrates or in situations where the resonance field is required in a distance very close to the magnets.

Among the different possible permanent magnet arrangements, the multicusp arrangement has been proven to be very efficient for the production and confinement of ECR plasmas [29–38]. Based on the original work on the multicusp array of magnets [39,40], much work has been done by applying the multicusp confinement to DC plasmas [41–44], either in the linear or checkerboard configuration. In the case of ECR plasmas, since the arrival of the first report on the application of multicusp magnets [45], there exist many works where linear multicusps [20–26] have been commonly used. The application of a checkerboard pattern of magnet arrangement was reported [27], which could produce uniform plasmas at low pressures. Thereafter, in the multicusp geometry, the plasma confinement was improved by magnetic-end plugging [28], which is an effective way to prevent axial particle loss. Most of these studies emphasized the production of large-diameter plasmas [46]. Although there had been some investigation in the development of ECR sources with a dimension smaller than cutoff or with a density above the geometrical cutoff of the waveguide, most of them suffer from such problems as described below.

An antenna was used earlier to launch the waves directly into the device [47]. However, an antenna cannot transmit microwaves efficiently when high-density plasma is formed around it. Most of the waves get absorbed in the plasma near the neighborhood of the antenna. This results in a decrease of the plasma density as we go away from the antenna, as the discharge becomes a downstream plasma. There have also been experiments on guiding the waves through a coaxial tube [48] (the coaxial ECR discharge), where the central conductor lies within the discharge volume. Although coaxial waveguides allow the propagation of transverse electromagnetic (TEM) mode microwaves in vacuum without any limitation of geometrical cutoff, depending upon the application, the central conductor lying within the discharge volume may become an obstruction. Due to sputtering of the central conductor, the plasma can become unclean and contaminated. The whistler mode has also been applied to study plasma production under the conditions of the geometrical [15,16] and plasma density cutoffs [2,48]. However, the plasma confined in the mirror field has a loss cone leading to a lower percentage of trapped particles and an instability, such as the drift instability, and axial particle losses are large. Moreover, depending upon the ECR magnetic field, it is difficult to make the device compact.

## Microwave Propagation in Waveguides

In this section, we briefly review [46,47] the derivation of the expressions for the geometrical cutoff condition in a rectangular and a circular waveguide kept in vacuum. The change in the propagation constants in the presence of a plasma and a magnetic field are discussed qualitatively for the sake of completeness and are of relevance to the experimental situation.

We consider the general wave equation for the case when the microwave electric field $E$ is entirely transverse ($E_z = 0$). We consider plane waves, where the variation with $z$ of the five components of $E$ and $H$ is given by $e^{-\gamma z}$, then the field components take the form $E_x(z) = e^{-\gamma z} E_x$, and similar expressions for the other components. We also consider harmonically varying fields of the form, $E = E_o e^{-j\omega t}$ and $H = H_o e^{-j\omega t}$. Thus, $E_x(z) = E_x e^{-\gamma z} e^{-j\omega t}$, where $\gamma$ is called the propagation constant of the wave given by

$$\gamma = (j\omega\mu\sigma - \omega^2\varepsilon\mu)^{1/2} = j(k_r - jk_i), \tag{2.1}$$

where $\varepsilon$, $\mu$, and $\sigma$ are the permittivity, the permeability, and the conductivity of the medium, respectively. The electric field $E$ is in V/m, and the magnetic field $H$ is in A/m. For free space, $\mu$ and $\varepsilon$ will be written as $\mu_o$ and $\varepsilon_o$, with the values $\mu_o = 1.257 \times 10^{-6}$ H/m and $\varepsilon_o = 8.854 \times 10^{-12}$ F/m. The velocity of light will be represented as $c = 1/\sqrt{\varepsilon_o\mu_o} = 2.998 \times 10^8$ m/s. A quantity called the wave impedance is given by $z_w = \sqrt{\mu/\varepsilon} = 377\ \Omega$ for free space. The real part of $\gamma$, $k_r$, is called the attenuation constant, and its imaginary part, $k_i$, is called the phase constant. The phase velocity of the wave is $v = \omega/k_r$ and the wavelength $\lambda = 2\pi/k_r$. If the conductivity of the medium $\sigma$ is small, $k_r \approx \omega\sqrt{\varepsilon\mu}$ and $v = 1/\sqrt{\varepsilon\mu}$. If $\sigma$ is not negligible, then the exact expressions for $k_i$ and $k_r$ are

$$k_i = \left[\frac{1}{2}\omega\mu\left(\sqrt{\sigma^2 + \omega^2\varepsilon^2} - \omega\varepsilon\right)\right]^{1/2}, \tag{2.2}$$

and

$$k_r = \left[\frac{1}{2}\omega\mu\left(\sqrt{\sigma^2 + \omega^2\varepsilon^2} + \omega\varepsilon\right)\right]^{1/2}. \tag{2.3}$$

Using Maxwell's equations in Cartesian coordinates, the wave equation for $H_z$ can be written as

$$\frac{\partial H_z}{\partial x^2} + \frac{\partial H_z}{\partial y^2} + \left(\gamma^2 + \omega^2\varepsilon\mu\right)H_z = 0, \tag{2.4}$$

and by employing cylindrical coordinates $r$, $\theta$, $z$ and choosing the z-axis as the axis of propagation,

$$\frac{\partial H_z}{\partial r^2} + \frac{1}{r}\frac{\partial H_z}{\partial r} + \frac{\partial^2 H_z}{\partial\theta^2} + \left(\gamma^2 + \omega^2\varepsilon\mu\right)H_z = 0. \tag{2.5}$$

Knowing $H_z$, other components of the field can be determined.

## Waveguide Geometrical Cutoff

### Rectangular Waveguide

Let us consider a waveguide of rectangular cross-section, as shown in Figure 2.1, which has dimensions $b$ in the $y$ direction and $a$ in the $x$ direction, and which has walls of infinite conductivity. The conductivity of the medium is omitted explicitly and will be assumed to be contained in the imaginary

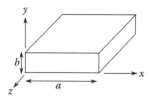

**FIGURE 2.1**
Coordinates for a rectangular waveguide.

part of the dielectric constant $\varepsilon$. We will consider the TE modes. Equation 2.4 for $H_z$ can be applied. $H_z$ is separable in rectangular coordinates and leads to simple sinusoidal solutions. Let

$$H_z = \cos k_x x \cos k_y y, \tag{2.6}$$

where the separation constants $k_x$ and $k_y$ are related by

$$k_c^2 = k_x^2 + k_y^2 = \gamma^2 + \omega^2 \varepsilon \mu. \tag{2.7}$$

The quantities $k_x$ and $k_y$ are wave numbers in the $x$ and $y$ directions, respectively. The field components $E_x$ and $E_y$ are given by

$$E_x = \frac{j\omega\mu k_y}{k_x^2 + k_y^2} \sin k_y y \cos k_x x, \tag{2.8}$$

and

$$E_y = -\frac{j\omega\mu k_x}{k_x^2 + k_y^2} \sin k_x x \cos k_y y. \tag{2.9}$$

We now apply the boundary conditions, when $y = 0$ or $y = b$, $E_x = 0$, therefore, $k_y = n\pi/b$, where $n$ is an integer. Further, when $x = 0$ or $x = a$, $E_y = 0$, therefore, $k_x = m\pi/a$, where $m$ is an integer. It is clear that both $m$ and $n$ may take any values including zero, except that both $m$ and $n = 0$ are excluded. Thus, the propagation constant $\gamma$ is given by

$$\gamma^2 = \left(\frac{m\pi}{a}\right)^2 + \left(\frac{n\pi}{b}\right)^2 - \omega^2 \varepsilon \mu. \tag{2.10}$$

To have propagation down the waveguide, $\gamma^2$ must be negative, hence no waves are propagated below a certain frequency. In the absence of any losses in the medium, that is, if $\varepsilon$ is purely real, there is a sharply defined critical frequency $\omega_{cr}$, which is given by

$$\omega_{cr}^2 = \frac{1}{\varepsilon\mu}\left[\left(\frac{m\pi}{a}\right)^2 + \left(\frac{n\pi}{b}\right)^2\right]. \tag{2.11}$$

In terms of the cutoff wavelength $\lambda_c$, and for free space $\varepsilon \to \varepsilon_0$ and $\mu \to \mu_0$, and $c = 1/(\varepsilon_0\mu_0)^{1/2}$,

$$\frac{1}{\lambda_c^2} = \left(\frac{m}{2a}\right)^2 + \left(\frac{n}{2b}\right)^2. \tag{2.12}$$

Thus, the value of the cutoff wavelength depends on the values of $m$ and $n$ and the dimensions of the waveguide. The eigenmodes are denoted by $TE_{mn}$. For the fundamental or the $TE_{10}$ mode (where $m = 1$, $n = 0$) and for a fixed-wave frequency, for example, 2.45 GHz as used in the experiments (free space wavelength, $\lambda_0 = 12.24$ cm), the cutoff width in a rectangular waveguide from Equation 2.12 is given by

$$a_c = \frac{\lambda_0}{2} = 6.12 \text{ cm}. \tag{2.13}$$

It may be noted that in the $TE_{10}$ mode the cutoff width does not depend upon the height $b$ of the waveguide.

### Circular Waveguide

Let us consider a waveguide of circular cross-section, as shown in Figure 2.2. The wave is considered to be launched along the $z$ direction. We solve the wave Equation 2.5 in cylindrical coordinates with the boundary conditions that at $r = R$ ($R$ is the radius of the waveguide),

$$\frac{\partial H_z}{\partial r} = 0. \tag{2.14}$$

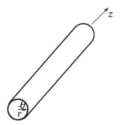

**FIGURE 2.2**
Coordinates for a cylindrical waveguide.

If the variables are separated,

$$H_z = \Theta(\theta)\xi(r). \tag{2.15}$$

The equation for $\Theta$ can be written as

$$\frac{\partial^2 \Theta}{\partial \theta^2} + m^2 = 0, \tag{2.16}$$

and the equation for $\xi$ can be written as

$$\frac{\partial^2 \xi}{\partial r^2} + \frac{1}{r}\frac{\partial \xi}{\partial r} + \left(k_c^2 - \frac{m^2}{r^2}\right)\xi = 0. \tag{2.17}$$

Here $m^2$ is a separation constant and $k_c^2 = \gamma^2 + \omega^2 \varepsilon \mu$. The solution for $\Theta$ is

$$\Theta = e^{im\theta}, \tag{2.18}$$

where $\theta$ can lie between 0 and an integral multiple ($m$) of $2\pi$. Thus, $m$ must be an integer or zero (= 0, ±1, ±2, ±3 ...). The complex form of $\Theta$ indicates that two solutions are possible—one is $\Theta = \cos m\theta$ and the other is $\Theta = \sin m\theta$. Thus, the modes are degenerate in pairs. The two modes may be interpreted as two states of polarization of the field.

In Equation 2.17, substituting $x = k_c r$, the equation becomes a Bessel's equation in the canonical form. Thus, the solution for $H_z$ (Equation 2.15) can be written as

$$H_z = e^{im\theta} J_m(k_c r), \tag{2.19}$$

where $J_m$ is the Bessel function of the first kind of order $m$. The solution $N_m$, the Bessel function of the second kind, is excluded because of the singularity at $r = 0$. For the satisfaction of the boundary condition given by Equation 2.14, the first derivative of $J_m(k_c r)$ is zero at $r = R$, giving

$$J_m'(k_c R) = 0, \tag{2.20}$$

and

$$k_c R = p_{mn}', \tag{2.21}$$

where $m = 0, 1, 2, 3 \ldots; n = 1, 2, 3, \ldots;$ and $\rho'_{mn}$ is the $n$th root of $J'_m$. The cutoff frequency can be calculated from above as

$$\omega_{cr}^2 = \frac{1}{\varepsilon\mu}\left(\frac{\rho'_{mn}}{a}\right)^2, \qquad (2.22)$$

and the cutoff wavelength is given by

$$\lambda_c = \frac{2\pi a}{\rho'_{mn}}. \qquad (2.23)$$

The modes are designated by $TE_{mn}$ and the fundamental mode for a round pipe is $TE_{11}$, for which $\rho'_{11} = 1.841$. Therefore, the cutoff radius for a circular waveguide is given by

$$R_c = \frac{\rho'_{11}\lambda_o}{2\pi} \cong 3.6 \text{ cm}. \qquad (2.24)$$

**Waveguide Filled with a Plasma**

Next, we review the propagation characteristics in a waveguide filled with a plasma. Although the problem can be formulated by considering a complex dielectric constant or a complex permeability, for a plasma the formulation in terms of conductivity has a more direct physical interpretation and is more useful. For simplicity, we study the case of a rectangular waveguide where the dominant mode, that is, the $TE_{10}$ mode, is considered to be propagating, then $k_c = \pi/a$. When the medium filling the waveguide has a conductivity $\sigma$, then the propagation constant $\gamma$ takes the form

$$\gamma^2 = \left(\frac{\pi}{a}\right)^2 + j\omega\mu\sigma - \omega^2\varepsilon\mu. \qquad (2.25)$$

If $\sigma$ is a complex quantity, it may be written as

$$\sigma = \sigma^r - j\sigma^c, \qquad (2.26)$$

where $\sigma^r$ and $\sigma^c$ denote the real and imaginary components, respectively. Substituting Equation 2.26 into Equation 2.25, $\gamma$ can be written as

$$\gamma^2 = \left(\frac{\pi}{a}\right)^2 - \omega^2\mu\left(\varepsilon - \frac{\sigma^c}{\omega}\right) + j\omega\mu\sigma^r. \qquad (2.27)$$

In Equation 2.27, $\sigma^c/\omega$ is the correction term to the dielectric constant in the presence of a medium with a conductivity $\sigma$. The classical definition of $\sigma$ in a high-frequency plasma is given by

$$\sigma = \frac{N_e e^2}{m(v_e + j\omega)},\tag{2.28}$$

separating into real and imaginary parts,

$$\sigma = \frac{N_e e^2 v_e}{m(v_e^2 + \omega^2)} - j\frac{N_e e^2 \omega}{m(v_e^2 + \omega^2)}.\tag{2.29}$$

If the wave frequency is much higher than the electron-neutral collision frequency $v_e(\omega^2 \gg v_e^2)$, then

$$\sigma = \frac{N_e e^2 v_e}{m\omega^2} - j\frac{N_e e^2}{m\omega}.\tag{2.30}$$

The imaginary portion of $\sigma$ ($\sigma^c$) is the expression for entirely free electrons. Since $\sigma^r$ is inversely proportional to $\omega^2$, and $\sigma^c$ is inversely proportional to $\omega$, at sufficiently high frequencies $\sigma^r \to 0$ and $\sigma \approx -j\sigma^c$. Then from Equation 2.27

$$\gamma^2 = \left(\frac{\pi}{a}\right)^2 - \omega^2\mu\left(\varepsilon - \frac{\sigma^c}{\omega}\right),\tag{2.31}$$

substituting the value of $\sigma^c$ from Equation 2.30 into Equation 2.31, the equation can be written as

$$\gamma^2 = \left(\frac{\pi}{a}\right)^2 - \omega^2\mu\left(\varepsilon - \frac{N_e e^2}{m\omega^2}\right).\tag{2.32}$$

The above equation can be rewritten in a convenient form so as to bring in the angular plasma frequency $\omega_p$, which can lead to the usual propagation constant in free space in the absence of the plasma ($\omega_p = 0$), where

$$\gamma^2 = \left(\frac{\pi}{a}\right)^2 - \frac{\omega^2}{c^2}\left(1 - \frac{\omega_p^2}{\omega^2}\right),\tag{2.33}$$

and

$$\omega_p^2 = \frac{N_e e^2}{m \varepsilon_o}. \tag{2.34}$$

Since $k_c^2 = \gamma^2 + \omega^2 \varepsilon \mu$, depending upon the value of $\gamma$, $k_c$ will be modified accordingly, and the new condition for cutoff radius can be obtained from $k_c R = p'_{mn}$.

## Effect of a Magnetic Field

For the waves launched perpendicular to the magnetic field ($K \perp B$), electromagnetic waves in the ordinary (O-wave) and extraordinary (X-wave) modes are known to be excited in the plasma [49], depending upon whether the wave electric field is parallel ($E \parallel B$) or perpendicular ($E \perp B$) to the static magnetic field, respectively.

Considering the plasma as cold and collisionless and under an infinite plasma approximation, the plasma appears as a dielectric with relative permittivity $\varepsilon_{ro}$ for the ordinary mode and $\varepsilon_{rx}$ for the extraordinary mode given by the relations

$$\varepsilon_{ro} = 1 - \frac{\omega_p^2}{\omega^2}, \tag{2.35}$$

and

$$\varepsilon_{rx} = 1 - \frac{\omega_p^2 \left( \omega^2 - \omega_p^2 \right)}{\omega^2 \left( \omega^2 - \omega_h^2 \right)}, \tag{2.36}$$

respectively. Taking $\mu$ to be the same as in free space ($\mu_o$), and that $\varepsilon = \varepsilon_r \varepsilon_o$, the propagation constants in the ordinary mode $\gamma_o$ and in the extraordinary mode $\gamma_x$ are given by

$$\gamma_o^2 = \left( \frac{\pi}{a} \right)^2 - \frac{\omega^2}{c^2} \left( 1 - \frac{\omega_p^2}{\omega^2} \right), \tag{2.37}$$

and

$$\gamma_x^2 = \left( \frac{\pi}{a} \right)^2 - \frac{\omega^2}{c^2} \left( 1 - \frac{\omega_p^2 \left( \omega^2 - \omega_p^2 \right)}{\omega^2 \left( \omega^2 - \omega_h^2 \right)} \right), \tag{2.38}$$

respectively, where $\omega_h(\omega_h^2 = \omega_p^2 + \omega_c^2)$ is known as the upper hybrid frequency. It may be noted that the propagation constant in the ordinary mode is similar to the case without a magnetic field.

## Plasma Density Cutoff

When the frequency of the wave $\omega$ equals the plasma frequency $\omega_p$, the wave is reflected from the plasma [50]. This property of plasma is utilized in an important diagnostic known as plasma reflectometry, which is primarily utilized to determine plasma density. The plasma frequency given by Equation 2.34 is proportional to the electron density. On substituting the value of the constants, Equation 2.34 can be written in a more readily applicable form as

$$f_p = 8.98 \times 10^3 \left(N_e\right)^{1/2}, \tag{2.39}$$

where $f_p$ is the plasma frequency in Hz and $N_e$ is the plasma (electron) density in $cm^{-3}$. Equating the value of $f_p$ to the wave frequency, we can readily calculate the critical density $N_c$. In the case of a magnetized plasma, the cutoff density depends upon the magnetic field. For the ordinary and the extraordinary modes, the relation of the plasma frequency to the wave frequency is shown below, from which the cutoff density for a particular wave frequency and magnetic field can be obtained.

### Ordinary Wave

The dispersion relation is given by $n^2 = 1 - X$, where $n = ck/\omega$ is the refractive index and $X = \omega_p^2/\omega^2$. The cutoff density given by the condition $X = 1$ leads to the relation

$$\omega_p^2 = \omega^2. \tag{2.40}$$

### Extraordinary Wave

The dispersion relation for the extraordinary wave is $n^2 = [(X - 1)^2 - Y]/[1 - X - Y]$, where $Y = \omega_c^2/\omega^2$. The X-mode cutoff condition is given by $Y = (X - 1)^2$, which leads to the relation

$$\omega_p^2 = \omega^2\left(1 + \frac{\omega_c}{\omega}\right). \tag{2.41}$$

# References

1. B.H. Quon and R.A. Dandl. 1989. Preferential electron–cyclotron heating of hot electrons and formation of overdense plasmas, *Phys. Fluids B* **1**: 2010–2017.

2. M. Tanaka, R. Nishimoto, S. Higashi, N. Harada, T. Ohi, A. Komori, and Y. Kawai. 1991. Overdense plasma production using electron cyclotron waves, *J. Phys. Soc. Jpn.* **60**: 1600–1607.

3. M. Sugimoto, M. Tanaka, and Y. Kawai. 1996. Electron cyclotron wave plasma production using a concave lens, *Jpn. J. Appl. Phys.* **35**: 2803–2807.

4. S.Y. Shapoval, V.T. Petrashov, O.A. Popov, M.D. Yoder Jr., P.D. Maciel, and K.C. Lok. 1991. Electrons cyclotron resonance plasma chemical vapor deposition of large area uniform silicon nitride films, *J. Vac. Sci. Technol.* **9**: 3071–3077.

5. J. Forster and W. Holber. 1989. Plasma characterization for a divergent field electron cyclotron resonance source, *J. Vac. Sci. Technol. A* **7**: 899–902.

6. M. Matsuoka and K. Ono. 1987. Low energy ion extraction with small dispersion from an electron cyclotron resonance microwave plasma stream, *Appl. Phy. Lett.* **50**: 1864–1866.

7. C. Charles. 2009. Plasmas for spacecraft propulsion, *J. Phys. D: Appl. Phys.* **42**: 163001.

8. C. Charles, K. Takahashi, and R. Boswell. 2012. Axial force imparted by a conical radiofrequency magneto-plasma thruster, *Appl. Phys. Lett.* **100**: 113504.

9. K. Takahashi, C. Charles, R. Boswell, and T. Fujiwara. 2011. Electron energy distribution of a current-free double layer: Druyvesteyn theory and experiments, *Phys. Rev. Lett.* **107**: 035002.

10. Z.Q. Xie and C.M. Lyneis. 1994. Plasma potentials and performance of the advanced electron cyclotron resonance ion source, *Rev. Sci. Instrum.* **65**: 2947–2952.

11. G.D. Alton and D.N. Smith. 1994. Design studies for an advance ECR ion source, *Rev. Sci. Instrum.* **65**: 775–787.

12. S. Fieder and H.P. Winter. 1994. Development of a compact 2.45 GHz ECR ion source, *Rev. Sci. Instrum.* 65: 1094–1096.

13. R. Geller. 1990. ECRIS: The electron cyclotron resonance ion sources, *Annu. Rev. Nucl. Part. Sci.* **40**: 15–43.

14. M. Delaunay. 1990. Compact ECR ion sources with permanent magnets in a cusp geometry and in a magnetic mirror structure, *Rev. Sci. Instrum.* **61**: 267–269.

15. M. Shimada and Y. Torii. 1993. Compact electron cyclotron resonance ion source with a permanent magnet, *J. Vac. Sci. Technol. A* **11**: 1313–1316.

16. O.A. Popov. 1989. Characteristics of electron cyclotron resonance plasma sources, *J. Vac. Sci. Technol. A* **7**: 894–898.

17. L. Mahoney and J. Asmussen. 1990. A compact, resonance cavity, five centimeter, multicusp, ECR broad-beam ion source, *Rev. Sci. Insrum.* **61**: 285–287.

18. W.D. Getty and J.B. Geddes. 1994. Size-scalable, 2.45-GHz electron cyclotron resonance plasma source using permanent magnets and waveguide coupling, *J. Vac. Sci. Technol.* **12**: 408–414.

19. T. Lagarde, J. Pelletier, and Y. Arnal. 1997. Influence of the multipolar magnetic field configuration on the density of distributed electron cyclotron resonance plasmas, *Plasma Sources Sci. Technol.* **6**: 53–60.

20. L. Pomathiod, R. Debrie, Y. Arnal, and J. Pelletier. 1984. Microwave excitation of large volumes of plasma at electron cyclotron resonance in multipolar confinement, *Phys. Lett. A* **106**: 301–304.
21. Y. Arnal, J. Pelletier, C. Pomot, B. Petit, and A. Durandet. 1984. Plasma etching in magnetic multipole microwave discharge, *Appl. Phys. Lett.* **45**: 132–134.
22. A. Hatta, M. Kubo, Y. Yasaka, and R. Itatani. 1992. Performance of electron cyclotron resonance plasma produced by a new microwave launching system in a multicusp magnetic field with permanent magnets, *Jpn. J. Appl. Phys.* **31**: 1473–1479.
23. L.A. Berry and S.M. Gorbatkin. 1995. Permanent magnet electron cyclotron resonance plasma source with remote window, *J. Vac. Sci. Technol. A* **13**: 343–348.
24. M. Pichot, A. Durandet, J. Pelletier, Y. Arnal, and L. Vallier. 1998. Microwave multipolar plasmas excited by distributed electron cyclotron resonance: Concept and performance, *Rev. Sci. Instrum.* **59**: 1072–1075.
25. H. Amemiya and S. Ishii. 1989. Electron energy distribution in multicusp-type ECR plasma, *Jpn. J. Appl. Phys.* **28**: 2289–2297.
26. H. Amemiya, K. Shimizu, S. Kato, and Y. Sakamoto. 1998. Measurements of energy distributions in ECR plasma, *Jpn. J. Appl. Phys.* **27**: 927–930.
27. M. Maeda and H. Amemiya. 1994. Electron cyclotron resonance plasma in multicusp magnets with a checkered pattern, *Jpn. J. Appl. Phys.* **33**: 5032–5037.
28. M. Maeda and H. Amemiya. 1994. Electron cyclotron resonance plasma in multicusp magnets with axial magnetic plugging, *Rev. Sci. Instrum.* **65**: 3751–3755.
29. M. Sadowski. 1967. Plasma confinement with spherical multipole magnetic field, *Phys. Lett. A* **25**: 695–696.
30. M. Sadowski. 1969. Spherical multipole magnets for plasma research, *Rev. Sci. Instrum.* **40**: 1545–1549.
31. R. Limpaecher and K.R. MacKenzie. 1973. Magnetic multipole containment of large uniform collisionless quiescent plasmas, *Rev. Sci. Instrum.* **44**: 727–731.
32. G.J. Brakenhoff and A. Goede. 1970. A warm quiescent plasma in a minimum B field obtained with permanent magnets, *Plasma Phys.* **12**: 815–817.
33. C. Koch and G. Matthieussent. 1983. Collisional diffusion of a plasma in multipolar and picket fence devices, *Phys. Fluids* **26**: 545–555.
34. C. Gauthereau and G. Matthieussent. 1984. Plasma density profiles in discharges surrounded by magnetic multipole walls, *Phys. Lett. A* **102**: 231–234.
35. K.N. Leung, T.K. Samec, and A. Lamm. 1975. Optimization of permanent magnet plasma confinement, *Phys. Lett. A* **51**: 490–492.
36. K.N. Leung, N. Hershkowitz, and K.R. MacKenzie. 1976. Plasma confinement by localized cusps, *Phys. Fluids* **19**: 1045–1053.
37. M. Katsch and K. Wiesemann, 1980. Relaxation of suprathermal electrons due to Coulomb collisions in a plasma, *Plasma Phys.* **22**: 627–638.
38. F.A. Hass, L.M. Lea, and A.J.T. Holmes. 1991. A hydrodynamics model of the negative ion source, *J. Phys. D: Appl. Phys.* **24**: 1541–1550.
39. M. Sadowski. 1967. Plasma confinement with spherical multipole magnetic field, *Phys. Lett. A* **25**: 695–696.
40. M. Sadowski. 1969. Spherical multipole magnets for plasma research, *Rev. Sci. Instrum.* **40**: 1545–1549.
41. C. Gauthereau and G. Matthieussent. 1984. Plasma density profiles in discharges surrounded by magnetic multipole walls, *Phys. Lett. A* **102**: 231–234.

42. K.N. Leung, T.K. Samec, and A. Lamm. 1975. Optimization of permanent magnet plasma confinement, *Phys. Lett. A* **51**: 490–492.

43. K.N. Leung, N. Hershkowitz, and K.R. MacKenzie. 1976. Plasma confinement by localized cusps, *Phys. Fluids* **19**: 1045–1053.

44. M. Katsch and K. Wiesemann. 1980. Relaxation of suprathermal electrons due to Coulomb collisions in a plasma, *Plasma Phys.* **22**: 627–638.

45. G.J. Brakenhoff and A. Goede. 1970. A warm quiescent plasma in a minimum B field obtained with permanent magnets, *Plasma Phys.* **12**: 815–817.

46. C.G. Montgomery and R.H. Dicke, 1948. E.M. Purcell (editors), *Principles of Microwave Circuits* (McGraw-Hill Book Company, Inc., New York).

47. M.A. Heald and C.B. Wharton. 1965. *Plasma Diagnostics with Microwaves* (John Wiley & Sons Inc., New York).

48. H. Amemiya and S. Ishii. 1989. Electron energy distribution in multicusp-type ECR plasma, *Jpn. J. Appl. Phys.* **28**: 2289–2297.

49. W.P. Allis, S.J. Buchsbaum, and A. Bers. 1963. *Waves in Anisotropic Plasmas* (MIT Press, Cambridge, MA).

50. Z. Sitar, M.J. Paisley, D.K. Smith, and R.F. Davis. 1990. Design and performance of an electron cyclotron resonance plasma source for standard molecular beam epitaxy equipment, *Rev. Sci. Instrum.* **61**: 2407–2411.

# 3

## Plasma and Wave Diagnostics

### Plasma Diagnostics with Langmuir's Probes

The Langmuir probes have been used as a simple measurement method for obtaining the local plasma parameters. The technique is well known and has been discussed in detail by several authors throughout the literature [1–5]. Here we briefly summarize the method, its usefulness, and the equations necessary for determining the plasma parameters. We also present results of an actual probe characteristic taken from the experiment. The probe measurement method consists of inserting into the plasma a small metallic electrode or probe of known area whose shapes are generally planes, cylinders, or spheres. In our experimental work, we mainly employ planar Langmuir's probes. The Langmuir probe measurement has the following advantages:

a. Local measurements can be made. This point is the most important because almost all other measurement techniques, such as spectroscopy or measurements based on microwave propagation, give information averaged over a large plasma volume.

b. The disturbance caused by the presence of the probe is localized under a wide range of conditions.

c. The measurement circuitry is simple as compared to other diagnostics.

d. Their usefulness is attributable to the fact that an analysis of the $I$–$V$ characteristics measured by the probe yields the electron and ion density, the electron temperature (assuming that the electron energy distribution is Maxwellian), and the space or plasma potential.

### Single Langmuir Probe

Figure 3.1 shows a basic circuit for single probe measurement. C is a vacuum chamber. One side of the probe P is inserted into the plasma, and the other side is connected electrically to a variable DC power supply, which in turn is connected to a second reference electrode immersed in the plasma. The reference electrode MC, for the case of the single probe diagnostics, has to

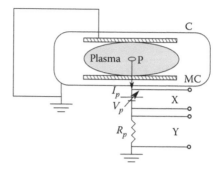

**FIGURE 3.1**
Simple circuit for determining *I–V* characteristics of the plasma.

be much larger than the probe itself and could typically be the anode or
cathode of a DC-generated plasma or the metal walls confining the plasma,
as in multicusp confinement devices. The probe current $I_p$ depends on the
probe voltage $V_p$ and is detected through a resistance $R_p$. The voltage across
$R_p$ can be applied to the vertical scale of an X–Y recorder represented by Y in
Figure 3.1, and the voltage $V_p$ to the horizontal scale represented by X in the
figure. This way, the *I–V* characteristics can be traced. The signal can also
be taken to an oscilloscope if it has an X–Y mode of plotting. A schematic
diagram of a typical planar Langmuir's probe is shown in Figure 3.2, along
with the details of the electronic measurement circuit. A digital photograph
of the probe is shown in Figure 3.3a, and the details of the current-to-voltage
converter (CVC) are shown in Figure 3.3b.

A slightly advanced probe measurement circuit is shown in Figure 3.2 and
consists of a bipolar power supply (BPS; Kepco BOP 100-1 M), a differential
amplifier (DA), National Instruments data acquisition card (DAQ; NIBNC-2110),

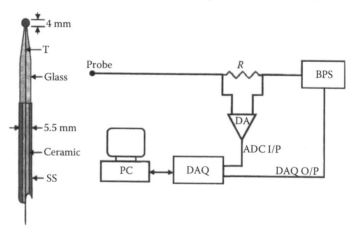

**FIGURE 3.2**
Schematic of Langmuir's probe and the measuring circuit.

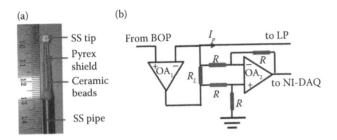

**FIGURE 3.3**
(a) Digital picture of Langmuir's probe showing the dimensions and assembly components.
(b) Detailed circuit of the current-to-voltage converter (CVC) for measurement of plasma current ($I_p$).

and a personal computer (PC). With the digital-to-analog converter (DAC) of DAQ, the probe potential, $V_p$, is swept from –80 to +80 V using the BPS. About 1000 data points are acquired in a single sweep of ~1.5 min. Each data point is averaged 100 times before acquiring. The voltage drop across the resistor R (~500 ohm) due to current drawn by the probe is fed back to the DA, which amplifies the difference between the two inputs. This is the input of the analog to digital converter (ADC) in DAQ. Control of the instrumentation and data storage is done using a PC with LabVIEW software (Version 8.10, National Instruments Corporation, Texas, the USA).

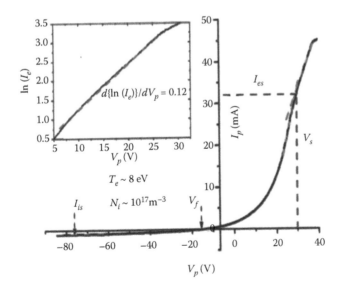

**FIGURE 3.4**
A typical Langmuir's probe characteristic obtained with 0.60 mTorr, 270 W, at the center of the MC. The inset shows the semilog plot of the electron current ($I_e$) for determination of electron temperature ($T_e$). The determined electron density and temperature are indicated in the figure.

A typical plane probe characteristic (curve of $I_p$ versus $V_p$) obtained from a microwave plasma experiment is shown in Figure 3.4. The qualitative behavior of the curve can be explained as follows. At the point $V_s$, the probe is at the same potential as that of the plasma (this is commonly called the space or plasma potential) and is conveniently chosen as the reference zero voltage. When the probe is biased positively with respect to the local plasma potential, the flux of all the negative particles to the probe is collected and current saturation occurs. Because the negative charge carriers are usually electrons, this region is called the *electron saturation region*. If now the probe is made negative relative to $V_s$, the probe begins to repel electrons and to accelerate the ions toward the probe. The probe can, however, collect those electrons that have energies large enough to overcome the energy barrier $e(V_s - V_p)$. This region is called the *electron retardation region*. If the electron energy distribution is Maxwellian, the shape of the curve in this region would be exponential.

Finally, for increasing negative probe potentials, a point is reached where the probe draws no net current because the flux of positive ions and electrons to the probe is equal. This point is called the floating potential $V_f$. This is the potential that would be acquired by an insulated substrate placed into the plasma.

At large negative values of $V_p$, almost all electrons are repelled and only positive ions are collected. The probe current is almost totally ion current $I_+$. This region is called the *ion saturation region*. The situation is similar to the electron saturation region, except for the difference in the absolute magnitude of the currents, which is due to their mass difference. At very large positive or negative values of $V_p$, breakdown occurs. Therefore, measurements can best be carried out between the electron and ion saturation regions.

The current drawn by a planar probe at a potential $V$ given by $V = V_s - V_p$, where $V_p$ is the probe potential and $V_s$ is the space or plasma potential lower than the space potential is given by

$$I(V) = (eN_eS_p/4)(8\kappa T_e/\pi m)^{1/2} \exp\left(-\frac{(V_s - V_p)}{\kappa T_e}\right),  \tag{3.1}$$

where $S_p$ is the probe area, $\kappa$ is the Boltzmann constant, and $T_e$ is the electron temperature.

### Electron Saturation Region

At the electron saturation region, when the probe potential is at the local space potential ($V_p = V_s$), the electron saturation current is given by

$$I_p = I_{es} = (eN_eS_p/4)(8\kappa T_e/\pi m)^{1/2},  \tag{3.2}$$

where the term $(8\kappa T_e/\pi m)^{1/2}$ is called electron thermal velocity, $v_e$. At larger positive probe potential, $V_p > V_s$, $I_e$ is almost constant.

### Ion Saturation Region

In the positive ion sheath, the electron concentration at the sheath edge can be assumed to be a Boltzmann distribution, $N_e \exp(-eV/\kappa T_e)$, where $V$ is the potential at the sheath edge. Each ion responds to $V$ to maintain quasi-neutrality. The potential $V$ at the sheath edge can be obtained by considering the charge ($\rho$) balance at the sheath edge from the following relation:

$$\rho = J_+/(2eV/M)^{1/2} - eN_e \exp(-eV/\kappa T_e) = 0, \tag{3.3}$$

$$d\rho/dV = 0, \tag{3.4}$$

where $J_+$ is the ion current density and $M$ is the positive ion mass. We therefore obtain $V = \kappa T_e/2e$.

From Equation 3.3, substituting the value of $V$ and assuming charge equality ($N_e = N_+$), the positive ion current can be obtained as

$$I_{is} = I_+ = eZN_+S_p \exp(-1/2)(\kappa T_e/M)^{1/2}, \tag{3.5}$$

where $Z$ is the electric charge number. It may be noted that the ion saturation current depends on the electron temperature rather than the ion temperature.

## Methods to Determine Plasma Parameters

### Electron Temperature ($T_e$)

Differentiating Equation 3.1, we have

$$\frac{d\ln\{I(V)\}}{dV} \propto \frac{1}{\kappa T_e}. \tag{3.6}$$

So, the electron temperature can be obtained from the gradient of the semi-log plot of the probe current versus probe potential. This has been shown in Figure 3.4.

### Electron Density ($N_e$)

From the electron saturation current $I_{es}$ (Equation 3.2), the electron density is obtained using the relation

$$N_e = \frac{4}{eS_p}\left(\frac{\pi m}{8\kappa T_e}\right)^{1/2} I_{es}. \tag{3.7}$$

## Ion Density ($N_+$)

Similar to the electron density, the ion density is obtained using the ion saturation current (Equation 3.5) and electron temperature as

$$N_+ = \frac{\exp(-1/2)}{eZS_p}\left(\frac{M}{\kappa T_e}\right)^{1/2} I_{is}. \tag{3.8}$$

## Space Potential ($V_s$) and Floating Potential ($V_f$)

The value of the potential of the plasma is important for two main reasons. First, knowledge of $V_s$ over a spatial region of interest facilitates the calculation of the electric fields present. Second, in probe theory, potentials are measured with respect to $V_s$. $V_s$ is determined as the point of a boundary between the electron retardation region and the electron saturation region, and can be measured directly from the knee of a single probe characteristics if it is clear, or by finding the intersection of the straight line used to determine the electron temperature and the linear extrapolation of the electron saturation current.

Another parameter, which can be determined from the probe characteristics, is the floating potential. This is the easiest and the most convenient potential to identify from probe characteristics, because it is the potential for which the net current to the probe is zero. In the event that electron saturation current cannot be measured or is distorted, $V_f$ is often useful in determining $V_s$. Since the difference between $V_f$ and $V_s$ is related to $T_e$ by the approximate relation

$$V_s - V_f = (\kappa T_e/2)\ln(M/2.3\,\mathrm{m}), \tag{3.9}$$

by knowing $V_f$ and $T_e$, we may be able to make an estimate of $V_s$.

The probe head used in the experiment is usually a planar probe made of a stainless-steel sheet of 0.1 mm thickness and has a diameter of 4 mm. Both surfaces of the probe are used for charge particle collection. The stainless-steel (SS) probe tip is spot welded on a Tungsten wire that is insulated using a tapered glass tube in the front part. The Tungsten wire is covered with ceramic beads both inside the glass tube and the SS tube behind it. The end of the probe is vacuum-sealed using Torr seal (see Figure 3.3a). Inside the multicusp, the probe surface is oriented parallel to the $k$ vector of the incident waves to avoid any possible disturbance by microwaves. In the center of the multicusp, where the $B$-field is almost zero, the probe measurements are not influenced by the magnetic field. In the peripheral region, the probe surface is perpendicular to $B$ so that charged particles can be collected by the probe. In the experiment, MC serves as a reference electrode, grounded with the chamber C. The ratio of the inner surface of MC to the probe area is ~5000, which could be considered adequate for obtaining reliable probe characteristics. The Debye length, $\lambda_D$, is much smaller than the probe radius and varies in the range of 0.07–0.1 mm, depending upon the plasma parameters obtained in the experiment.

## Practical Difficulties Encountered in Determination of the Plasma Parameters

The above-discussed theoretical determination of the plasma parameters is often difficult to execute from the experimentally determined probe characteristics. For example, for a planar probe, the ion saturation current $I_{is}$ is theoretically a constant value at deeper negative bias voltage; however, often the measured probe current shows that it is not constant but linearly increasing (see Figure 3.4). It thus becomes difficult to decide the precise value of $I_{is}$. This ambiguity has been avoided by taking the $I_{is}$ as the average value of the currents at a voltage where the electron current starts to appear, and the current at a deeper negative voltage, for example, at $V_p = -70$ V.

The second problem is that it is often difficult to determine $V_s$ distinctly because, as predicted by theory, the knee region is not so sharp. Practically, the meeting point of the tangential lines in the electron retardation region and the electron saturation region is used as the location of the space potential. Further, this problem may be overcome by plotting ln $(I_e)$ versus $V_p$, and a sharp kink may be discernable (see Figure 3.4).

For the probe and the reference electrode system to work properly, there are additional considerations. These have been applied in one of the transversely magnetized plasma experiments employing a rectangular wave guide with a dimension below cutoff [6]. The waveguide had permanent magnets on two sides and could be rotated so as to make the wave electric field ($E$) parallel or perpendicular to the static magnetic field ($B$). The condition is that [6]

$$\frac{N_p e S_p (\kappa T_e / \epsilon M)^{1/2}}{N_r e S_r (\kappa T_e / 2\pi m)^{1/2}} < 1 \tag{3.10}$$

or

$$\eta = S_p / S_r < (N_r / N_p) S_c, \tag{3.11}$$

where $N_p$ and $N_r$ are the plasma densities near the probe and the reference electrode, $\epsilon$ is the base of natural logarithm, and $S_p$ and $S_r$ are the areas of the probe and the reference electrode, respectively. The critical area limit is $\eta_c = (\epsilon M / 2\pi m)^{1/2}$. For Ar, $\eta \cong 178$. The area ratio between the reference electrode and the plane probe $S_r / S_p$ is $5.5 \times 10^3$, and that with the cylinder probe is about $1.1 \times 10^4$. Thus, for the system to work correctly, it is needed that $\eta \ll S_c$, which is satisfied for our probes.

Calculation of the electron-neutral ($\lambda_e$) and ion-neutral ($\lambda_i$) mean free paths give $\lambda_e = 90-340$ cm for the $E \perp B$ case, $\lambda_e = 60-190$ cm for the $E \parallel B$ case,

$\lambda_i = 20$–90 cm for $E \perp B$, and $\lambda_i = 15$–50 cm for $E \parallel B$ case, respectively [6]. The Debye length ($\lambda_D$) ranges from 0.02 to 0.05 cm. That is, $\lambda_i, \lambda_e \gg R_{cy}, R_{pl}$ and $\lambda_D$, where $R_{cy}$ and $R_{pl}$ are the radius of cylindrical and planar probes, respectively. These results indicate that the probe–plasma situation is effectively collisionless. Let us consider the ion and electron Larmor radii $R_{li}$ and $R_{le}$. For $B = 1000$ G, a typical value, we have $R_{li}/R_{pl} = 20$–50 and $R_{li}/R_{cy} = 300$–700, whereas $R_{le}/R_{pl} = 0.06$–0.8 and $R_{le}/R_{cy} = 0.9$–1.2. Therefore, ions are almost unaffected by the field while the electrons could be affected. Hence, the plasma density was derived from the ion saturation current [6].

In determining the plasma density from the ion saturation current, an assumption of a Maxwellian plasma is made, which will introduce a certain error in the determination of the plasma parameters if the electron energy distribution is non-Maxwellian. Therefore, it is important to determine the electron energy distribution of the plasma and compare it to standard distributions such as Maxwellian or Druyvesteyn and consider corrections to the plasma parameters.

## Measurement of the Electron Energy Distribution

The probe characteristics can be employed to measure the electron energy distribution of the plasma. It is useful to know the degree of deviation from Maxwellian, the extent of the high-energy tail, and the value of the average electron energy $E_{av}$. The Druyvesteyn method [7] can be applied for this purpose. However, there are limitations with this method that can be overcome if an electron energy analyzer (EEA) can be applied for this purpose. This will be discussed later.

According to Druyvesteyn's method, the electron energy distribution function (EEDF) $f(E)$ is related to the second derivative $I_p''$ of the probe characteristics through the relation

$$f(E) = \left[ (e^3 S_p N_+)/(8m)^{1/2} \right] I_p''(V_p)(V_s - V_p)^{1/2}, \tag{3.12}$$

where $E = e(V_s - V_p)$. $f(E)$ is calculated from the measured $I_p''$ and calibrated as

$$\sum_0^{E_{max}} f(E)\Delta E = 1, \tag{3.13}$$

where $\Delta E = e\Delta V_p$; $\Delta V_p$ is the probe voltage division. $E_{max} = e(V_s - V_{pm})$ and $V_{pm}$ is defined as the lowest $V_p$ where $I_p''$ reached a noise level when measured with the highest available sensitivity. Experimentally, $I_p''(V_p)$ is obtained by the beat method [8], which can be improved by phase-locked trigger sampling [9]. The average energy $E_{av}$ is obtained from the measured energy distribution $f(E)$ as

$$E_{av} = \sum_{0}^{E_{max}} Ef(E) / \sum_{0}^{E_{max}} f(E). \tag{3.14}$$

$E_{av}$ can be used to obtain the equivalent temperature $T_{eq}$ ($T_e$ when the plasma is Maxwellian) from the relation $\kappa T_{eq} = (2/3)E_{av}$, and using $T_{eq}$, the plasma (ion) density $N_+$ can be found out from Equation 3.8 and the Maxwellian $F_M(E)$ and Druyvesteyn $F_D(E)$ distributions from

$$f_M(E) = \left[ 2\pi E^{1/2} / (\pi \kappa T_{eq})^{3/2} \right] \exp(-E/\kappa T_{eq}), \tag{3.15}$$

and

$$f_D(E) = \left[ 2wE^{1/2} / (E_{av})^{3/2} \right] \exp(-xE^2/E_{av}^2), \tag{3.16}$$

where $w = \Gamma(5/4)^{3/2}/\Gamma(3/4)^{5/2} = 0.519$, $x = \{\Gamma(5/4)/\Gamma(3/4)\}^2 = 0.547$, and $\Gamma$ is the complete Gamma function. The experimentally obtained $f(E)$ can be compared with standard distributions such as the Maxwellian and Druyvesteyn distributions $f_M(E)$ and $f_D(E)$, respectively. One can also employ a numerical technique to obtain the second derivative of the probe characteristics. At first, Savitzky–Golay smoothing is applied to the probe data obtained experimentally. Then $I_P''(V_p)$ can be obtained by numerical differentiation of the smoothed characteristics. However, even a small fluctuation of the measured data of probe $I$–$V$ characteristics can result in large distortion during differentiation and regularization procedures. Moreover, we have found that the electron energy probability function (EEPF) results are limited in energy. The results for Argon using the numerical technique in a pulsed plasma at 1.4 mTorr and peak power 3.2 kW are shown in Figure 3.5a and b.

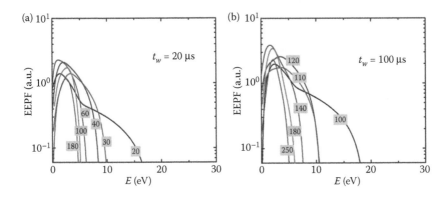

**FIGURE 3.5**
Time evolution of EEPF after the end of the pulse ($t = 0$) for pulse duration, $t_w$ (a) 20 µs, and (b) 100 µs, using Druyvesteyn's method.

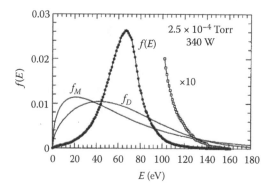

**FIGURE 3.6**

Electron energy distribution function $f(E)$ for the circular tube at a pressure of $2.5 \times 10^{-4}$ Torr and a power of 340 W. $f_M$ and $f_D$ indicate the Maxwellian and Druyvesteyn distribution with the same average energy. (Reprinted with permission from S. Bhattacharjee and H. Amemiya. 1999. Production of microwave plasma in narrow cross-sectional tubes; effect of the shape of cross section, *Rev. Sci. Instrum.* **70**: 3332. Copyright 1999, American Institute of Physics.)

Also, a result using the *experimental* beat method is shown in Figure 3.6 for a plasma in a multicusp magnetic field at a pressure of $2.5 \times 10^{-4}$ Torr and a power of 340 W.

## Retarding Field Analyzer Probe (or Electron Energy Analyzer)

The electron and ion energy distribution functions are extremely important parameters of a discharge, and indicate how the delivered power is coupled with the charged particles. The EEDF helps in understanding the wave–plasma interaction, especially in microwave plasmas [10], and is the parameter of interest in the investigations. The EEDF may be obtained by analyzing the Langmuir probe characteristics by the Druyvesteyn method, which gives the average EEDF over all directions [10–13]. However, it might be interesting to know the EEDF for electrons that have their velocities parallel or perpendicular to the axis of the MC. In that case, a retarding field analyzer probe must be used, which restricts the measurement to one direction [14–16]. The design of the EEA probe for our system is based on the works of Bohm et al. [15], Aanesland et al. [14], and Amemiya et al. [16].

Figure 3.7a shows the schematic of the EEA probe employed in the system. It consists of four electrodes of outer diameter 7 mm, enclosed by an aluminum jacket of outer diameter ~12 mm. The electrodes are separated by machinable ceramic (macor) rings of 1 mm thickness with tiny holes in them to carry out the wires from the jacket to the 5.5 mm stainless-steel pipe, welded to the jacket.

The outer jacket has a 2-mm orifice through which plasma penetrates up to the first electrode, which has an orifice of 1 mm. It is covered by a fine stainless-steel mesh of 66% transparency [14], having wire spacing of 200 μm and wire diameter of 75 μm. Since the wire spacing is less than the Debye length,

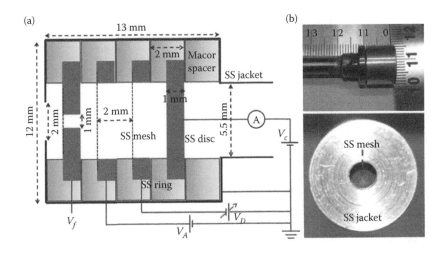

**FIGURE 3.7**
(a) Schematic of the EEA showing the biasing. (b) Longitudinal (top) and cross-section (bottom) view of the fabricated EEA.

a nearly uniform sheath forms on the mesh near the orifice. This electrode is floated so that it acquires the floating potential $V_f$ (usually negative) of the plasma. The second electrode is held at a positive potential of $V_A \sim 40$ V ($>V_s$), which attracts the electrons by a potential difference of ($V_A - V_f$) and repels the ions. The third electrode acts as a discriminator, and is swept between 0 and $-100$ V to retard the electrons so that only electrons having energy greater than the discriminator voltage ($V_D$) can pass through it. The energy-discriminated electrons are collected by the last electrode held at $V_c \sim 40$ V, and the current ($I_c$) of the order of a few $\mu A$ is measured by a GPIB-interfaced Keithley 2001 multimeter, and stored in a PC. The second and third electrodes are in the form of rings with inner diameters of 3 mm, with the mesh welded to them as shown in Figure 3.7. The meshes are primarily used to maintain uniform and curvature-free equipotentials between the electrodes, so that there is no radial component of the field [13–15] that may cause unequal acceleration of the electrons. It also helps in reducing the electrons in the beam so that space charge repulsion effect, and hence the beam spread, is minimized [13–16]. Figure 3.7b shows a digital photograph of the fabricated probe.

Figure 3.8 shows a typical $I$–$V$ characteristic obtained in the Ar plasma at 0.35 mTorr, 180 W. The characteristic is fitted with a smooth Boltzmann fit, which is then differentiated to obtain ($dI_c/dV_D$) versus $V_D$ plot, which is proportional to the electron energy distribution variation [13–16].

## Ion Energy Analyzer Probe

An electrostatic ion energy analyzer (IEA) probe is used to extract information regarding the ion energy distribution and ion temperature of a plasma

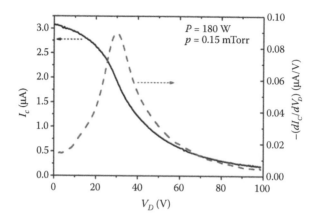

**FIGURE 3.8**
Collector current ($I_C$) versus discriminator bias ($V_D$) (solid black line). The derivative is shown by the dashed line in the same plot.

source [17–23]. The construction is similar to an EEA and consists of a series of grids maintained at various potentials. The grids separate out the electron and ion components from the plasma, so that all the electrons are cut off and ions with sufficient energy will be collected [17,21]. The design of IEA probes for our system is based on the works of Bohm and Perrin [15] and Aanesland and Fredriksen [14]. The schematic of the probe with the electrical connections is shown in Figure 3.9.

The probe essentially consists of a grounded front plate (P), three metal grids ($G_1$–$G_3$), and a collector (C). The aperture of P is 0.8 mm and is covered with a fine mesh with 60 lines/cm. All the grids are made of the same mesh.

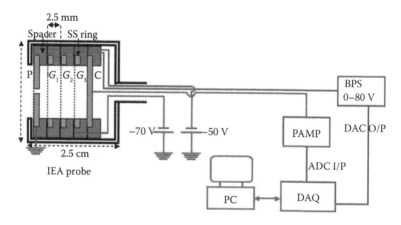

**FIGURE 3.9**
Schematic of ion energy analyzer probe along with the circuit used for the probe measurement.

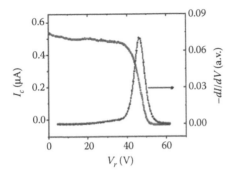

**FIGURE 3.10**
A typical *I–V* characteristics for IEA probe obtained in a microwave-generated Ar plasma in one of our experiments at 0.25 mTorr and 200 W power. The derivative plot of the *I–V* characteristics is also shown.

The mesh is spot-welded on SS rings of 1-mm thickness, and they are insulated by machinable ceramic (Macor) spacers. The total geometric transmission for the grid is given by $\delta_T = (l - d)/l^2 = 66\%$, where $l$ is the line spacing and $d$ is the wire diameter.

The front plate along with the probe body is grounded and is in contact with the plasma. Hence the sheath that forms in front of P accelerates the ions and decelerates the elections. Grid $G_1$ is slightly biased to the negative to repel the electrons that could enter the probe. The ion discriminator grid, $G_2$, is swept from 0 to 80 V. Only the ions with axial energy greater than the applied energy pass through $G_2$. A constant negative bias is applied to the third grid, $G_3$, again to repel any more electrons that are able to pass through the potential barrier of P and $G_1$. The collector, C, is held at a higher potential than $G_3$ for reducing the secondary electron contribution. The ion current at C, $I_c$, is measured as a function of the sweep potential $V_r$.

Figure 3.9 also shows the schematic of the circuit used for probe measurement. The sweep potential, $V_r$, is given by BPS using the DAC or DAQ. The collector current, $I_c$ (in microamperes), is taken to the current preamplifier (PAMP; SRS SR570) and the amplified voltage signal is fed to the input of the ADC in DAQ. Figure 3.10 shows a typical ion energy distribution obtained by taking the derivative of the current–voltage characteristics.

## Design of Electromagnetic Probes

Measurement of electromagnetic wave field intensity in the plasma is carried out by the application of thin "pick-up" antenna probes so that the ambient plasma and the permeating wave are least disturbed [11,24,25]. These probes

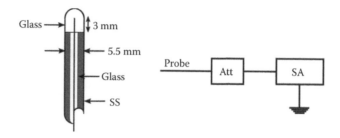

**FIGURE 3.11**
Schematic of electric field probe along with the circuit. Att, attenuator; SA, spectrum analyzer.

have a small active area (area in which the oscillating electric or magnetic field may induce voltage) compared to the incident wavelength. This ensures detailed spatial profiling, and also utilizes a very small fraction of the wave power. This is in contrast to receiver or transmitter antennas (quarter-wave, dipole, etc.), which are designed to actually absorb or emit as much of the wave power as possible, and are also larger in dimension [25]. The probes as shown in Figure 3.11, and a digital photograph is shown in Figure 3.12, are constructed with high-grade micro-coax cables having low loss up to 18 GHz; the shielding and the dielectric are removed from a small portion, ~2–3 mm to use it as the active area for construction of an electric field probe [11,24] for the measurement of the wave electric field, or for a B-dot probe [26,27] for the measurement of the wave magnetic field. The active area is enclosed within a quartz bulb to isolate it from the plasma, and the rest of the coaxial cable is enclosed within a thin (~3 mm OD) stainless-steel pipe that is attached to a 5.5 mm OD pipe by an adapter, and vacuum-sealed. The other end of the coax is terminated in a vacuum-compatible 50 Ω BNC female connector,

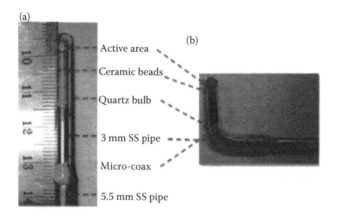

**FIGURE 3.12**
Digital snap of a (a) linear antenna probe and (b) bent antenna probe showing the dimensions and assembly components.

from where the detected signal is carried by a microwave-grade BNC cable to spectrum analyzer (SA), via a 30 dB attenuator (Att), all having matched impedances of 50 $\Omega$ at their terminals. The SA measures the power dissipated in an equivalent matched impedance of 50 $\Omega$ by the voltage induced in the active element of the probe, as a function of frequency. From the obtained spectrum, the peak power corresponding to 2.45 GHz is recorded. Most of the experiments are performed with an Agilent N1996A (0–3 GHz) SA, while, later on, a more sophisticated Agilent E4408B (0–26 GHz) SA is used.

## Wave Electric Field Probe

In an electric field probe, from the definition of electric potential ($\Phi$) corresponding to an electric field ($\vec{E} = -\vec{\nabla}\Phi$), the oscillating wave electric field induces a voltage in the exposed element of the micro-coax, which is given by $\Phi_E = -\int_0^L \tilde{\vec{E}} \cdot \vec{dl}$, where $\Phi_E$ is the induced voltage due to the oscillating electric field $\tilde{\vec{E}}$ acting on a conductor of length $L$. The electric field intensity designated by $\langle|E|^2\rangle$ is then proportional to the power dissipated, that is, $\Phi_E^2/Z$, where $Z$ is the equivalent matched load ($Z$ = input impedance of SA ~50 $\Omega$).

Since a conductor will respond to the electric field that is tangential to its surface, a small spherical metallic blob coated with silver paste is used as the tip of the active element so that it picks up field oscillating in any direction, and hence the total electric field intensity $\langle|E_{tot}|^2\rangle$ at any point is recorded. Figure 3.12a shows the picture of a linear antenna probe for axial and radial measurement, and Figure 3.12b shows a bent antenna probe used for polar measurement of the electric field intensity.

## B-Dot Probe

The B-dot probe operates on the principle of Faraday's law, where an oscillating magnetic flux enclosed by a conducting loop induces an oscillating electromotive force (EMF) in the loop given by

$$\Phi_B = -\frac{\partial}{\partial t}\left[\int_S \tilde{\vec{B}} \cdot \vec{dS}\right] = -\omega N S \tilde{B}.$$

Here, $\omega$ is the frequency of the wave field, $S$ is the area enclosed by the loop, $N$ is the number of turns of the loop, and $\tilde{B}$ is the component of the oscillating magnetic field, perpendicular to the plane of the loop. Depending on the orientation of the loop with respect to the axis of the MC, wave magnetic field intensity $\langle|B_{r,\theta,z}|^2\rangle \propto \Phi^2_{B_r, B_\theta, B_z}/Z$ corresponding to the $r$, $\theta$, and $z$ components of the wave magnetic field can be measured. The loop is constructed by 4-turn coiling of a 0.2-mm-diameter enameled copper wire.

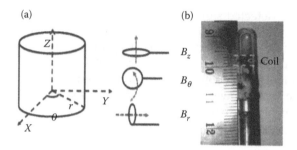

**FIGURE 3.13**
(a) Schematic showing the loops orientation scheme with respect to the multicusp. (b) Picture of a fabricated B-dot probe.

Initial measurements are performed with a 3-mm-diameter coil, and for advanced experiments a 1-mm coil is used. One end of the coil is soldered to the 2-mm exposed micro-coax, while the other end is wrapped around the exposed portion tightly to prevent pickup of the electric field, and then soldered to the outer ground shielding of the micro-coax. Figure 3.13a shows the schematic of the loop orientations with respect to the MC for measurement of $\langle |B_z|^2 \rangle$, $\langle |B_\theta|^2 \rangle$, and $\langle |B_r|^2 \rangle$, and Figure 3.13b gives a digital photograph of one of the fabricated probes used for measuring the radial variation of $\langle |B_z|^2 \rangle$ and $\langle |B_\theta|^2 \rangle$.

## References

1. J.D. Swift and M.J.R. Schwar. 1969. *Electric Probes for Plasma Diagnostics* (Elsevier, New York).
2. P.M. Chung, L. Talbot, and K.J. Touryan. 1975. *Electric Probes in Stationary and Flowing Plasmas* (Springer, Berlin).
3. F.F. Chen. 1965. *Plasma Diagnostic Techniques* R.H. Huddlestone, and S.L. Leonard (editors), (Academic, New York).
4. J.E. Heidenreich III, J.R. Paraszczak, M. Moisan, and G. Sauve. 1987. Electrostatic probe analysis of microwave plasmas used for polymer etching, *J. Vac. Sci. Technol. B* **5**: 347–354.
5. R.M. Clements. 1978. Plasma diagnostics with electric probes, *J. Vac. Sci. Technol.* **15**: 193–198.
6. S. Bhattacharjee and H. Amemiya. 1997. Transversely magnetized microwave plasma ion a waveguide with a dimension below cutoff, *Rev. Sci. Instrum.* **68**: 3061–3067.
7. M.J. Druyvesteyn. 1930. Der Niedervoltbogen, *Z. Phys. A: Hadrons Nuclei* **64**: 781–798.
8. K. Wiesemann. 1969. Der Einfluß einer Blende auf die Verteilungsfunktion der Elektronen in einem Gasentladungsplasma, I. Die Verteilungsfunktion der

Elektronen hinter einem Potentialsprung (Theorie), *Annalen Der Physik* **478**: 104–112.

9. H. Amemiya, K. Shimizu, S. Kato, and Y. Sakamoto. 1988. Measurements of energy distributions in ECR plasma, *Jpn. J. Appl. Phys.* **27**: 927–930.

10. V.A. Godyak, R.B. Piejak, and B.M. Alexandrovich. 2002. Electron energy distribution function measurements and plasma parameters in inductively coupled argon plasma, *Plasma Sources Sci. Technol.* **11**: 525–543.

11. S. Bhattacharjee and H. Amemiya. 2000. Production of pulsed microwave plasma in a tube with a radius below the cut-off value, *J. Phys. D: Appl. Phys.* **33**: 1104–1116.

12. B. Lipschultz, I. Hutchinson, B. LaBombard, and A. Wan. 1986. Electric probes in plasmas, *J. Vac. Sci. Technol. A* **4**: 1810–1815.

13. I.H. Hutchinson. 2002. *Principles of Plasma Diagnostics* 2nd edition (Cambridge University Press, Cambridge) pp. 55–98.

14. A. Aanesland and A. Fredriksen. 2001. Pressure dependent mode transition in an electron cyclotron resonance plasma discharge, *J. Vac. Sci. Technol. A* **19**: 2446–2452.

15. C. Bohm and J. Perrin. 1993. Retarding-field analyzer for measurements of ion energy distributions and secondary electron emission coefficients in low pressure radio frequency discharges, *Rev. Sci. Instrum.* **64**: 31–44.

16. H. Amemiya, K. Shimizu, S. Kato, and Y. Sakamoto. 1998. Measurements of energy distributions in ECR plasma, *Jpn. J. Appl. Phys.* **27**: L927–L930.

17. I.H. Hutchinson. 1987. *Principles of Plasma Diagnostics* (Cambridge University Press, UK).

18. J.E. Allen. 1974. *Plasma Physics*, B.E. Keen (editor), (Institute of Physics, London and Bristol).

19. P.M. Chung, L. Talbot, and K.J. Touryan. 1975. *Electric Probes in Stationary and Flowing Plasmas* (Springer-Verlag, New York).

20. V.I. Demidov, S.V. Ratynskaia, and K. Rypdal. 2002. Electric probes for plasmas: The link between theory and instrument, *Rev. Sci. Instrum.* **73**: 3409–3439.

21. R.L. Stenzel, R. Williams, R. Agüero, K. Kitazaki, A. ling, T. McDonald, and J. Spitzer. 1982. Novel directional ion energy analyzer, *Rev. Sci. Instrum.* **53**: 1027–1031.

22. M. Sugawara. 1998. *Plasma Etching Fundamentals and Applications* (Oxford University Press, Oxford, England).

23. H.E.M. Peres and F.J. Ramirez-Fernandez. 1997. High resistivity silicon layers obtained by hydrogen ion implantation, *Braz. J. Phys.* **27A**: 237–239.

24. T.H. Stix. 1990. Waves in plasmas: Highlights from the past and present, *Phys. Fluids B* **2**: 1729–1743.

25. S. Silver. 1997. *Microwave Antenna Theory and Design* (Peter Peregrinus Ltd. On behalf of IEE, London).

26. R.C. Phillips and E.B. Turner. 1965. Construction and calibration technique of high frequency magnetic probes, *Rev. Sci. Instrum.* **36**: 1822–1825.

27. G.S. Eom, G.C. Kwon, I.D. Bae, G. Cho, and W. Choe. 2001. Heterodyne wave number measurement using a double B-dot probe, *Rev. Sci. Instrum.* **72**: 410–412.

28. S. Bhattacharjee and H. Amemiya. 1999. Production of microwave plasma in narrow cross-sectional tubes; effect of the shape of cross section, *Rev. Sci. Instrum.* **70**: 3332.

# 4

## Genesis of a Wave Induced Discharge: A New Perspective

### Introduction

With increased use of intense short-pulsed lasers or high-frequency electromagnetic waves (EWs) for basic and applied plasma research, an understanding of the predischarge electron dynamics has become crucial, and for this a knowledge of the dependence of $N$ over $\Lambda/\lambda$ in the presence of the high-frequency EWs and magnetostatic ($B$) fields (in case of magnetically assisted discharges) is required, where $N$ is the number of collisions suffered by an electron in diffusing out of the container, $\lambda$ is the mean-free path, and $\Lambda$ is the characteristic diffusion length. This is essentially a problem where the classical field-free random walk needs to be evaluated in the presence of constraints imposed on the electron motion by the EW and $B$ fields. We show that there is an intricate connection of the physics of this problem to the problem of finding out the plasma initiation time (breakdown time) and critical threshold electric fields [1–8].

One such example belongs to a recent observation of quasi-steady-state interpulse plasmas realized by the application of short-pulse (~1 μs pulse width), high-power (~100 kW) microwaves [1], where the dynamics of the initial seed electrons is considered important for predicting plasma growth. The electron random walk in a gaseous medium is a subject of considerable interest in many other branches of physics as well, such as wave transmission through gas-filled *rf* cavities employed in particle accelerators; high-power microwave (HPM) window breakdown at high pressures (a major limiting factor in radiating HPM); and microwave plasma interaction, including propagation of powerful EWs through the atmosphere where they lead to the formation of artificially ionized layers, which is of significance in physics studies of the upper atmosphere. At high pressures, the collision of electrons with the background gas in the presence of EWs can significantly alter a multipactor discharge, or even render it irrelevant, thereby making it similar to a volume breakdown process. The coupled phenomena of electron random walk, energy gain, and, in particular, knowledge of the predischarge electron dynamics and breakdown time are important in the aforesaid physical processes and applications.

The objective is to investigate the process of electron transport in a background gas, in the presence of EWs in the microwave regime, before the discharge fully develops by ionization processes. First-time comparisons are made with the classical random walk in the absence of electromagnetic fields, and a new dependence of $N$ with $\Lambda/\lambda$ is found. To investigate the purely statistical process, a Monte Carlo (MC) simulation technique is employed, which considers scattering and randomization of the phase of individual electrons, electron–neutral collisions, and the effect of wave amplitude and frequency. The actual experimental cross-section of the test gas (argon) is taken into account.

## Historical Development of Breakdown Studies

Brown and MacDonald [7,8] employed the classical relation for explaining optimum breakdown in high-frequency gas discharges ($uN = U_{ins}$, where $u$ is the energy gained per electron per collision and $U_{ins}$ is the ionization energy); however, the "field-free" relation explains experimental breakdown curves over a limited region (low pressure side of the Paschen-like breakdown curve). An accurate and consistent prediction of critical threshold fields requires knowledge of the dependence of $N$ over $\Lambda/\lambda$ in the presence of electromagnetic and magnetostatic fields for magnetically assisted discharges.

Historically, the determination of breakdown time received numerous interpretations both in theory and experiments [2–7]. The Boltzmann equation was expanded by spherical functions in the mean-free path and moderate amplitude limits, providing a regime of breakdown for weak to moderate fields [8]. An analytical criterion with particle balance equations is more related to the discharge maintenance criterion because it assumes that breakdown and steady state occur almost at the same time [9]. In pulsed discharges, the use of averaged fields has some limits of validity because the statistical phenomenon is smeared out. Moreover, breakdown actually occurs much earlier than the time taken to reach the critical (cutoff) density ($\omega = \omega_p$) [4,5].

In DC discharges, the exponential growth criterion like the "Townsend criterion" is used for obtaining breakdown time, where the number of electrons multiply by an arbitrary factor say, $10^8$ from its initial value, that is, $n = n_0 e^{\langle v_i \rangle t}$, where $n_0$ is the initial density and $<v_i>$ is the average ionization rate [9]. In experiments, often the output of an optical signal (flashover) is employed to identify the breakdown time [10]. In reality, a sufficient density of electrons is required to produce a detectable level of optical output, implying that breakdown has already occurred much earlier in time. Therefore, despite several works on the subject, the interpretation of the breakdown time has remained debatable. Thus, a unified definition of the breakdown criterion and knowledge of the breakdown time is desirable. A unified approach requires the knowledge of the electron collision processes with the neutrals in the presence of the high-frequency electromagnetic (EM) waves.

The objective of the research is twofold: (i) to determine an empirical relationship between $N$, $\lambda$, and $\Lambda$ in the presence of polarized EW. A field-free square law was obtained by Kennard as $N \propto (\Lambda/\lambda)^2$. In this work, we determine how $N$ is modified in the presence of EW; (ii) we demonstrate that the solution to the random walk problem (obtained in (i)) provides a unified approach to the determination of the breakdown time and breakdown electric fields in both the high- and low-pressure regions of the Paschen-like breakdown curve. (iii) Thereafter, we extend the investigation to magnetically assisted wave-induced discharges.

## Modeling of the Electron Dynamics

The dynamics of a classical electron (point charge) in space can be described by Newton's second law, namely

$$m_e \frac{d^2\vec{r}}{dt^2} = \vec{F}_{tot},$$
(4.1)

where $m_e$ is the electron mass, $\vec{r} = x\hat{i} + y\hat{j} + z\hat{k}$ is the position vector at any instant of time $t$ defined with respect to a right-handed coordinate system shown in Figure 4.1, and $\vec{F}_{tot}$ is the total force acting on the electron, which may include electrostatic, magnetostatic, electromagnetic waves, thermal gradient, or pressure gradient forces. For an ensemble of electrons moving in the background of neutral atoms, collisional interaction comes into the

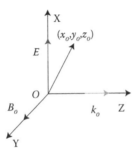

**FIGURE 4.1**
Reference coordinate system.

picture. The dynamics of such an ensemble of electrons is conventionally addressed in the kinetic theory with the Boltzmann transport equation, which incorporates the collision term [11]. Particle number and momentum averaging of the Boltzmann equation lead to the fluid equation, where the collisions occur as the Langevin drag force [11,12]. In this work, in contrast to the fluid-average method, a discrete approach is taken where the motion of each electron in the ensemble is followed by solving Equation 4.1 numerically, where the phenomenon of random collisions is incorporated explicitly by considering actual random collisions of the electrons with neutrals by employing MC methods [13,14]. All the electron collision and ionization cross-sections are taken into account [15,16].

A number of seed electrons $n$ is taken to be present initially in a certain volume of the test gas (argon). These electrons may arise from cosmic rays or other stray electron and radiation sources and usually lie in the range $10^3$–$10^4$ per $cm^3$. Typically, a spherical volume defined by a radius $\Lambda = 1.5$ cm is taken as the test volume, estimated from the characteristic diffusion length $[1/\Lambda^2 = (2.4/a)^2 + (\pi/L)^2]$ [15,17,18] calculated for a cylindrical chamber of radius, $a = 36$ mm, and length, $L = 300$ mm; $n$ is determined to be ~$10^4$.

The electrons and neutrals follow the assumptions of kinetic theory of gases, and obey Maxwell's velocity and energy distribution laws. The ratio of electron mass ($m_e$) to neutral (Ar) mass ($M_n$) is $m_e/M_n$ ~ $10^{-4}$; therefore, the energy loss in the elastic collision process is ~$10^{-2}$ times smaller compared to the thermal energy (~0.03 eV). At the same energy, the velocity ratio of an electron ($v_e$) to a neutral ($v_n$) is $v_e/v_n$ ~ $10^3$ hence, in the timescale of the simulation, which is of the order of a few nanoseconds, the neutrals do not move appreciably from their positions as compared to the electrons and may be considered a static background. An estimate of the internal electric field due to the $10^4$ randomly distributed electrons is calculated at $10^4$ random field points in the volume of interest, and the most probable field ($E_{int}$) is determined to be ~$3 \times 10^{-4}$ V/cm. With the magnitude of this field being much smaller than the typical electric field, ~$10^3$ V/cm, applied in experiments, the space-charge effect due to the electrons is neglected.

The electrons ionize as they gain energy from the waves, and the density of electrons as well as ions increases. The increment in the number of electrons is exactly the same as the increment in the number of ions, so the most probable electric field does not change, but the ions do not respond to the EM wave because of their heavy mass and the electrons shield out the applied wave electric field. The above assumptions would become invalid when the electron density reaches a point where the shielding starts to become effective. Until the initiation time, the charged particle density is of the order of $10^6$–$10^7$ $cm^{-3}$. A calculation for the skin depth of the EM wave in the volume indicates that, for the growing electron density, at a density ~$10^9$ $cm^{-3}$, the wave amplitude is damped by about 10%. Therefore, it may be said that an electron density of ~$10^9$ $cm^{-3}$ may be taken as the limiting density when the assumptions of the kinetic theory for low-density gases does not remain valid and the space-charge effects must be taken into account. However, since the regime of interest in the current model is limited to the predischarge regime, where the number density of the charged particles is low, that is, $n_0 \ll 10^9$ cm$^{-3}$, the effect of the space charge is neglected, and the electrons can be treated as point particles that follow the assumptions of the kinetic theory for low-density gases.

The process of collision between the electrons and neutrals is implemented by considering their collision cross-sections as a function of electron energy [15]. The mean-free path ($\lambda$), which an average electron traverses between successive collisions, is given by the relation $\lambda = 1/n_g\sigma$, where $n_g$ is the neutral gas density given by the relation $p = n_g k_B T$; $p$ is the neutral gas pressure, $T \sim 298$ K (room temperature); $k_B$ is *Boltzmann's* constant; and $\sigma$ is the electron–neutral collision cross-section and depends on the energy $U$ of the electrons. The variation of $\sigma$ with energy $U$ [15,16] for argon is given in Figure 4.2. Vahedi and Surendra [19] considered differential cross-section and energy-dependent scattering angles to model collisional plasmas and self-sustained discharges. However, such models can lead to the overestimation of the collision cross-section at low energies (~5 eV), as observed by Hans Rau [13]. Since the electron energy in the current model does not go to very high numbers, as will be seen later, an isotropic cross-section has been assumed and implemented from data available in the literature [16]. The interelectron collisions are neglected due to their lower collision cross-section compared to the neutrals. Also, since the density of electrons and ions are quite small, the chances of electron–electron, and electron–ion collisions are negligible compared to electron–neutral collisions.

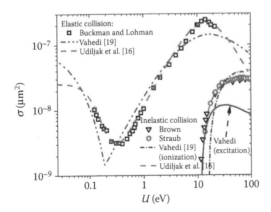

**FIGURE 4.2**
Experimental variation of elastic and inelastic collision cross-sections with electron energy for argon [16,17]. (Reprinted with permission from I. Dey et al. 2008. Sub-nanosecond electron transport in a gas in the presence of polarized electromagnetic waves, *J. Appl. Phys.* **103**: 083305–083306. Copyright 2008, American Institute of Physics.)

The electrons gain energy from the waves until the energy growth starts to saturate, as it crosses the elastic limit and inelastic collisions start to dominate. During the course of its motion, the electron satisfies any one of the loss criterion described below. The electrons may gain energy and

a. Escape from the volume purely by elastic collisions.
b. They may undergo inelastic collisions after traveling a certain distance, in the volume determined by $r < \Lambda$.

In case (a), the electrons escape from the volume purely by elastic collisions, and the variation of $N$ with $\Lambda/\lambda$ is readily obtained. In case (b), the electrons undergoing an inelastic process (viz., excitation or ionization) satisfy the following energy conservation relations:

i. Excitation: $U_{pe} + U_{NA} \cong U'_{pe} + U^*_{NA}$
ii. Ionization: $U_{pe} + U_{NA} \cong U''_{pe} + U_{ee} + U_{ion} - U_{inz}$,

where $U_{pe}$, $U_{NA}$, $U'_{pe}$, $U^*_{NA}$, $U_{ee}$, $U''_{pe}$, $U_{ion}$, and $U_{inz}$ are the energies of the primary electron, neutral atom (negligible), slow primary electron after inducing excitation, excited neutral atom, slow-ejected electron, slow primary electron after inducing ionization, ion (negligible), and the ionization energy, respectively. In both cases, the excitation or ionization energy is subtracted from the energy of the primary electron and the slow electron(s) is allowed to undergo the process of elastic collision and energy gain all over again, and the value of $N$ is determined as described earlier. The variation of $N$ with $\Lambda/\lambda$ is not affected by the increase in electron number because it is determined by a statistical average method over all the involved electrons.

In the simulation, the implementation of $\lambda$ is achieved by considering a randomized Gaussian spread with $w \sim 10\%$ in the value of $\lambda$ calculated at a particular neutral density and electron energy, to incorporate the fact that the electron may travel any length of free path from 0 to $\infty$, which when averaged over a large number of collisions will give the mean-free path $\lambda$ at that energy and density. A collision is counted if the distance traveled by the electron exceeds the free path determined from the randomized Gaussian distribution [17,18]. This mean-free path is compared with the one determined by Hans Rau [13] with the null method and no commendable difference was found.

## Development of the General Algorithm

The outline of the algorithm of the program for studying the predischarge electron dynamics is as follows. The program starts at some particular pressure, with one electron out of the $10^4$ seed electrons available in the spherical test volume of radius $\Lambda = 1.5$ cm. The initial position $(x_0, y_0, z_0)$ of the electron is chosen arbitrarily by random numbers in the volume such that $x_0^2 + y_0^2 + z_0^2 \leq \Lambda^2$. In the initial calculations [17], the electrons were assumed to be monoenergetic, having a kinetic energy (0.03 eV) corresponding to the room temperature (~300 K). The electrons were allowed to evolve with velocity (~9.5 × 10⁴ m/s) corresponding to this energy. Bhattacharjee et al. [18] improved the model further by considering the electrons to have a Maxwell–Boltzmann distribution of energies, with the mean energy corresponding to room temperature. It may be noted that during the birth of a seed electron, it might have part of the energy corresponding to the process of creation, that is, either by photons or by any other particle or radiation. However, the created electrons equilibrate in the ambient gas, which is at room temperature, and as a result of collisions have a distribution of velocities with the mean at room temperature, which is taken as the starting distribution.

The electron, after being designated a random origin $\langle x_0, y_0, z_0 \rangle$, is then randomly designated an initial velocity $v_0$ utilizing a Maxwell–Boltzmann velocity distribution function given by [18]

$$Fdv = \frac{dN_v}{N} = 4\pi \left( \frac{m_e^2}{2\pi k_B T} \right)^{3/2} e^{-m_e v^2/2k_B T} v^2 dv, \qquad (4.2)$$

where $Fdv$ is the probability that the velocity of an electron lies between $v$ and $v + dv$, $N$ is the total number of electrons, $dN_v$ is the number of electrons having velocity between $v$ and $v + dv$, $T$ is the equilibrium temperature, and $k_B$ is the Boltzmann constant, with the electron mean energy $(3k_B T/2)$ taken as room temperature ($T = 300$ K). Since all directions are equally probable, the velocity components $v_x$, $v_y$, and $v_z$ are designated arbitrary directions using random numbers with the magnitude constraint $v_x^2 + v_y^2 + v_z^2 = v_0^2$.

The electron is then allowed to evolve with time, in steps of $dt = 0.001$ ns, subject to the total external force of $\vec{F}_{tot}$ (Equation 4.1). Three conditions are investigated depending on the nature of $\vec{F}_{tot}$, namely, (i) field-free case, (ii) high-frequency EM field, and (iii) high-frequency EM field with its wave vector perpendicular to a static magnetic field and with electric field polarization (a) parallel and (b) perpendicular to the magnetostatic field. Depending on the nature of the force field, the electron evolves in space and may collide elastically or inelastically (in the presence of the EM field) with the neutrals in the background. The frequency of collision is decided by the free path obtained from the mean-free path $\lambda(p, U)$ as described above. On each collision, the velocity directions are randomized and the electron motion is allowed to continue until the electron meets an exit criterion, like escape from the test volume. Parameters like the average number of collisions for the 10⁴ electrons ($N$), mean-free path traveled, energy gained (in case of EM field), and so on, are recorded. Details of the studies are presented in the subsequent sections.

From the above description of the algorithm formulation, it is evident that the stochastic nature of the processes involved must be implemented properly with the use of random numbers, so that the generated random variables are uniformly distributed without any bias. The most important parameter that needs uniform randomization is the initial velocity and subsequent velocity of the electrons in the event of elastic or inelastic collisions. Initially [18], the velocity randomization was implemented by considering the components to be given by $v_x = v\sin\theta\cos\phi, v_y = v\sin\theta\sin\phi$, and $v_z = v\cos\theta$, where $v = \sqrt{v_x^2 + v_y^2 + v_z^2}$ is the magnitude of either the initial velocity obtained from the Maxwell–Boltzmann distribution or the velocity obtained from the time evolution of the electron at the instant of collision. The azimuthal angle $\theta(0 \le \theta \le \pi)$ and the polar angle $\phi(0 \le \phi \le \pi)$ were randomly chosen by two independent random numbers $r_1$ and $r_2$, such that $\theta = r_1\pi$ and $\phi = 2r_2\pi$, with $r_1$ and $r_2$ having uniform distribution from 0 to 1. However, it was later observed that this does not result in a uniform distribution of $v_x$, $v_y$, and $v_z$ in the velocity space. Bhattacharjee et al. [18] rectified the above shortcoming by modifying the velocity components to $v_z = (2r_1 - 1)v$, $v_x = v\sqrt{1 - (2r_1 - 1)^2}\cos\phi$, and $v_y = v\sqrt{1 - (2r_1 - 1)^2}\sin\phi$, where $0 \le r_1 \le 1$ and $\phi$ is implemented by $r_2$ as above. In this case, the distribution was found to be more uniform. The initial position of the electron was implemented in a similar manner, namely, $z_0 = (2r_1 - 1)\Lambda$, $x_0 = \Lambda\sqrt{1 - (2r_1 - 1)^2}\cos\phi$, and $y_0 = \Lambda\sqrt{1 - (2r_1 - 1)^2}\sin\phi$, where $\Lambda$ is the radius of the sphere defined by the characteristic diffusion length. The histogram plots in Figure 4.3a and b show the difference in distribution between the old and the new formulation of the randomly designated velocity components.

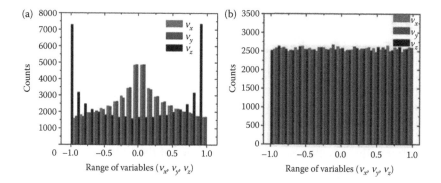

**FIGURE 4.3**
Histogram plot of the normalized velocity components generated by random numbers using (a) old formulation (Reprinted with permission from I. Dey et al. 2008. Sub-nanosecond electron transport in a gas in the presence of polarized electromagnetic waves, *J. Appl. Phys.* **103**: 083305–083306. Copyright 2008, American Institute of Physics.) and (b) formulation by Bhattacharjee et al. (S. Bhattacharjee, I. Dey, and S. Paul. 2008. *Physics of Plasmas* **20**: 042118. Copyright 2013, American Institute of Physics.)

The algorithm was initially implemented by using an Euler method [17,18], which is easy to implement and leads to fast-running algorithm. However, the discretization and round-off errors accumulate, and lead to large errors in the computed output. In contrast, the fourth-order Runge–Kutta (RK4) method is more accurate, but requires careful implementation and a slower-running algorithm since the number of discretization steps increases.

The difference between the accuracy of the two methods is demonstrated below in Figure 4.4a and b with two phase plots, wherein the motion of an electron is evaluated in the presence of an oscillating electric field, without any collisions. The phase plot for the RK4 method (Figure 4.4a) is a closed ellipse, as is expected for the motion of the electron moving only under the influence of an oscillating electric field, since the electron gains no energy on an average. In contrast, the phase plot for the Euler method (Figure 4.4b) is not closed, and violates the conservation of energy. It may be noted that initially (at small time durations) the difference between the methods is small, but increases rapidly with time.

The formulated algorithm is implemented in FORTRAN 90 and compiled using the gfortran compiler. RK4 method is used to solve the differential equation for the electron. The MC method is implemented by using the internal random number generator of gfortran, which generates uniformly distributed random numbers. The time-step size used is 0.001 ns $\ll$ 0.41 ns, the time period corresponding to 2.45 GHz EM waves. The time step is very small to increase the accuracy of the result and to study the electron motion in fine steps, under the influence of various fields (electromagnetic, magnetostatic). The roundoff error is reduced by using double precision variables.

The following sections describe the physics behind the algorithm formulation under various conditions of the external force $\vec{F}_{tot}$, where all the improvements and considerations discussed above, namely, (i) initial velocity designation by the Maxwell–Boltzmann velocity distribution, (ii) uniform

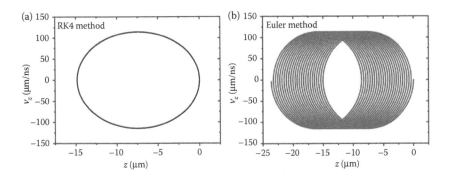

**FIGURE 4.4**
Velocity ($v_z$) versus displacement ($z$) phase plots for (a) the RK4 method and (b) the Euler method for the solution of the equation of motion of an electron under the influence of an oscillating electric field in absence of collisions.

randomization in the velocity space, and (iii) RK4 method with small time steps, have been implemented.

### Field-Free Case: Three-Dimensional Random Walk

In the conventional field-free problem, $\vec{F}_{tot} = 0$ and the electrons perform three-dimensional random walk freely in the volume consisting of neutral atoms (or molecules) under equilibrium conditions, where there are no gradients of density or temperature. The total external force $\vec{F}_{tot} = 0$ in this case. The motion of the electrons is dictated by their initial velocity obtained from a Maxwell–Boltzmann velocity distribution (Equation 4.2). The electrons do not gain any energy in this case and the collisions are elastic.

The variation of number of collisions $N$ with $\Lambda/\lambda$ averaged over $10^4$ electrons is studied in the simulation by varying $\lambda$ (by changing the gas pressure $p$), and $\sigma$ is obtained from the elastic electron–neutral collision cross-section data, shown in Figure 4.2, for the energy corresponding to the velocity of the electron. The electrons are followed until they escape from the test volume. The total number of collisions and the average of the free paths (i.e., the mean-free path) that the electron travels before leaving the volume are recorded. The variation is shown in Figure 4.5, where the conventional field-free variation, $N \propto (\Lambda/\lambda)^2$, as obtained by Kennard, is reproduced [20]. The fitting equation is given by

$$N = A(\Lambda/\lambda)^2 = 1.22. \tag{4.3}$$

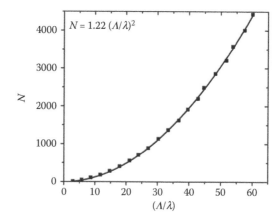

**FIGURE 4.5**
Field-free variation of $N$ with $(\Lambda/\lambda)$.

The coefficient $A$ obtained is less than the value of 1.5 obtained by Kennard, and can be attributed to the fact that Kennard [20] assumed all the electrons originate from the center of the reference coordinate system with the same thermal velocity (this agrees with the simulation), in contrast to the more realistic situation of random origin of the electrons with a Maxwell–Boltzmann velocity distribution considered in the MC algorithm used. Bhattacharjee et al. [18] obtained a factor of 1.25, which is about 2% higher than the one obtained here. The discrepancy may be attributed to the use of the Euler method with a coarser time step (0.01 ns).

The reproduction of the conventional square law verifies the assumptions made in the study and the MC recipes based on the random numbers used in the algorithm. Hence, the program can be advanced to include the effect of a high-frequency EM field, which is discussed in the next section.

## Electron Transport in the Presence of an EM Field

Let a linearly polarized plane EM field with the electric field vector of the form $\vec{E} = E_0 \cos(k_0 z - \omega t)\hat{i}$ be imposed on the seed electrons in the same test volume. The primary frequency used in the simulation is 2.45 GHz, which provides $\omega = 1.54 \times 10^{10}$ rad/s and $k_0 = 0.05141$ cm$^{-1}$. The wave electric field $(E_0)$ amplitude and the power $(P_0)$ are related, as $E_0 = \sqrt{2P_0/(\varepsilon_0 Sc)}$, where $S$ is the cross-sectional area on which the power is incident and $c$ is the velocity of light. $E_0$ is varied from 5 to 100 kV/m, corresponding approximately to a power range of 0.1–50 kW imposed on a circular cross-sectional area of radius 3.6 cm. Equation 4.1 can then be written in terms of component equations as

$$\frac{d^2x}{dt^2} = C_0 \cos(k_0 z - \omega t), \tag{4.4a}$$

$$\frac{d^2y}{dt^2} = 0, \tag{4.4b}$$

$$\frac{d^2z}{dt^2} = 0, \tag{4.4c}$$

where $C_0 = -eE_0/m_e$ is the maximum acceleration. The electrons are allowed to evolve in time, similar to the field-free case, but with a constraint along the axis of electric polarization ($x$ axis) and random along the $y$ and $z$ axes. Therefore, the motion is not truly random but may also be referred to as a constrained random walk [17,18].

In this case, at a particular neutral pressure, the velocity and energy change after each time step $dt$, and hence the free path is evaluated from $\lambda(p,u)$ at each time step and the collision criterion is checked. The collisions randomize the phase of the electrons and they gain energy, as explained in Chapter 2. The collisions remain elastic up to the inelastic threshold energy of 13.66 eV for Ar gas, which is calculated by taking the mean of the excitation energy (11.56 eV) and the ionization energy (15.76 eV) (cf. Figure 4.2), after which the electron may undergo an inelastic collision. The chance of an inelastic collision is decided randomly, weighted by the ratio of elastic to inelastic cross-section at that energy. It may be noted that elastic collisions are dominant up to about 100 eV, after which the inelastic processes dominate (Figure 4.2). In the event of an inelastic collision, the inelastic threshold energy is subtracted from the energy of the electron, and it is allowed to evolve in time and gain energy. The electron is followed until it escapes from the test volume.

The variation of $N$ with $\Lambda/\lambda$ is plotted in Figure 4.6, where it is observed that the variation is no longer quadratic, but has a strong linear component at higher values $\Lambda/\lambda$. The variation may be fitted with an equation of the form

$$N = A(\Lambda/\lambda) + B(\Lambda/\lambda)^2, \tag{4.5}$$

where $A$ and $B$ are coefficients, which is a function of the applied electric field. Typically for $E_0 = 30$ kV/m, $A = 84$ and $B = 0.33$. It is observed that the average number of collisions is an order higher than the field-free case due to the fact that the electrons oscillate in the volume and hence more collisions are probable. With increase in the maximum field intensity, $N$ decreases since the electrons are accelerated to higher velocities and hence their escape probability increases.

Bhattacharjee et al. [18] had obtained a slightly different dependence, showing a saturation tendency of $N$ with $\Lambda/\lambda$. The variation was fitted with

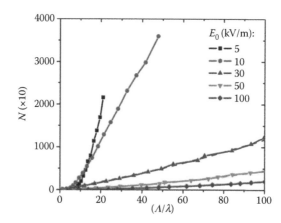

**FIGURE 4.6**
Variation of $N$ versus $\Lambda/\lambda$ in the presence of a linear electromagnetic field of 2.45 GHz with maximum field intensity as the parameter.

an equation of the form $N = C(\Lambda/\lambda)^2/(1 + D(\Lambda/\lambda))^2$. The discrepancy may be related to the use of the Euler method with coarse time steps.

The inelastic collisions have the probability of causing ionization either directly or by providing excited atoms that are ionized easily by low-energy electrons. The process of ionization introduces a new electron (secondary electrons) and an ion in the system, which is recorded. Since this investigation is aimed toward the study of the dynamics of the initial seed electrons, before an actual discharge fully develops due to the avalanche of inelastically generated electrons, it is essential to know the crossover when the inelastic processes start to take over the elastic processes. This, in principle, would indicate the initiation of the discharge and is studied by comparing the percentage of electrons out of the initial seed electrons, involved in elastic and inelastic collisions as a function of time and shown in Figure 4.7a. It is observed that the percentage of the number of elastic electrons decreases and those of inelastic ones increase with time as expected. The two curves meet at a crossover point $\tau_c$ and give as new definition of the breakdown time. Typical $\tau_c$ for 50, 75, and 100 kV/m at a pressure of 1 Torr is 25, 2.8, and 1.4 ns, respectively. The values are similar to the ones obtained by Bhattacharjee et al. [18] (at 105 kV/m; $\tau_c \sim 1.6$ ns). The discrepancy is much less in this case since the electron dynamics is studied for a short period of time <10 ns. The crossover at 30 kV/m may occur at a much later time compared to that of the higher fields.

Figure 4.7b shows the variation of $\tau_c$ with neutral pressure at three different electric field strengths. The variation demonstrates an optimum pressure for discharge formation, where $\tau_c$ is minimum. The minimum lies in the range 1–2 Torr for 75 and 100 kV/m, similar to what was obtained in Ref. [18]. The minimum value of $\tau_c$ increases with decrease in field strength as expected. The variation can be explained from the fact that, as the number of collisions is less in the lower pressure regime, the ionization and excitation rates are rather small, so the breakdown time is longer. As the neutral

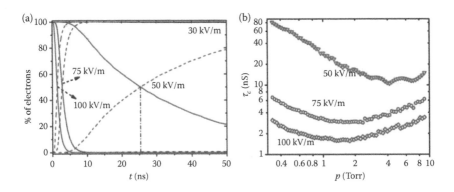

**FIGURE 4.7**
(a) Plot of the percentage of elastic (solid lines) and inelastic (dashed lines) electrons with time at $p = 1$ Torr. (b) Variation of crossover time ($\tau_c$) with neutral pressure.

density increases, an optimum condition occurs, where there are just enough neutrals to cause efficient ionization. In the higher pressure regime (>1 Torr), the collision frequency is rather high ($v_c \gg \omega/2\pi$), and the electrons do not get sufficient time to gain enough energy required for inelastic collisions; therefore, a longer time is required for breakdown.

Figure 4.8a shows the variation of average energy gained by an electron per nanosecond versus pressure for the 50, 75, and 100 kV/m case. The variation demonstrates an optimum pressure, where energy gain is at maximum. At low pressures, the electrons are highly accelerated and most of them escape from the volume without undergoing many collisions, and hence not gaining enough energy. As the neutral density increases, the number of collisions increases, the electrons gain energy between the collisions, and changes occur due to phase randomization. At higher pressures, the collisions are quite frequent and the electrons do not get enough free paths between successive collisions to gain energy; hence, the energy gain decreases.

Figure 4.8b shows the variation of mean energy ($<U>$) possessed per electron with time, at a pressure of 1 Torr for the same field strengths of Figure 4.7a. It can be seen that at 30 kV/m the mean energy saturates at ~0.2 eV, indicating that the electrons have very low energies. Hence, discharge initiation is not expected at this field, which is also evident from Figure 4.7a, where there is no crossover for 30 kV/m at $p = 1$ Torr. At higher field strengths, the mean energies increase, going up to ~35 eV for 100 kV/m. At 50 kV/m, the growth of $<U>$ is slow, and is reflected in the crossover point, which occurs at ~25 ns compared to the 75 and 100 kV/m cases, where $\tau_c \sim 2$ ns (Figure 4.7a). It may be noted that a mean energy of ~4 eV has enough population of energetic tail electrons in the electron energy distribution to cause ionization (50 kV/m case). Thus, it is seen that in the presence of an EM field alone, electric fields of the order of at least 50 kV/m are required to initiate discharge comfortably in the optimum pressure regime of 1–2 Torr.

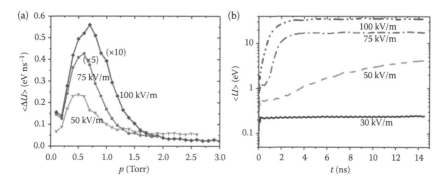

**FIGURE 4.8**
(a) Average energy gain per electron per nanosecond ($<\Delta U>$) versus the gas pressure ($p$).
(b) Mean energy per electron ($<U>$) versus time ($t$).

## Magnetostatically Assisted Electron Transport in the Presence of an EM Field

When a static magnetic field ($B_0$) is imposed upon the seed electrons, perpendicular to the wave vector of the high-frequency EM field, their dynamics becomes all the more interesting due to the cyclotron motion of the electrons, which increases the probability of collisions. The external force term now becomes the Lorentz force $\vec{F}_{tot} = e(\vec{E} + \vec{v} \times \vec{B}_0)$, and two principal situations may arise depending on the orientation of the $B_0$ field lines with the wave electric field ($E$) polarization, namely, $E \parallel B_0$ and $E \perp B_0$ cases. For the parallel case, that is, both $E$ and $B_0$ along the $x$ axis, the equations of motion become

$$\frac{d^2x}{dt^2} = C_0 \cos(k_0 z - \omega t), \tag{4.6a}$$

$$\frac{d^2y}{dt^2} = -\omega_{ce} v_z, \tag{4.6b}$$

$$\frac{d^2z}{dt^2} = \omega_{ce} v_y, \tag{4.6c}$$

where $\omega_{ce} = eB_0/m_e$ is the electron–cyclotron frequency.

The magnetic field affects the motion along the $y$ and $z$ axes, further destroying the random nature of the phenomenon; however, the dynamics is dominated by the electric field along the $x$ axis, and the variations are similar to the ones obtained for the only EM field case. For an individual electron, since the magnetic field is parallel to the electric field, the magnetic field will cause no gyration in the motion of the electron, which would perform an oscillatory motion predominantly along the electric field vector ($x$ axis). There will be no effect even if the EM wave frequency becomes equal to the cyclotron frequency corresponding to the magnetic field.

For the perpendicular case, with $B_0$ along the $y$ axis (Figure 4.1), the equations become

$$\frac{d^2x}{dt^2} = C_0 \cos(k_0 z - \omega t) + \omega_{ce} v_z, \tag{4.7a}$$

$$\frac{d^2y}{dt^2} = 0, \tag{4.7b}$$

$$\frac{d^2z}{dt^2} = -\omega_{ce} v_x. \tag{4.7c}$$

The electron motion in this case is strongly affected by the magnetostatic field, and when the electron–cyclotron frequency equals the wave frequency, resonant transfer of energy occurs from the EM electric field to the electrons, accelerating the electrons to high velocities and hence bringing about the onset of discharge much earlier in time.

Four cases of $B_0$ are studied, namely, 500 G (below resonance), 875.5 G (at resonance), 1000 G (above resonance), and 875.5 G with $E \parallel B_0$. Figure 4.9 shows the $N$ versus $\Lambda/\lambda$ for the four cases at $E_0 = 10$ kV/m. Another plot for 875.5 G (resonance) at 30 kV/m is also shown for comparison. The number of average collisions $N$ is of the same order as obtained in Figure 4.6, and the variations can be fitted with the equation $N = A(\Lambda/\lambda) + B(\Lambda/\lambda)^2$. As resonance is approached, the $N$ value decreases since the electrons are highly accelerated and may leave the volume without enough collisions. The magnetic field has very little effect when it is parallel to $E$, and resembles the case of EM wave-induced dynamics in the absence of a magnetic field.

Figure 4.10a shows the crossover, the four magnetic field conditions for $E_0 = 50$ kV/m, where inelastic collisions are more probable at 1 Torr. The crossover for the parallel case is about an order higher, ~26 ns as expected. Another interesting observation is the occurrence of the crossover for $E_0 = 10$ kV/m at resonance (~57 ns), which was not obtained in any of the other cases. This indicates that at resonance the threshold electric field required for discharge is significantly lower. The crossover for the 500, 875.5, and 1000 G for $E_0 = 50$ kV/m is very close (~3 ns) and overlaps each other. Figure 4.10b gives the expanded view of the crossover for the overlapping cases, where it is observed that the crossover time ($\tau_c$) decreases from 500 G (~3.56 ns) to 1000 G (~3.28 ns).

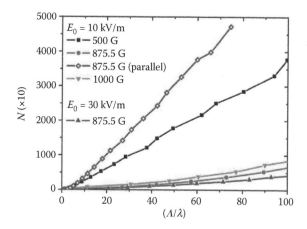

**FIGURE 4.9**

$N$ versus ($\Lambda/\lambda$) plot in the presence of a magnetostatic field. (Reprinted with permission from S. Bhattacharjee, I. Dey, and S. Paul. 2008. *Physics of Plasmas* **20**: 042118. Copyright 2013, American Institute of Physics.)

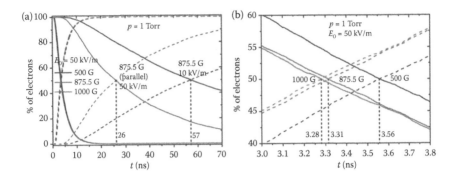

**FIGURE 4.10**
(a) Plot of the percentage of elastic (solid line) and inelastic (dashed line) electrons with time at 1 Torr. (b) Expanded view of (a) in the time range 3–3.8 ns for 500, 875.5, and 1000 G. (Reprinted with permission from S. Bhattacharjee, I. Dey, and S. Paul. 2013. Electron random walk and collisional crossover in a gas in presence of electromagnetic waves and magnetostatic fields, *Phys. Plasmas* **20**: 42118. Copyright 2013, American Institute of Physics.)

Figure 4.11a shows the variation of $\tau_c$ with neutral pressure for the conditions shown in Figure 4.10a, except for the parallel case. It is observed that the minima of the curves lie between 0.1 and 1 Torr. The variation at resonance is quite interesting and different from the other variations. At lower pressures ($10^{-5}$–$10^{-2}$ Torr), where inelastic collisions are usually less due to unavailability of adequate number of neutrals, $\tau_c$ (ECR) is much lower than $\tau_c$ (no ECR). This is because at lower pressures the electrons acquire a large energy ($U > 100$ eV) by resonant coupling, which increases the probability of inelastic collisions (Figure 4.2) and induces a faster crossover. As the neutral

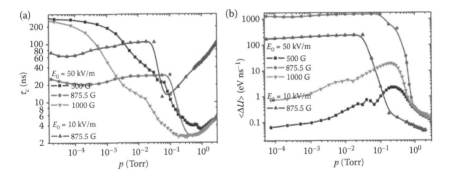

**FIGURE 4.11**
(a) Variation of crossover time ($\tau_c$) with gas pressure ($p$). (b) Average energy gain per electron per nanosecond ($<\Delta U>$) versus $p$ in the presence of magnetostatic field. Symbols: 500 G (black square), 875.5 G (circle), 1000 G (inverted triangle), at 50 kV/m and 875.5 G (triangle) at 10 kV/m. (Reprinted with permission from S. Bhattacharjee, I. Dey, and S. Paul. 2013. Electron random walk and collisional crossover in a gas in presence of electromagnetic waves and magnetostatic fields, *Phys. Plasmas* **20**: 42118. Copyright 2013, American Institute of Physics.)

density increases, an optimal condition is realized ~0.1–0.5 Torr, where there are just enough neutrals for efficient ionization leading to the steep minima. After the minima, $\tau_c$ (ECR) increases and converges with $\tau_c$ (no ECR) values, since at higher neutral densities (Figure 4.11a), the resonance mechanism is rendered less effective due to increased elastic collisions.

Figure 4.11b shows the variation of the average energy gained by an electron per nanosecond ($<\Delta U>$) versus pressure for the conditions shown in Figure 4.11a. For the nonresonant cases, the maxima occur between 0.1 and 1 Torr, complementary to the plot in Figure 4.11a, as expected. In the resonant case, it is seen that $<\Delta U>$ is about two orders higher in the low-pressure range ($10^{-5}$–$10^{-2}$ Torr), confirming the fact that the electrons gain higher energies due to resonant coupling. At higher pressures, due to enhanced collisions, the variations converge with the nonresonant cases as expected.

Figure 4.12a shows the variation of the mean energy $<U>$ over $10^4$ electrons with time, given the same conditions as in Figure 4.10a. In contrast to Figure 4.8b, for only the EM field, where the mean energy at 30 kV/m is very small, indicating no crossover and hence no discharge initiation, in the magnetically assisted case, the electron possesses a mean energy of ~4 eV at 10 kV/m, with $B_0$ = 875.5 G (ECR). At 50 kV/m, in the presence of a magnetic field (500–1000 G), the mean energy is higher than what was obtained for the same field strength in the no-magnetic-field case (Figure 4.8b). Thus, the discharge probability at lower field strength is enhanced in the presence of a magnetic field. The effect of ECR is studied at lower pressures at a low field strength of 6 kV/m (~200 W). It is observed that the electrons possess very high mean energies (~100 eV) at 0.1–10 mTorr, enabling the initiation of discharge at low pressures, which is the prime advantage of magnetic resonance-based sources such as ECR. With increase in pressure, the mean energy decreases, and the mechanism becomes less effective at high pressures (>100 mTorr).

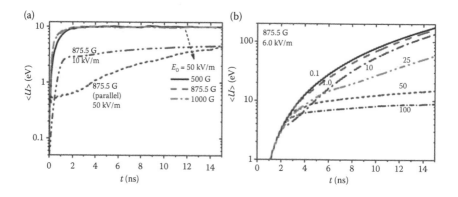

**FIGURE 4.12**
Variation of electron mean energy $<U>$ with time ($t$) for (a) 50 kV/m with $B_0$ = 500, 875.5, 1000 G, and 875.5 G with $E\|B$; 10 kV/m with $B_0$ = 875.5 G; at 1 Torr and (b) for 875.5 G with $p$ = 0.1, 1, 10, 25, 50, and 100 mTorr.

## Electric Field Effects and the Random Walk Parameter

In order to understand the dependence of $N$ as shown in Equation 4.5, we determined the effect of the electric field on $\lambda$ and the average velocity $\langle v \rangle$. Figure 4.13 shows the variation of the mean-free path $\lambda$ and $\langle v \rangle$ acquired by an electron with electric field, at a fixed frequency and pressure as the parameter. We see that $\langle v \rangle$ increases with increase in field strength and saturates at higher values. This is because at a fixed pressure and frequency, the electrons gain higher energy at higher fields; therefore, their velocity increases until they undergo collisions so frequently that they are not able to gain any more energy. The mean-free path is related to the electron energy (hence velocity) via the collision cross-section ($\sigma$) and hence tends toward saturation. The dependence $\lambda$ and $\langle v \rangle$ over the electric field $E$ can be expressed as $\lambda = c_1 + c_2 \exp(-E/E_{th})$ (cm) and $\langle v \rangle = [c_3 + c_4 \{1 - \exp(-E/E_{th})\}] \times 10^8$ (cm/s), where $c_1 = 0.3$ cm, $c_2 = 0.4$ cm, $c_3 = 0.1$ cm/s, $c_4 = 2.8$ cm/s, and $E_{th} = 3.5 \times 10^2$ V/cm at 0.25 Torr. At $E = 0$, $\lambda$ and $\langle v \rangle$ attain their field-free values of 0.7 cm and $0.1 \times 10^8$ cm/s, respectively.

The nature of the variation of $\lambda$ and $\langle v \rangle$ upon the electric field (Figure 4.13) is found to have a nonlinear dependence because of the interrelationship between quantities; the field changes the velocity of the particle,

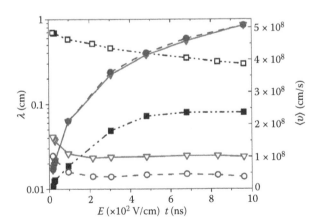

**FIGURE 4.13**

Variation of mean-free path $\lambda$ and average velocity $\langle v \rangle$ with electric field. Solid symbols: $\langle v \rangle$, and open symbols: $\lambda$. Solid and open symbols of the same shape have identical conditions, that is, square: 0.25 Torr, inverted triangle: 12.5 Torr, and circle: 25 Torr. (Reprinted with permission from I. Dey et al. 2008. Sub-nanosecond electron transport in a gas in the presence of polarized electromagnetic waves, *J. Appl. Phys.* **103**: 083305–083306. Copyright 2008, American Institute of Physics.)

which changes the collision cross-section $\sigma$, and this, in turn, changes the mean-free path. As may be noted from Figure 4.2, $\sigma$ is a nonlinear function of electron energy, for example, undergoing minima and maxima at around 0.3 and 11 eV, respectively. In such a varied study over several parameters, a Monte Carlo simulation helps ascertain the true dependence.

In Figure 4.14, we study the temporal evolution of the random walk parameter $\langle \chi^2 \rangle$. The quantity $\langle \chi^2 \rangle \left[ = \langle (\Delta x)^2 \rangle, \langle (\Delta y)^2 \rangle, \text{or} \langle (\Delta z)^2 \rangle \right]$ is the mean square displacement of an electron along the $x$, $y$, and $z$ axes in a time interval $\Delta t$. The time period corresponds to about a quarter cycle of the wave (3 GHz in this case), where the electric field is a maximum. The parameter $\langle (\Delta z)^2 \rangle$ remains linear until ~0.04 ns when the field is low, and then rapidly increases with the field. The temporal variation of $\langle (\Delta z)^2 \rangle$ is dominated by the influence of the field, and the effect of scattering by collisions is negligibly small, as can be seen from the corresponding variation of $\langle (\Delta x)^2 \rangle$ and $\langle (\Delta y)^2 \rangle$ (Figure 4.14). For the field-free case as shown in Figure 4.15, the averaging has been performed over 100 electrons and a 100-point adjacent averaging has been done to indicate a close to smooth variation for a large number of electrons. We see that $\langle \chi^2 \rangle$ is 0 for $t = 0$ and is a constant on an average for $t > 0$ as expected, since there is no force field for accelerating the electrons. The field-free dynamics are determined solely by random scattering, and no preference in $\langle \chi^2 \rangle$ along any one of the directions is observed.

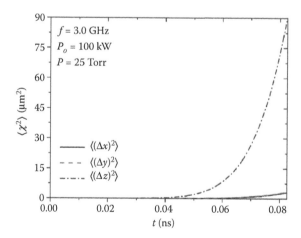

**FIGURE 4.14**
(Color online) Variation of the random-walk parameters $\langle \chi^2 \rangle$, with time for the electric field case at 25 Torr. (Reprinted with permission from I. Dey et al. 2008. Sub-nanosecond electron transport in a gas in the presence of polarized electromagnetic waves, *J. Appl. Phys.* **103**: 083305–083306. Copyright 2008, American Institute of Physics.)

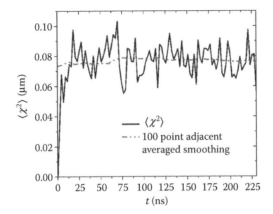

**FIGURE 4.15**

Variation of random-walk parameter $\langle\chi^2\rangle$ with time for the field-free case at 25 Torr. (Reprinted with permission from I. Dey et al. 2008. Sub-nanosecond electron transport in a gas in the presence of polarized electromagnetic waves, *J. Appl. Phys.* **103**: 083305–083306. Copyright 2008, American Institute of Physics.)

## Linear and Circularly Polarized Waves

In this section, we consider the analysis of linearly (LP) and circularly (CP) polarized waves given by $\vec{E}(y, t) = E_0 \cos(ky - \omega t)\hat{z}$ and $\vec{E}(y, t) = E_0\left[\cos(ky - \omega t)\hat{z} + \sin(ky - \omega t)\hat{x}\right]$, at a wave frequency of 2.45 GHz. The power $P$ and electric field $E_0$ for LP and CP waves are related by $P_{LP} = (1/2)\varepsilon_0 E_0^2 \, \alpha c$ and $P_{CP} = \varepsilon_0 E_0^2 \, \alpha c$, respectively, where $\alpha$ is the area of the cross-section, and $c$ is the velocity of the wave. From the average electron energy gain per unit time for the wave polarizations (linear [LP] and circular [CP]), it is possible to make predictions for the breakdown time over the collision frequencies.

Figure 4.16 shows the variation of $\tau_c$ with pressure for different microwave electric fields and for LP and CP cases. A minimum is observed in the variation of $\tau_c$ with pressure as expected. As the number of collisions is less in the lower-pressure regime, the ionization and excitation rates are rather small, so the breakdown time is longer. In the higher-pressure regime (>~1 Torr), the collision frequency is rather high ($v_c \gg \omega/2\pi$) and the electrons do not get sufficient time to gain enough energy required for inelastic collisions; therefore, a longer time is required for breakdown. It is seen that a minimum value of $\tau_c$ occurs ~1 Torr for both the LP and CP wave polarizations. Also, for the same peak electric field $E_o$, the crossover happens at an earlier time for CP EW as compared to LP. This is because the intensity $\langle\vec{S}\rangle$ of the wave launched into the volume for CP waves ($\langle\vec{S}_{CP}\rangle = \varepsilon_0 c E_0^2 \hat{y}$) is two times greater than that of the LP waves ($\langle\vec{S}_{LP}\rangle = \varepsilon_0 c E_0^2)/2\hat{y}$, where the time average is taken over the Poynting vector $\vec{S}$.

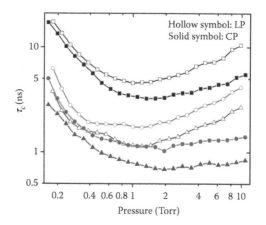

**FIGURE 4.16**

Variation of crossover time with pressure for different microwave electric fields and for linear and circular polarization. Square symbols: $6.08 \times 10^4$ V/m. circular symbols: $1.05 \times 10^5$ V/m. Triangular symbols: $1.36 \times 10^5$ V/m. (Reprinted with permission from S. Bhattacharjee and S. Paul. 2009. Random walk of electrons in a gas in the presence of polarized electromagnetic waves: Genesis of a wave induced discharge, *Phys. Plasmas* **16**: 104502–104504. Copyright 2009, American Institute of Physics.)

Figure 4.17 shows the average energy gain of an electron per nanosecond ($\langle \Delta E \rangle$) at two different wave electric fields ($1.36 \times 10^5$ and $1.05 \times 10^5$ V/m) and for LP and CP EW. It is seen that the energy gain is maximum at an optimum pressure and is larger for CP waves. The energy gain has a maximum (~1 Torr, $1.36 \times 10^5$ V/m LP microwave electric field) that corresponds to the minimum $\tau_c$ (cf. Figure 4.17).

## Summary and Conclusion

The dynamics of the initial seed electrons distributed randomly in space with background-neutral gas atoms providing collision centers is studied, subject to various external applied forces, by a Monte Carlo based simulation algorithm, by considering the problem as a random walk phenomenon. The conventional field-free dynamics as predicted by Kennard [20] is verified, and the square law variation relating the number of collisions to the mean-free path is confirmed. In the presence of a high-frequency EM field, the motion is found to be constrained and a new variation between $N$ and $\Lambda/\lambda$ is proposed. A new discharge initiation criterion is defined as the crossover time of dominance of inelastic collisions over elastic collisions, and is aimed at solving the ambiguities in the proper evaluation of the discharge onset time. The study has shown that the dependence of the mean electron velocity and the free

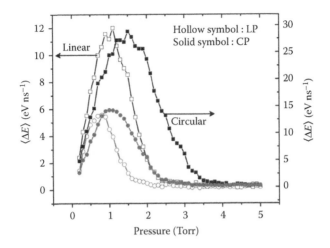

**FIGURE 4.17**
Average energy gain of electron per nanosecond for linear and circular polarizations. The electric fields are square symbols: $1.36 \times 10^5$ V/m, circular symbols: $1.05 \times 10^5$ V/m. (Reprinted with permission from S. Bhattacharjee and S. Paul. 2009. Random walk of electrons in a gas in the presence of polarized electromagnetic waves: Genesis of a wave induced discharge, *Phys. Plasmas* **16**: 104502–104504. Copyright 2009, American Institute of Physics.)

path when an oscillating electric field is imposed on the system is a nonlinear function of the electric field amplitude (see Figure 4.13). The modification of $v$ and $\lambda$ with electric field results in the modified relation between $N$ and $\Lambda/\lambda$, as indicated by the simulations. The electron energy, number of collisions, and their dependence upon wave power, frequency, and electron–neutral collisionality have been studied. As observed in the results, in a purely random process, the mean square displacement $\langle \chi^2 \rangle$ tends to be a fixed value for all the three directions (Figure 4.14), whereas in the constrained random process, $\langle \chi^2 \rangle$ is a nonlinear function of the wave amplitude, and in accordance to the explanation above, there is asymmetry in its values along the three directions (Figure 4.14). The random-walk parameter is a true indicator of the physical process and shows distinct differences of a true versus constrained random process. In the presence of a magnetic field, the cyclotron motion of the electrons is found to enhance the probability of collisions. At resonance, the discharge formation is found to occur at much lower pressures and also at lower electric fields, confirming the observations of electron–cyclotron resonance.

# References

1. S. Bhattacharjee, I. Dey, A. Sen, and H. Amemiya. 2007. Quasisteady state interpulse plasmas, *J. Appl. Phys.* **101**: 113311–113318.

2. B. Lax, W.P. Allis, and S.C. Brown. 1950. The effect of magnetic field on the breakdown of gases at microwave frequencies, *J. Appl. Phys.* **21**: 1297–1304.

3. V.A. Lisovskii. 1999. Criterion for microwave breakdown of gases, *Tech. Phys.* **44**: 1282–1285.

4. M.J. Mulbrandon, J. Chen, P.J. Palmadesso, C.A. Sullivan, and A.W. Ali. 1989. A numerical solution of the Boltzmann equation for high-powered short pulse microwave breakdown in nitrogen, *Phys. Fluids B: Plasma Phys.* **1**: 2507–2515.

5. A. Lacoste, L.L. Alves, C.M. Ferreira, and G. Gousset. 2000. Simulation of pulsed high-frequency breakdown in hydrogen, *J. Appl. Phys.* **88**: 3170–3181.

6. H. Hammen, D. Anderson, and M. Lisak. 1991. A model for steady-state breakdown plasmas in microwave transmit-receive tubes, *J. Appl. Phys.* **70**: 93–98.

7. S.C. Brown and A.D. MacDonald. 1949. Limits for the diffusion theory of high frequency gas discharge breakdown, *Phys. Rev.* **76**: 1629–1633.

8. A.D. MacDonald. 1966. *Microwave Breakdown in Gases* (Wiley, New York).

9. Y.Y. Lau, J.P. Verboncoeur, and H.C. Kim. 2006. Scaling laws for dielectric window breakdown in vacuum and collisional regimes, *Appl. Phys. Lett.* **89**: 261501–261503.

10. A.A. Neuber, J.T. Kline, G.F. Edminston, and H.G. Krompholz. 2007. Dielectric surface flashover at atmospheric conditions under high-power microwave excitation, *Phys. Plasmas* **14**: 057102–057107.

11. F.F. Chen. 1984. *Introduction to Plasma Physics and Controlled Fusion* (vol. 1, 2nd edition, Plenum Press, New York) p. 125.

12. W.P. Allis, S.J. Buchsbaum, and A. Bers. 1963. *Waves in Anisotropic Plasmas* (MIT Press, Cambridge, MA) pp. 133, 179.

13. H. Rau. 2000. Monte Carlo simulation of a microwave plasma in hydrogen, *J. Phys. D: Appl. Phys.* **33**: 3214–3222.

14. C.O. Hwang, J.A. Given, and M. Mascagani, 2001. The simulation-tabulation method for classical diffusion Monte Carlo, *J. Comput. Phys.* **174**: 925–946.

15. M.A. Lieberman and A.J. Lichtenberg. 1994. *Principles of Plasma Discharges and Material Processing* (Wiley-Interscience, New York) p. 73.

16. R. Udiljak, D. Anderson, M. Lisak, V. Semenov, and J. Puech. 2004. Improved model for multipactor in low pressure gas, *Phys. Plasmas* **11**: 5022–5031.

17. I. Dey, J.V. Mathew, S. Bhattacharjee, and S. Jain. 2008. Sub-nanosecond electron transport in a gas in the presence of polarized electromagnetic waves, *J. Appl. Phys.* **103**: 083305–083306.

18. S. Bhattacharjee and S. Paul. 2009. Random walk of electrons in a gas in the presence of polarized electromagnetic waves: Genesis of a wave induced discharge, *Phys. Plasmas* **16**: 104502–104504.

19. V. Vahedi and M. Surendra. 1995. A Monte Carlo collision model for the particle-in-cell method: Applications to argon and oxygen discharges, *Comp. Phys. Comm.* **87**: 179–198.

20. E.H. Kennard. 1938. *Kinetic Theory of Gases* (McGraw-Hill, New York), Chap. 7, pp. 272–273.

21. S. Bhattacharjee, I. Dey, and S. Paul. 2013. Electron random walk and collisional crossover in a gas in presence of electromagnetic waves and magnetostatic fields, *Phys. Plasmas* **20**: 42118.

# 5

## Plasma Production in Continuous-Mode Microwaves: Near-Circular Multicusp

### Introduction

This chapter describes the experimental set-up used for experiments with continuous-mode microwaves. The plasma configuration including the microwave system, the vacuum chamber, the pumping system, and other experimental accessories is described in detail. Thereafter, the production of microwave plasma in a circular waveguide with the transverse dimension smaller than the cutoff value is discussed. The construction of the waveguide makes it slightly polygonal, so its cross-section is referred to as near circular.

In this experiment, a circular conducting tube was inserted into a multicusp (12 poles) to reduce the cross-section to below cutoff. Results indicated that high-density plasma could be produced in the tube. However, the finite width of the inserted cylinder would have disturbed the ECR action occurring near the walls of the multicusp, in the path of bounce motion of the electrons. The current experiment improves the previous one by having designed the waveguide with a radius smaller than the cutoff value. Additionally, a detailed study of the plasma characteristics has been made, including measurements of the axial and radial plasma density variation in the narrow waveguide.

The construction of the waveguide will be described in the section "Waveguide Design." The experimental results will be presented in the section "Experimental Results." Some possible mechanisms of plasma production in a narrow waveguide and a method for obtaining an overdense plasma will be discussed in the "Discussion" section.

### Apparatus

Figure 5.1 shows a schematic view of the experimental apparatus. The apparatus consists of a vacuum chamber C inside which the waveguide Multicusp

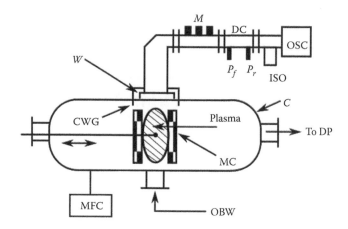

**FIGURE 5.1**

Top view of the experimental apparatus. (Copyright 1998, The Japan Society of Applied Physics.)

(MC; near-circular, rectangular, and square) designed for the study is placed. The chamber is evacuated by a diffusion pump (DP) backed by a rotary pump to pressures below $2 \times 10^{-6}$ Torr. In later experiments, the DP and the rotary pump have been replaced with a turbomolecular and a diaphragm pump. A variety of gases such as Ar, Ne, He, $H_2$, $N_2$, and $O_2$ are used to investigate the plasma production possibilities. However, the results in Ar will be presented. Ar is introduced into the chamber through a mass flow controller (MFC), and the pressure is maintained at a constant level in the range of $10^{-3}$–$10^{-4}$ Torr. Microwaves in the continuous mode are launched into the waveguide through a vacuum quartz window W. The expanded view of the window section is shown in Figure 5.2. The standard rectangular waveguide (WG) has been terminated at W, mounted in a flange (cf. Figure 5.2). Plasma parameters are measured with planar and/or cylinder Langmuir's probes. Observation and photographs of the plasma cross-section are made through the observation window (OBW).

## Vacuum Chamber

The vacuum chamber C (Figure 5.1) is made of stainless steel with a diameter of 60 cm and a length of 120 cm. The chamber is much larger as compared to MC, and this is somewhat helpful to avoid wall contamination during the discharge. There are six large ports, each of 24 cm diameter—four ports are on the sides and two are on the top. The ports on the sides were used for launching the microwaves, evacuation by vacuum pumps, for visual observation of the plasma, and for plasma diagnostics. Among the ports on the top, one was usually used for mass spectroscopy measurements and

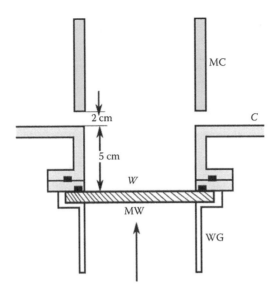

**FIGURE 5.2**
Schematic of the window section. MC, waveguide; C, chamber; W, quartz window; MW, micro-wave input; WG, standard waveguide. (Reprinted with permission from S. Bhattacharjee and H. Amemiya. 1997. Transversely magnetized microwave plasma in a rectangular waveguide under cutoff conditions, *Rev. Sci. Instrum.* **68**: 3061–3067. Copyright 1997, American Institute of Physics.)

the other was blind, with smaller ports that could be used for insertion of Langmuir's probes and the MFC. The chamber has a vacuum leak, and four ports of 6 mm diameter which could be used for probe measurements or optical measurements either in the axial or radial direction of MC. Two other ports of 15 mm diameter on the top were used for the measurement of the gas pressure in the chamber using an ionization and a Baratron's gauge.

## Microwave System

The microwave system consists of a *cw* magnetron oscillator OSC (Figure 5.1) at a frequency of 2.45 GHz. The forward power was varied from 200 to 360 W. Forward and reflected powers, $P_f$ and $P_r$, were monitored through a directional coupler DC through power monitors $P_f$ and $P_r$. An isolator and a dummy load ISO protected the magnetron from the reflected power $P_r$. A three-stub tuner $M$ was used to impedance-match the applied microwave power into the plasma. Microwaves from the magnetron oscillator were guided by a standard rectangular WG (Toshiba type, which has a reduced size in the direction of the $E$ vector) in the $TE_{10}$ mode to the quartz window $W$

at the entrance into the vacuum chamber C. An expanded view of the region is shown in Figure 5.2. Inside the vacuum chamber, the electromagnetic wave was let to propagate into the nonstandard magnet containing WG, MC after a short gap of 2–4 cm. In the experiment with the circular WG, the wave mode was changed to the $TE_{11}$ mode using a circular WG, CWG. In the case of the rectangular and square WG, CWG was not used. Although the experiments have been designed to pass predominantly the $TE_{11}$ or the $TE_{10}$ mode as described above, there could be some other multimode due to the gap at the transition region. Such a small gap is, however, inevitable because of the joint of the standard WG outside the chamber C. Furthermore, if the WG is very tightly attached to the window, it could damage the quartz window W due to the heat caused by the plasma.

## Vacuum Pumps

Figure 5.3 shows a schematic of the circuit of the pumping system. DP and RP represent the diffusion and the rotary pumps, respectively. The valves M, BP, L, and D denote the main, the bypass, the leak, and the one connecting the DP to the rotary pump. The chamber C is initially evacuated through the bypass valve BP using the rotary pump RP to a pressure of about $10^{-4}$ Torr. Thereafter, BP is closed and evacuation is continued with the DP backed by RP through the main valve M, until a base pressure of about $2 \times 10^{-6}$ Torr or lower is attained. The valve D is used to evacuate the DP oil chamber to pressure below $10^{-3}$ Torr to avoid any contamination of the oil before the heater is set on. The leak valve L can be used to maintain gas flow conditions during the experiment. The pumping speed of the rotary pump is about 20 L/min and that of the DP is about 1500 L/s. The gas flow into the chamber could be adjusted using the MFC.

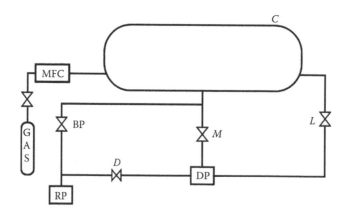

**FIGURE 5.3**
A schematic to show the circuit of the pumping system.

## Pressure Gauges

In our experiment, an ionization gauge and Baratron's gauge were used for the gas pressure measurements. The thermionic ionization gauge consists of a filament, a grid, and the collector, and operates on the basis of gas cooling balanced by the heated coil. Pressures below $10^{-4}$ Torr have been measured using the ionization gauge. The Baratron gauge is an absolute pressure transducer that uses a dual-electrode Inconel transducer design, coupled with a low-impedance, fixed frequency bridge signal conditioner. The Baratron gauge can be used to make pressure measurements of four decades in the range 0.1–1 K Torr.

## Multicusp Design

Different types of multicusp geometries have been extensively studied for the generation of microwave plasmas such as square [1], rectangular [2], and, more commonly, circular [3,4]. In a circular multicusp, magnets are arranged in a cylindrical geometry with neighboring magnets having opposite polarity, resulting in a minimum $B$ field at the center, and strong field near the wall of the multicusp [3,5,6–14]. The magnetic field varies radially, as $B(r) = B_0 r^{n/2-1}$ [1], where $B_0$ is the surface magnetic field, $r$ is the radius of the multicusp, and $n$ is the number of poles of the multicusp. Sub-cutoff dimensional multicusp WGs were studied for the generation of high-density plasmas in narrow cross-sections by overcoming the WG geometrical and plasma density cutoffs [1,3,4,14].

The multicusp is fabricated using rectangular pipes (SS, 1 mm thick), each containing six permanent magnets arranged in a cylindrical geometry, attached to two SS rings at the ends. In each pipe, the polarity of the end magnets is reversed for magnetic-end plugging. In some experiments, the dimension is brought below the cutoff value by putting a stainless-steel cylinder of appropriate thickness inside the multicusp. For a multicusp with $n$ poles, the effective radius $R_e$ is defined by $R_e = R/3\left\{1 + \sqrt{[1 + (3n\tan(\pi/n)/\pi)]}\right\}$, where $R$ is the radius of the inscribed circle [3].

The charged particles are confined in the multicusp by magnetic mirror effect [8,15,16]. The density of the plasma is dictated by the balance between production and loss processes. Energetic electrons, which cause ionization, are more easily lost to the chamber walls than the slower ions unless steps are taken to return the fast electrons to the plasma. The strong multipole magnetic field surrounding the plasma volume meets these requirements [3,5,11,12].

The radial $B$ field profile of the multicusps is shown in Figure 5.4 with the octupole MC radii at cutoff ($r = 3.6$ cm) and above cutoff ($r = 4$ cm)

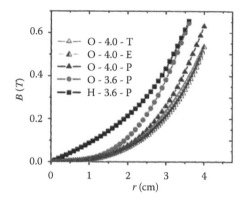

**FIGURE 5.4**

Radial magnetic field profile for hexapole (H) and octupole (O) multicusps at cutoff ($r = 3.6$ cm) and an octupole multicusp above cutoff ($r = 4$ cm) using Poisson simulation (P), fitting equation (T) and measured experimental data (E).

dimensions. For the above cutoff dimension, the experimental profile and the fitting curve are also plotted. The octupole geometry has a flatter minimum $B$ field region than hexapole, and it increases with the increase in radius of the MC. The magnetic field in the WGs is measured with a Hall effect Gaussmeter. The Hall probe is moved in the radial direction in the central plane of the WGs to map the transverse magnetic field.

## Waveguide Design

Figure 5.5 shows detailed views of the multicusp waveguide MC similar to the one constructed in reference [17]. Black and white regions represent north and south poles of the magnets as indicated by N and S. The

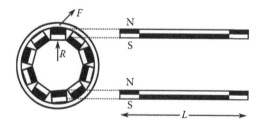

**FIGURE 5.5**

Transverse and longitudinal sections of the multicusp waveguide MC. The inner radius $R$ is 3.4 cm, and the length $L$ of the waveguide is 37 cm. Black and white parts indicate N and S poles of the magnets, respectively. (Copyright 1998, The Japan Society of Applied Physics.)

minimum radius $R = 3.4$ cm and the length $L = 37$ cm. $R$ is smaller than the cutoff radius $R_c$ (= 3.6 cm) of the $TE_{11}$ mode. The multicusp WG has 10 poles and is constructed using hollow rectangular pipes made of stainless steel of dimension $2 \times 1.75 \times 37$ cm$^3$ and of 1 mm thickness. Pieces of permanent magnets made of samarium cobalt (SmCo) or NdFeB of size $1.8 \times 1.5 \times 4.5$ cm$^3$ were inserted into these pipes in a linear arrangement, that is, the same poles on one surface. The poles of the magnets at both ends of the pipe were constructed opposite to the central six magnets. The reversed field at the end sections prevents axial particle loss [17]. Ten such magnet-filled pipes were then arranged to form the circular multicusp in a way that neighboring pipes have opposite polarity. Finally, in order to tighten the pipes in the circular geometry, they were fixed to circular aluminum rings $F$ at both ends of the WG.

## Experimental Results

Figure 5.6 shows the variation of $N_+$ with the microwave power $P_f$ at various pressures $p$. At higher pressures, the density increases with $P_f$ in two steps ($p = 4.6 \times 10^{-4}$ Torr), or in a single step ($p = 3.4 \times 10^{-4}$ Torr). As the pressure is lowered ($p = 3.0 \times 10^{-4}$ Torr), the steps disappear and $N_+$ tends toward a maximum ($p = 2.5 \times 10^{-4}$ Torr). When the pressure is further decreased ($p = 1.6 \times 10^{-4}$ Torr), $N_+$ decreases. It should be noted that $N_+ > N_c$ in a relatively

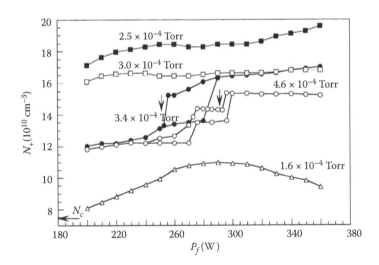

**FIGURE 5.6**
Variation of the plasma (ion) density $N_+$ with power $P_f$ for some pressures $p$. Arrows denote the return path. (Copyright 1998, The Japan Society of Applied Physics.)

wide $p$ region, where $N_c$ (= $7.4 \times 10^{10}$ cm$^{-3}$ for 2.45 GHz, cf. Equation 2.39) is the plasma cutoff density.

At $p = 4.6 \times 10^{-4}$ Torr, besides being formed inside the WG, the plasma was also formed around the outer surface. This corresponded to a state before the first step in $N_+$, where the reflected power $P_r$ was finite (<5 W). As $P_f$ was gradually increased and the matching adjusted, the first step of $N_+$ occurred where the outside plasma suddenly disappeared with $P_r \sim 0$. The second step in $N_+$ was due to the fact that the internal plasma mode changed to a higher density (single step in $N_+$ for $p = 3.4 \times 10^{-4}$ Torr). As $p$ was decreased, the plasma was sustained inside the WG (the state at $p = 3.0 \times 10^{-4}$ Torr) with jumps disappearing and $N_+$ showing only a weak dependence on $P_f$. These values of $p$ were slightly variable over different runs and hysteresis was seen in the steps.

Figure 5.7 shows the variations $N_+$, $T_e$, and $V_s$ with $p$ for $P_f = 330$ W. $N_+$ is seen to increase to a peak value at $(2.5 - 3.5) \times 10^{-4}$ Torr and then decrease with $p$. The existence of an optimum $p$ where the density has a peak is notable. It may be noted that in spite of a small discharge size, the optimum density occurs at a relatively lower $p$. $N_+ > N_c$ was realized for the operating $p$. $V_s$ and $T_e$ showed a monotonic decrease with increase in $p$. Below $1.0 \times 10^{-4}$ Torr, it was difficult to maintain the discharge.

Figure 5.8 shows the variations of $V_s$ and $T_e$ with $P_f$ for some $p$ values. Generally, $V_s$ and $T_e$ decrease with $p$ except for $V_s$ at a lower $p$, while their dependence on $P_f$ is relatively weak in spite of some irregularities.

Figure 5.9 shows the radial variation of $N_+$ and magnetic field $B$ from the center ($r = 0$) to the magnet surface ($r = 3.4$ cm), which is indicated as the "Wall."

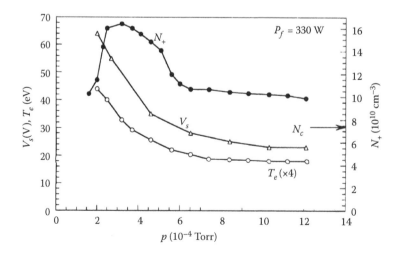

**FIGURE 5.7**

Variations of the plasma (ion) density $N_+$, space potential $V_s$, and electron temperature $T_e$ with pressure $p$ for $P_f = 330$ W. (Copyright 1998, The Japan Society of Applied Physics.)

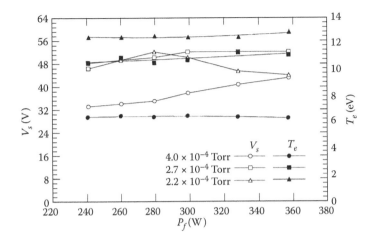

**FIGURE 5.8**
Variation of the space potential $V_s$, and electron temperature $T_e$ with power $P_f$ for some pressures $p$. (Copyright 1998, The Japan Society of Applied Physics.)

The radial variation of $B$, $B(r)$, exhibits a magnetic bottle where the region $r < 0.9$ cm is almost magnetic-field-free. $B(r)$ fits well to $18.0 \, r^4$ dependence for $0 < r < 2.8$ cm (thin-dotted curve). The ECR magnetic field $B_{ECR}$ ($= 875$ G) lies at $r = 2.65$ cm. $N_+$ is rather uniform for $r < 2.0$ cm, and rises slightly at $r = 2.0$ cm and then decreases. $N_+ > N_c$ is fulfilled for $r < 20.0$ cm.

Figure 5.10 shows the axial variations of $N_+$ from one end of the WG ($z = 0$) up to the center $z = L/2$. At a higher pressure ($p = 4.0 \times 10^{-4}$ Torr), some peaks

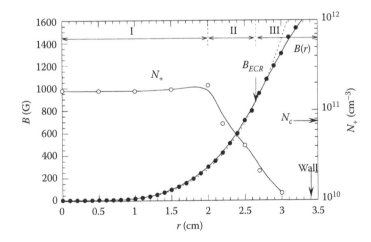

**FIGURE 5.9**
Radial variations of the magnetic field $B(r)$, plasma ion density $N_+$ at power $P_f = 300$ W, and pressure $p = 2.8 \times 10^{-4}$ Torr. (Copyright 1998, The Japan Society of Applied Physics.)

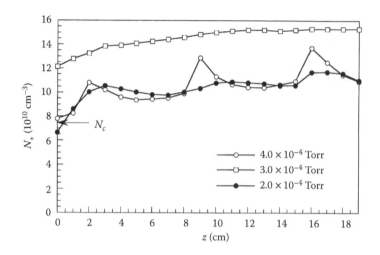

**FIGURE 5.10**

Axial variation of the plasma ion density $N_+$. Measurements for pressures $p = 2.0 \times 10^{-4}$ and $3.0 \times 10^{-4}$ Torr were made at power $P_f = 300$ W and for $p = 4.0 \times 10^{-4}$ Torr at $P_f = 260$ W. (Copyright 1998, The Japan Society of Applied Physics.)

were observed. The pressure range at which such peaks appear coincides with that of the steps shown in Figure 5.6. However, with decreasing $p$ they disappear, and $N_+$ tends to be uniform around the optimum pressure ($p = 3.0 \times 10^{-4}$ Torr). With a further decrease of $p$, $N_+$ decreases. $N_+ > N_c$ is almost satisfied along $z$.

## Spatial Profiles of the Plasma

The spatial profiling of the plasma is intended to provide information about the plasma density and temperature as a function of radial and axial distances inside the MC, with input pressure and power as parameters. Langmuir's probes are used for spatially scanning the plasma, and obtaining the probe characteristic at each point. Figure 5.11a and b shows the result of the measurements.

Figure 5.11a shows the axial variation of $T_e$ and $N_i$ at $P_{in} = 180$ W and pressures of 0.15 and 0.45 mTorr [18]. Along the $z$ direction, both $T_e$ and $N_i$ are uniform in the middle of the multicusp and fall off at the entrance and exit (30 cm). The decrease is about 6% for $T_e$ over a length of 22.5 cm from the entrance, and about 17% for $N_i$ over a length of 20.0 cm extending from 2.5 to 22.5 cm. In the axially uniform region, the average $T_e \sim 12$ eV and average $N_i \sim 1.0 \times 10^{11}$ cm$^{-3}$.

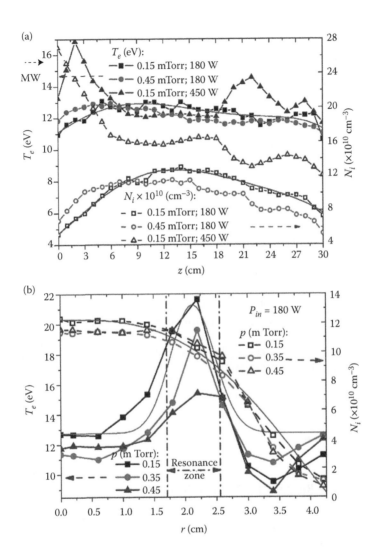

**FIGURE 5.11**

(a) Axial and (b) variation of electron temperature ($T_e$; solid lines with solid symbols) and ion density ($N_i$; dashed lines with hollow symbols) at 0.15 mTorr (squares) and 0.45 mTorr (circles) with 180 W input power. Radialfitting on the 0.15 mTorr $N_i$ data is shown by a solid line. (Reprinted with permission from I. Dey and S. Bhattacharjee. 2008. Experimental investigation of standing wave interactions with a magnetized plasma in a minimum – B field, *Phys. Plasmas* **15**: 123502–123507. Copyright 2008, American Institute of Physics.)

Figure 5.11b shows the radial variation of $T_e$ and $N_i$ at $P_{in} = 180$ W, given the same conditions as Figure 5.8 [18]. The data are taken at the center (15 cm from the entrance). $T_e$ shows a peak around 23 mm from the center, which may be attributed to wave—plasma resonances, whereby the electrons gain a large amount of energy in this region.

In the central region ($r = 0$–28 mm), the plasma is overdense (supercritical) (i.e., $>7.4 \times 10^{10}$ cm$^{-3}$ for $\omega_p/2\pi = 2.45$ GHz), as shown by the $N_i$ data (Figure 5.11b). Near the periphery (35–42 mm), $T_e$ shows an increase since the plasma density is much below cutoff and wave propagation can take place through this region [18–22], thereby accelerating the electrons. However, it may be noted that the $T_e$ in the central region is higher by about ~1 eV (180 W; 0.15 mTorr) compared with the peripheral value, though the plasma is overdense and minimally magnetized, thereby prohibiting any wave propagation (cf. Figure 5.7).

The axial and radial variations of $T_e$ and $N_i$ show similar trends at other powers and pressures [18]. From the results, it is observed that the radial plasma density inhomogeneity scale length $\left| n_e/(\partial n_e/\partial r) \right|_{r=30\,\text{mm}}$ ~1 cm is much less than the wavelength of the microwave, $\lambda_0$ ~ 12.2 cm, and hence the radial gradient of the density will affect the wave dispersion properties. Bhattacharjee et al. [20] solved the radial diffusion equation and found the radial variation of the plasma density to be of the form $N_i(X) = N_0 J_0(CX^q)$, where $X = r/a$ is the normalized radius in the MC, $J_0$ is the zeroth-order Bessel function, $N_0$ is the density at $r = 0$, $C = 2.405$ (1st root of the zeroth-order Bessel function) and $q$ is the exponent that depends on the number of poles of the MC, the ionization rate, and the mean free time of collision. For the case of the octupole having radius 41 mm, $N_0$ ~ $1.2 \times 10^{11}$ cm$^{-3}$ and $q = 2$ obtained by fitting the 180 W, 0.15 mTorr $N_i$ data in Figure 5.11b [18].

---

## Discussion

### On the Possible Reasons for Plasma Production

In Figure 5.9, we define for the sake of convenience regions I ($r < 2.0$ cm), II ($2.0 < r \le 2.65$ cm), and III ($2.65 < r < 3.4$ cm) as the core, resonance, and boundary regions, respectively. Theoretically, if the radial dimension of the WG is smaller than the cutoff value, waves do not propagate through it. However, ignition can occur near the WG entrance due to field penetration. If microwaves pass through the narrow region II, the microwave power density would increase considerably and favor plasma production in this region. At low pressures, the plasma can diffuse into the WG, where it is confined by the minimum-$B$ configuration and the radial loss is suppressed. Furthermore, because $V_s$ in region I is higher than the ionization potential (see Figures 5.7 and 5.8), electrons can be accelerated from the boundary region III toward the core region I and cause ionization. The high $V_s$ would also help to confine electrons in radial and axial directions. These effects would make $N_+ > N_c$ in the core region.

If a macroscopic approach is considered in the pressure range concerned, the effective radial diffusion coefficient $D_r$ would be governed by the Larmor radius in region III. Denoting the longitudinal diffusion coefficient by $D_z$, the density $N_+(z)$ would follow, $N_+(z) = N_{+0} \exp\{-(c/R)\sqrt{D_r/D_z}z\}$, where $N_{+0}$ is the density at the entrance ($z = 0$) and $c$ is a constant determined by the radial distribution of $N_+$ (for the lowest mode of Bessel $J_0$-type distribution, $c = 2.4$). Considering the classical or Bohm-type diffusion, $D_r/D_z \sim 1/(\omega_c\tau)^2$ or $0.06/(\omega_c\tau)$, where $\tau$ is the mean collision time, for $p = 4 \times 10^{-4}$ Torr, $T_e = 6$ eV, and $B = 10^3$ Gauss we obtain $\omega_c\tau \sim 6 \times 10^3$. $N_+(z)/N_{+0}$ at the exit ($z = 37$ cm) at ~0.99 and ~0.92 for the respective cases. Measured density profiles (Figure 5.10) indicate a trend of a slight rise along $z$ toward the center. This suggests that besides the plasma diffusing through the WG an additional increment of charged particles occurs due to the process as mentioned above, in connection with the high $V_s$ and wave absorption in region II. Thus, the multicusp field may be considered to assist diffusion, confinement, and production of charged particles.

## On the Produced Plasma

Experimental results have suggested two plasma regimes. I: Ignition regime at a higher $p$, where plasma exhibits steps and hysteresis (Figure 5.6), and some peaks in $N_+$ ($z$) (Figure 5.10) where the density was comparatively lower. II: Uniform and high-density regime by which $N_+ > N_c$ was obtained. This regime is reached only through the transition from the ignition regime by adjustment of plasma impedance matching, $p$ and $P_f$. As a result of parameter adjustment, the optimum plasma with the highest density is stabilized in the narrow WG.

In deriving $N_+$, an assumption of a Maxwellian plasma was made and this will introduce a certain error in $N_+$, the electron energy distribution, is non-Maxwellian. According to a recent paper [23] that compared the ion current for different electron energy distributions with the same mean energy, the error in $N_+$ in an ECR plasma was estimated to be ~5%. Therefore, the result of $N_+ > N_c$ is considered to be reliable.

The existence of an optimum $p$ (Figure 5.7) can be explained by the particle balance between the production and the loss. In the steady state, assuming a uniform plasma density, we have

$$N_e N_g \langle v_e\sigma_i \rangle \pi R^2 L = 2v_B N_e \pi R_0^2 + D_r \frac{N_e}{\Lambda} 2\pi RL + a_r N_e N_+ \pi R_0^2 L, \quad (5.1)$$

where $N_g$ is the gas density, $N_e$ is the plasma (electron) density, $\sigma_i$ is the ionization cross-section, $a_r$ is the recombination coefficient, $v_e$ is the electron mean thermal velocity ($v_e = 6.69 \times 10^7 T_e (\text{eV})^{1/2}$ cm/s), $v_B$ is the velocity at the sheath edge (or ion acoustic velocity, $v_B = 1.6 \times 10^5 T_e (\text{eV})^{1/2}$ cm/s), $D_r$ is the radial diffusion coefficient, and $R$ and $L$ are the radius and length or the multicusp, respectively. $R_o$ is the radius of the core overdense plasma ($R_o = 2.0$ cm). $\Lambda$ is the characteristic diffusion length assumed to be $R - R_o$. Although the present WG is nearly circular, the equivalent radius $R_e$ is given by [24]

$$R_e = \frac{R}{3}\left[1 + \sqrt{1 + \frac{3n\tan(\pi/n)}{\pi}}\right],\tag{5.2}$$

where $n$ is the number of sides of the polygon and $R$ is the inscribed circle ($R_e \to R$ as $n \to \infty$). For $n = 10$, the correction is 0.85%. Assuming charge neutrality ($N_e = N_+$), we can obtain the maximum $N_+$ under the condition of maximum $P - L_p$, where $P$ and $L_p$ are production and loss rates, respectively, that is,

$$P = N_g\langle v_e\sigma_i\rangle, \quad L_p = \frac{2v_B}{L}\left(\frac{R_o}{R}\right)^2\frac{2D_r}{(R - R_o)R}.\tag{5.3}$$

The first term of $L_p$ represents the axial loss and the second term the radial loss, respectively. In the current experiment with $\omega\tau \gg 1$ ($\omega$ is the angular cyclotron frequency, and $\tau$ is the mean collision time), it may be assumed that the radial loss is governed by the Bohm diffusion: $D_r = 6.25 \times 10^6 T_e$ (eV)/$B(G)$ cm$^2$/s, where the average value of $B$ is taken to be 1000 G.

Figure 5.12 shows the production rate $P$, the loss rate $L_p$, and their difference $P - L_p$, which are calculated for the parameters of Figure 5.7. $T_e$ below

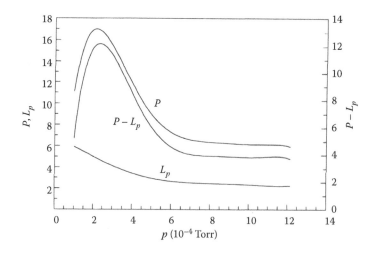

**FIGURE 5.12**
Variation of $P$, $L_p$, and $P - L_p$ with $p$ for the parameters of Figure 5.7 in units of $10^4$ s$^{-1}$. (Copyright 1998, The Japan Society of Applied Physics.)

$2 \times 10^{-4}$ Torr was extrapolated to increase with $1/p$ up to $1 \times 10^{-4}$ Torr. The ionization cross-section $\sigma_i(E)$ was fitted by $\sigma_i(E) = C(E - E_i)\exp\{-D(E - E_i)\}$, where $E_i$ is the ionization energy for Ar (= 15.76 eV), $C = 2.5 \times 10^{-17}$ cm$^2$/eV, and $D = 2.8 \times 10^{-2}$ (eV)$^{-1}$, respectively. Then, $P$ may be expressed in analytical form as $P = N_g C \, \kappa T_e v_e \exp(-x)\{2 + x(1 + y)\}/(1 + y)^3$, where $x = E_i/\kappa T_e$ and $y = D\kappa T_e$. The existence of an optimum $p$ as observed in Figure 5.7 can be understood from the profile of $P - L_p$ versus $p$, which shows a maximum nearly at the same $p$ as in Figure 5.7.

## Conclusions

A microwave plasma is produced in a near-circular WG with a cross-section smaller than the cutoff value. Plasma density above the cutoff value of about $2 \times 10^{11}$ cm$^{-3}$ is obtained in the pressure range of $10^{-4}$ Torr at a power density of 6–10 W/cm$^2$. The electron temperature is about 6–14 eV. In spite of the narrow cross-section of the WG, an optimum pressure exists at a lower-pressure regime $(2.5–3.5) \times 10^{-4}$ Torr, where the density becomes a maximum. Mode jumps and hysteresis have been observed before the plasma is stabilized in the narrow WG.

## References

1. S. Bhattacharjee and H. Amemiya. 1999. Production of microwave plasma in narrow cross sectional tubes: Effect of the shape of cross section, *Rev. Sci. Instrum.* **70**: 3332–3337.
2. S. Bhattacharjee and H. Amemiya. 1997. Transversely magnetized microwave plasma in a rectangular waveguide under cutoff conditions, *Rev. Sci. Instrum.* **68**: 3061–3067.
3. S. Bhattacharjee and H. Amemiya. 1998. Microwave plasma in a multicusp circular waveguide with a dimension below cutoff, *Jpn. J. Appl. Phys.* **37**: 5742–5745.
4. S. Bhattacharjee and H. Amemiya. 2000. Production of pulsed microwave plasma in a tube with a radius below the cut-off value, *J. Phys. D: Appl. Phys.* **33**: 1104–1116.
5. M. Moissan and J. Pelletier (editors). 1992. *Microwave Excited Plasmas* (Elsevier, Amsterdam).
6. R. Limpaecher and K.R. MacKenzie. 1973. Magnetic multipole containment of uniform collisionless quiescent plasmas, *Rev. Sci. Instrum.* **44**: 726–731.
7. K.N. Leung, R.D. Collier, L.B. Marshall, T.N. Gallaher, W.H. Ingham, R.E. Kribel, and G.R. Taylor. 1978. Characteristics of a multidipole ion source, *Rev. Sci. Inst.* **49**: 321–325.

8. F.K. Azadboni, M. Sedaghatizade, and K. Sepanloo. 2010. Design studies of a multicusp ion source with FEMLAB simulation, *J. Fusion Energ.* **29**: 5–12.
9. H. Amemiya and S. Ishii. 1989. Electron energy distribution in multicusp-type ECR plasma, *Jpn. J. Appl. Phys.* **28**: 2289–2297.
10. H. Amemiya, S. Ishii, and Y. Shigueoka. 1991. Multicusp type electron cyclotron resonance ion source for plasma processing, *Jpn. J. Appl. Phys.* **30**: 376–384.
11. H. Amemiya and M. Maeda. 1996. Multicusp type machine for electron cyclotron resonance plasma with reduced dimensions, *Rev. Sci. Instrum.* **67**: 769–774.
12. C.E. Hill. 1996. Ion and electron sources, *Proc. CERN Accelerator School on Cyclotrons, Linacs and their Applications*, La Hulpe, Belgium, CERN-96-02.
13. S. Bhattacharjee, H. Amemiya, and Y. Yano. 2001. Plasma buildup by short-pulse high-power microwaves, *J. Appl. Phys.* **89**: 3573–3579.
14. S. Bhattacharjee. 1999. *Production of microwave plasma in a waveguide with a dimension below cutoff*, Doctoral dissertation, Graduate School of Science and Engineering, Saitama University, Japan.
15. J.A. Bittencourt. 2004. *Fundamentals of Plasma Physics* (3rd edition, Springer, New York).
16. F.F. Chen. 1984. *Plasma Physics and Controlled Fusion* (Plenum Press, New York).
17. M. Maeda and H. Amemiya. 1994. Electron cyclotron resonance plasma in multicusp magnets with axial magnetic plugging, *Rev. Sci. Instrum.* **65**: 3751–3755.
18. I. Dey and S. Bhattacharjee. 2008. Experimental investigation of standing wave interactions with a magnetized plasma in a minimum-B field, *Phys. Plasmas* **15**: 123502–123507.
19. H. Amemiya and M. Maeda. 1996. Multicusp type machine for electron cyclotron resonance plasma with reduced dimensions, *Rev. Sci. Instrum.* **67**: 769–774.
20. M. Katsch and K. Wiesemann. 1980. Relaxation of suprathermal electrons due to Coulomb collisions in a plasma, *Plasma Phys.* **22**: 627–638.
21. R. Talman. 2006. *Accelerator X-Ray Sources* (Wiley-VCH Verlag GmbH Co. KGaA, Weinheim).
22. F.A. Hass, L.M. Lea, and A.J.T. Holmes. 1991. A hydrodynamics model of the negative ion source, *J. Phys. D: Appl. Phys.* **24**: 1541–1550.
23. H. Amemiya. 1997. Sheath formation criterion and ion flux for non-Maxwellian plasma, *J. Phys. Soc. Jpn.* **66**: 1335–1338.
24. X. Zhou and Y. Wang. 1996. Approximate formula for cutoff wave-number of lowest-order TM mode of a hollow metallic waveguide of arbitrary cross-section, *IEE Proc.-Microwave Antennas Propag.* **143**: 454–456.

# 6

## Plasma Production in Continuous-Mode Microwaves: Rectangular Nonmulticusp Waveguide

### Introduction

In Chapter 5, we have seen that a plasma could be produced in a near-circular multicusp waveguide with a dimension smaller than the cutoff value. In a waveguide with a multicusp magnetic field, the field lines are arched along the circumference connecting poles of opposite polarity, as shown schematically in Figure 6.1a. On the other hand, the electric field of the circular mode ($TE_{11}$) is peaked at the center and curved radially outward as shown in Figure 6.1b. Therefore, when waves are launched into a circular multicusp waveguide, the electromagnetic coupling gives rise to locations inside the waveguide where the electric field can be perpendicular or parallel to the static magnetic field (ordinary ($E \parallel B$)) and extraordinary $E \perp B$ modes. It is of interest to make an independent study of the dependence of the plasma parameters and the power absorption in these two principal modes. This will help us to understand the plasma production and maintenance mechanisms in a multicusp waveguide with regard to the effectiveness of the wave modes.

In this chapter, we will study the plasma properties and power absorption in the aforesaid electromagnetic modes, which can be experimentally realized in a rectangular waveguide with a linear magnetic field as shown schematically in Figure 6.2a and b. Additionally, the plasma production and its properties are investigated for the case when the dimension of the rectangular waveguide is below cutoff, and the magnetic field is of nonmulticusp type.

The next section describes the experimental procedures followed by a section on experimental results. Finally, the results are discussed and the particle motions are described from a single-particle theory.

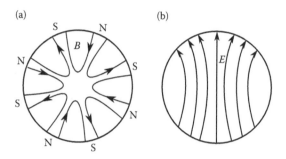

**FIGURE 6.1**
A schematic diagram to show (a) static magnetic field $B$ and (b) wave electric field $E$ ($TE_{11}$ mode) in a multicusp circular waveguide. N and S represent the north and south poles of the magnets.

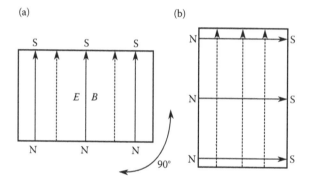

**FIGURE 6.2**
Schematic diagram to show (a) magnetic field $B$ (solid lines) and electric field $E$ (dashed lines) in the ordinary mode and (b) the same in extraordinary mode. The waveguide can be rotated to make $B$ parallel or perpendicular to $E$.

## Experimental Procedures

### Waveguide Design

A rectangular waveguide has been considered because of the following advantages:

1. It is easier to construct, and its clear geometry helps us to identify the fundamental vacuum waveguide mode ($TE_{10}$) as compared to a polygonal waveguide.

2. A rectangular waveguide does not suffer from any cutoff limitations in height $b$, allowing us an additional advantage of selecting $b$ according to our requirements (cf. Figure 6.3).

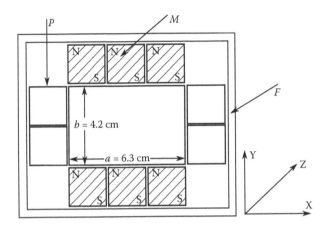

**FIGURE 6.3**
Schematic of the vertical cross-section of the rectangular waveguide. M, magnets; P, steel pipes; F, aluminum frame. Z coordinate is the direction of the microwave input. (Reprinted with permission from S. Bhattacharjee and H. Amemiya. 1997. Transversely magnetized microwave plasma in a rectangular waveguide under cutoff conditions, *Rev. Sci. Instrum.* **68**: 3061–3067. Copyright 1997, American Institute of Physics.)

3. The higher the electric field intensity, the better the electron cyclotron resonance (ECR) action, so a waveguide with smaller $b$ can lead to better ECR action.

4. A rectangular waveguide allows us the possibility of studying the two principal waves $E \parallel B$ and $E \perp B$ in a transversely magnetized plasma.

5. A rectangular plasma waveguide can be directly coupled with the standard rectangular waveguide from the magnetron oscillator, helping us to reduce power losses.

A schematic of the waveguide cross-section is shown in Figure 6.3. It is characterized by its cross-sectional width $a$ and height $b$. The rectangular waveguide was constructed in a similar manner as the near-circular waveguide in Chapter 4.

The pipes were fixed with screws to rectangular aluminum rings $F$ at both ends, which hold the pipes tightly in a rectangular shape. Three pipes each have been arranged on the top and bottom, as well as two pipes on each side, resulting in a final dimension of $6.3 \times 4.2 \times 37$ cm$^3$.

Considering the cross-sectional dimension of the waveguide, for the O-like mode propagation $E \parallel B$, we take $a = 6.3$ cm and $b = 4.2$ cm. On the contrary, for the X-like case, we take $a = 4.2$ cm and $b = 6.3$ cm. In both the cases, the principal mode that can be transmitted is the TE$_{10}$, and for this mode the cutoff wavelength is solely governed by the length $a$, which for microwaves of 2.45 GHz is 6.12 cm (cf. Equation 6.13). Therefore, for the "O-like" mode propagation, the waveguide dimension is made very close to cutoff and for the "X-like" mode propagation it is much below cutoff. The waveguide was rotated by 90° to make the constant magnetic field parallel or perpendicular to the microwave electric field (cf. Figures 6.2 and 6.3).

## Magnetic Field Design

The schematic of the arrangement of the permanent magnets can be seen in Figure 6.3. The types of permanent magnets are the same as those used in the near-circular waveguide, which has a dimension $1.8 \times 1.5 \times 4.5$ cm$^3$ and a surface magnetic field of abut $2 \times 10^3$ G. Eight such magnets were arranged in a row and inserted into top and bottom pipes. The pipes at the sides were kept hollow. This arrangement gave the desired transverse magnetization with an option or rotation of the waveguide to study either the "O-like" or the "X-like" mode.

The origin of measurements was the center of the waveguide cross-section. Figure 6.4 shows the mapping of the magnetic field distribution with respect to horizontal ($X$) direction and vertical direction ($Y$) as a parameter. Measurements along $X$ could be done from 0 to 25 cm and along $Y$ from 0 to 1.6 cm. Figure 6.4 shows that the field is almost uniform at the center of the waveguide ($y = 0$–1.0 cm), with values lying between 1125 and 1250 G, and it dropped at the edge. The field is higher at $y \geq 1.4$ cm, owing to its proximity to the magnets. At the join of the two pipes ($x = 1.2$ cm), the field is weaker because of fringe effect. It is important to note that the magnetic field for ECR

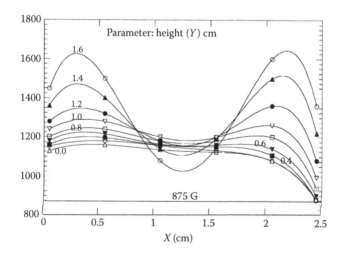

**FIGURE 6.4**
Magnetic field distribution inside the waveguide with the origin taken as the center of the waveguide cross-section and height $Y$ as the parameter. (Reprinted with permission from S. Bhattacharjee and H. Amemiya. 1997. Transversely magnetized microwave plasma in a rectangular waveguide under cutoff conditions, *Rev. Sci. Instrum.* **68**: 3061–3067. Copyright 1997, American Institute of Physics.)

action (875 G) is realized very close to the wall, at about $x = 2.45$ cm and at smaller values of $y$.

### Measurement of the Reflected Power and the Ion Current

To understand the characteristics of wave propagation through the wave-guide, the reflected power from the plasma and the ion current to the plane probe have been measured simultaneously as a function of the input power and the pressure. These measurements were done both for the "O-like" and the "X-like" modes. Marked differences were observed. For the measurement of the ion current, the plane probe was set at a fixed voltage of –50 V. The input power and the pressure were varied in the range 100–260 W and 0.12–0.3 mTorr. Data at powers above 260 W and pressures above 0.3 mTorr were not reliable, as the probe glowed and the plasma expanded outside the waveguide.

---

## Experimental Results

Plasma parameters were measured with a planar and cylinder Langmuir's probe. The probes were inserted through the narrow space between adja-cent magnet carrying pipes, at $z = 18.5$ cm from one end of the waveguide, roughly at the waveguide center. The probes were 2 cm apart and occupied identical positions inside the waveguide, which helped us to compare the plasma parameters measured through them.

A fairly dense plasma was obtained in the "O-like" mode. The plasma den-sity was higher than the cutoff density in the "X-like" mode. Photographs of the plasma taken in the two cases showed a relatively uniform plasma except for bright regions near the waveguide sides and near the join of two pipes. These bright regions could be locations for efficient electron acceleration and heating taking place. It is interesting to note that the plasma production was possible even though the waveguide dimension was much below cutoff in the $E \perp B$ case. Through some confinement mechanisms, the plasma could be main-tained inside the waveguide. Experimental results obtained from the "X-like" and the "O-like" cases are discussed separately in the following sections.

### "O-like" Mode Propagation ($E \parallel B$)

The reflected power (Figure 6.5) showed a marked dependence on pres-sure ($p$), with the power reflection as high as 40% of the input power at 0.15 mTorr. However, excellent absorption of microwave power has been observed around 0.21 mTorr, where the reflected power was within 2%. Such enhanced absorption has not been observed in the "X-like" mode propaga-tion in the range of operating pressures. Figure 6.6 shows the variation of the

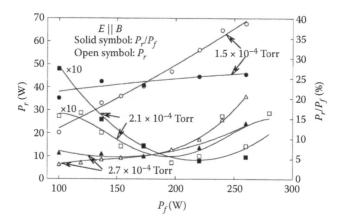

**FIGURE 6.5**

Relation between the input power $P_f$ and the reflected power $P_r$ as a function of pressure $p$ in the ordinary mode ($E \parallel B$). (Reprinted with permission from S. Bhattacharjee and H. Amemiya. 1997. Transversely magnetized microwave plasma in a rectangular waveguide under cutoff conditions, *Rev. Sci. Instrum.* **68**: 3061–3067. Copyright 1997, American Institute of Physics.)

plasma density ($N_e$) with pressure as a function of three different powers. For all powers, the "O-like" mode showed a bump in $N_e$. The maximum $N_e$ at 200 W was $2 \times 10^{10}$ cm$^{-3}$. No such peaks were observed in the "X-like" mode at any pressure. Figure 6.7 shows the space potential ($V_s$) measured as a function of $p$ for different powers. The $V_s$ also has a fairly sharp increase around 0.2 mTorr, with values in the range 55–65 V.

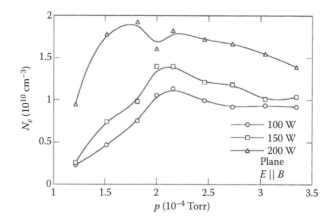

**FIGURE 6.6**

Variation of the plasma density $N_e$ with pressure $p$ at different powers in the ordinary mode ($E \parallel B$). (Reprinted with permission from S. Bhattacharjee and H. Amemiya. 1997. Transversely magnetized microwave plasma in a rectangular waveguide under cutoff conditions, *Rev. Sci. Instrum.* **68**: 3061–3067. Copyright 1997, American Institute of Physics.)

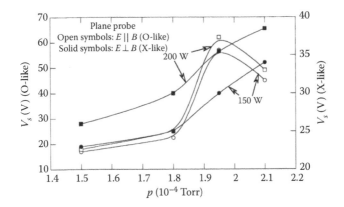

**FIGURE 6.7**

Variation of the space potential $V_s$ with pressure $p$ at 150 and 200 W in the ordinary ($E \parallel B$) and the extraordinary ($E \perp B$) modes. (Reprinted with permission from S. Bhattacharjee and H. Amemiya. 1997. Transversely magnetized microwave plasma in a rectangular waveguide under cutoff conditions, *Rev. Sci. Instrum.* **68**: 3061–3067. Copyright 1997, American Institute of Physics.)

The plane and the cylinder probe could give us the electron temperature in a direction parallel ($T_{e\parallel}$) and perpendicular ($T_{e\perp}$) to the magnetic field, respectively. Figures 6.8 and 6.9 show the variation of the electron temperature with respect to $p$ for different powers, as measured with a planar and a cylinder probe, respectively. $T_{e\parallel}$ and $T_{e\perp}$ show quite different variations with $p$.

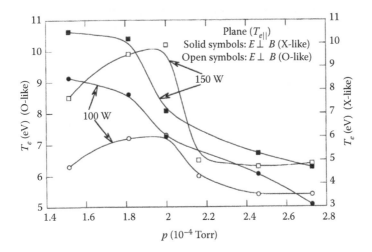

**FIGURE 6.8**

Parallel electron temperature $T_{e\parallel}$ as a function of pressure $p$ for 100 and 150 as measured in the ordinary and extraordinary modes with a plane probe. (Reprinted with permission from S. Bhattacharjee and H. Amemiya. 1997. Transversely magnetized microwave plasma in a rectangular waveguide under cutoff conditions, *Rev. Sci. Instrum.* **68**: 3061–3067. Copyright 1997, American Institute of Physics.)

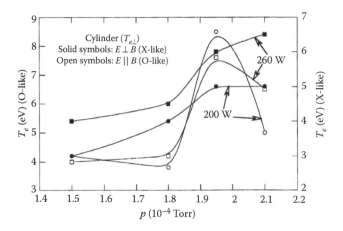

**FIGURE 6.9**

Perpendicular electron temperature $T_{e\perp}$ as a function of pressure $p$ for 200 and 260 W as measured in the ordinary and extraordinary modes with a cylinder probe. (Reprinted with permission from S. Bhattacharjee and H. Amemiya. 1997. Transversely magnetized microwave plasma in a rectangular waveguide under cutoff conditions, *Rev. Sci. Instrum.* **68**: 3061–3067. Copyright 1997, American Institute of Physics.)

## "X-Like" Mode Propagation ($E \perp B$)

In the "X-like" case, the reflected power (Figure 6.10) was about 10–20% of the input power, and its dependence on pressure was small. This implies that the input power was effectively absorbed by the plasma. Figure 6.11 shows the variation of $N_e$ with respect to $p$ for different powers. $N_e$ increased with both

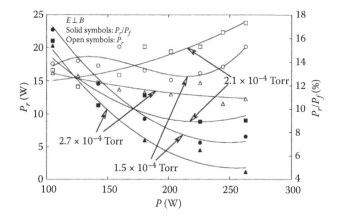

**FIGURE 6.10**

Relation between the input power $P_f$ and the reflected power $P_r$ as a function of pressure $p$ in the extraordinary mode ($E \perp B$). (Reprinted with permission from S. Bhattacharjee and H. Amemiya. 1997. Transversely magnetized microwave plasma in a rectangular waveguide under cutoff conditions, *Rev. Sci. Instrum.* **68**: 3061–3067. Copyright 1997, American Institute of Physics.)

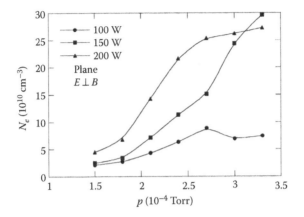

**FIGURE 6.11**
Variation of the plasma density $N_e$ with pressure $p$ at different powers in the extraordinary mode ($E \perp B$). (Reprinted with permission from S. Bhattacharjee and H. Amemiya. 1997. Transversely magnetized microwave plasma in a rectangular waveguide under cutoff conditions, *Rev. Sci. Instrum.* **68**: 3061–3067. Copyright 1997, American Institute of Physics.)

power and pressure, and at higher powers it increased nearly in proportion to the pressure. This proves that there is efficient ionization at all experimental conditions, and much higher current densities could be obtained by increasing the microwave power. $N_e$ of about $3 \times 10^{11}$ cm$^{-3}$ has been recorded in this case, which is about an order higher than that of the "O-like" mode. Figure 6.7 shows the variation of $V_s$ with $p$. $V_s$ increases almost in proportion to both power and pressure. Figures 6.8 and 6.9 show that $T_{e\parallel}$ and $T_{e\perp}$ as measured by the plane and the cylinder probe, respectively. $T_{e\parallel}$ decreased with pressure whereas $T_{e\perp}$ increased with pressure.

## Discussion

In summary, the following was demonstrated.

1. Plasma could be created in a waveguide with dimensions below cutoff.

2. Plasma density for the $E \perp B$ case is greater than the plasma density for $E \parallel B$ case.

3. Plasma density much above the cutoff value of $7.4 \times 10^{10}$ cm$^{-3}$ for microwaves of 2.45 GHz could be obtained in the $E \perp B$ case.

4. The "O-like" mode showed higher temperature, higher plasma density, and higher space potential around 0.2 mTorr.

5. The microwave-generated high-density plasmas obtained in such small volume waveguides look promising for industrial applications.

The experimental results observed in the two cases will be discussed separately. Understanding of the motion of the particles near the center of the waveguide will be made from a single particle theory [1]. It is worthwhile to study the effect of the microwave electric field on the electrons, as in the central region of the waveguide, owing to a field of about 1200 G, the electron cyclotron frequency is about 1.4 times higher than the applied microwave frequency of 2.45 GHz. E and B are assumed to be uniform in space, which is true to a good degree of approximation at the center of the waveguide. The system of coordinates is defined as shown in Figure 6.3.

### "O-Like" Mode Propagation (E ∥ B)

In the "O-like" mode propagation, the electric field E is parallel to the magnetic field. Therefore, the E cannot accelerate the electrons gyrating around the magnetic field lines. However, near the join of two pipes and at the waveguide sides owing to the fringing of the field lines, it is very likely that a component of the electric field be perpendicular to the magnetic field lines. It is at these locations that electron acceleration can efficiently occur to bring about ionization. Moreover, as described earlier, the magnetic field required for ECR action is also met at these sites. The existence of an optimum pressure in microwave discharges is well known to be where efficient ionization occurs, leading to higher plasma densities. This effect of optimum pressure becomes even more pronounced if there are ECR regions. Moreover, a temperature of about 11 eV at around 0.2 mTorr favors a higher rate of ionization. Another important finding is a peaked space potential at this particular pressure. This high space potential could lead to an increase in the ionization by potential acceleration of the charged particles. The temperature measurements by the cylinder probe also showed that the perpendicular electron temperature, otherwise relatively cold, became high at this definite pressure, proving that pronounced electron motion in the perpendicular direction occurs only at this pressure. Thus, we believe that although in this case the electric field is parallel to the magnetic field, peaks of $N_e$ and $T_e$ could be explained by: (1) when efficient electron acceleration along with ECR action occurs at a particular pressure at the field-fringing locations, and (2) through enhanced potential acceleration at the particular pressure.

To study the motion of a single particle, for the E ∥ B case, we assume the electric and the magnetic field to be of the form $E(t) = E_0 e^{i\omega t} \hat{x}$ and $B = B_0 \hat{x}$, and both of them to lie in the x direction. The x, y, and z components of the equation of motion are

$$m\dot{v}_x = q\tilde{E}_x \quad m\dot{v}_y = qv_z B_0 \quad m\dot{v}_z = -qv_y B_0.$$

$$(6.1)$$

The $y$ and $z$ components have simple harmonic solution, which can be written as

$$v_y, v_z = G\sin\omega_c t + H\cos\omega_c t, \tag{6.2}$$

where $G$ and $H$ are arbitrary constants and $\omega_c = |q|B_0/m$. Thus, in the $y$–$z$ plane, the particle describes a circular motion at the cyclotron frequency. The guiding center drifts in the $x$ direction, and the drift is oscillatory. The charge-dependent drift is opposite for electrons and ions. This motion of the charge particles helps plasma confinement because as the particles drift in the $x$ direction they are caught in the magnetic field lines.

The appearance of a bump in $N_e$ could be explained as follows. As $p$ decreases, the ECR action becomes stronger. On the other hand, due to the decrease of gas density, the ionization rate decreases. Therefore, there should be an optimum pressure at which the ionization rate is at maximum. As for the loss mechanism, there would be a wall effect by which the particles are lost. However, as the electron and the ion mean-free paths are much longer than the height $b$ of the waveguide, the wall loss is independent of pressure, and hence the overall effect of plasma production and loss could explain the existence of an optimum pressure at which the plasma density reaches its maximum. Our data have been interpreted as a function of $p$, where the height $b$ of the waveguide is fixed. However, the dimensionally correct parameter for the peak would be $pb$ as in Paschen's law, which is mostly used in DC discharges.

## "X-Like" Mode Propagation Case ($E \perp B$)

In the "X-like" case, as the dimension of the waveguide is below cutoff, so it is very hard for the electromagnetic waves to penetrate the waveguide and produce a plasma inside it. The wave penetration aspect in a cylindrical geometry slightly below cutoff has been qualitatively discussed elsewhere [2]. However, there seems to be no quantitative theory concerning the present rectangular geometry. A previous work [3] treated the plane wave propagation in a magnetized plasma along and across the magnetic field, where the plasma density could be made above the cutoff value at the sacrifice of matching. However, no discussion about geometrical effects was made of the cylindrical tube used. This work explores the geometrical effect of the waveguide, which would arise in some applications. We have demonstrated that this geometrical limitation can be overcome by the aforementioned arrangement of magnets and that the plasma production is possible even though the dimension of the waveguide is smaller than the cutoff value for the initially launched wave.

In the "X-like" case, it is understood that full use of the electric field has to be made and also the electron temperature should be sufficiently high for discharge initiation. The greatest advantage of this mode is that the electric field being perpendicular to the magnetic field, electron acceleration, leading to ionization, and hence higher densities can occur at all ranges of pressures and power (cf. Figure 6.11). It is because of this reason that the density and temperature peaks occurring at a particular pressure in the ordinary mode are not seen here. Added to this fact, here the space potential is an increasing function of both power and pressure. In the range of 0.15–0.21 mTorr (cf. Figure 6.7), the space potential increases by as much as 8 V for 150 W to 11 V for 200 W. This increase in space potential at the axis of the waveguide further helps the acceleration of electrons to the center and prevents particle loss to the wall. Once the electrons gain more than 15 eV of energy, the ionization potential for Ar, they can ionize additional Ar atoms. With increase in either power or pressure, it has been noticed that the cylinder probe recorded an increase in temperature, implying the actual increase in the cyclotron motion of the electrons at higher powers. For all the aforesaid reasons, we observe a steady increase in the plasma density with both power and pressure. The characteristic system size can be taken as $b$ (4.2 cm) because the magnetic field is parallel in this direction for both mode of operation. The estimated ratio of the mean-free path to $b$ is about 15–80 for electrons and 4–20 for ions.

To study the single-particle motion, in the extraordinary mode, we take the magnetic field in the $y$ direction, where $B = B_0 \hat{y}$, and the electric field as usual in the $x$ direction. Then, the $x$, $y$, and $z$ components of the equation of motion are

$$m\dot{v}_x = q\left[\tilde{E}_x - v_z B_o\right] \quad m\dot{v}_y = 0 \quad m\dot{v}_z = qv_x, \tag{6.3}$$

The principal difference between the ordinary and extraordinary particle motions is that here the $y$ component of the velocity of the particles is a constant. This helps to increase the plasma density and prevent particle loss to the wall. In the central region of the waveguide, as the field lines are aligned in the $y$ direction, the particles can get trapped in these field lines. Moreover, as the fields exert no force on the electrons, it is easy for them to get accumulated. Increase in power brings about ionization, and more and more particles get trapped in the field lines, leading to an increase in their number density near the vicinity of the probes. The equation for $v_s$ and $v_z$ on further simplification tells us that the $z$ component perpendicular to $B$ and $E$ is the usual $E \times B$ drift, except that the drift oscillates. The $x$ component has the accelerating effect by the electric field.

## Conclusions

In this chapter, it has been shown that a plasma can be produced in a rectangular $TE_{10}$ mode waveguide with a cross-sectional area (6.3 cm width × 4.2 cm height) much smaller than standard waveguides (12.4 cm width × 5.6 cm height) under a nonmulticusp magnetic field. The characteristics of the plasma and the wave propagation have been studied for the perpendicularly launched waves ($K \perp B$) in the ordinary like $E \parallel B$ and extraordinary like ($E \perp B$) modes. The maximum electron density as measured by a plane probe in the $E \parallel B$ and $E \perp B$ cases are $2 \times 10^{10}$ cm$^{-3}$ (underdense) and $3 \times 10^{11}$ cm$^{-3}$ (overdense), respectively, while the electron temperatures ranged from 3 to 11 eV.

## References

1. F.F. Chen. 1984. *Introduction to Plasma Physics and Controlled Fusion* (Vol. 1: Plasma Physics, 2nd edition, Plenum, New York, Chap. 2).
2. H. Amemiya and M. Maeda. 1996. Multicusp type machine for electron cyclotron resonance plasma with reduced dimensions, *Rev. Sci. Instrum.* **67**: 769–774.
3. E.G. Bustamante, M.A.G. Calderon, J.M. Senties, and E. Anabitarte. 1989. Absorption of high-frequency electromagnetic waves by a transversely magnetized cold plasma waveguide, *J. Phys. D: Appl. Phys.* **22**: 408–412.
4. S. Bhattacharjee and H. Amemiya. 1997. Transversely magnetized microwave plasma in a rectangular waveguide under cutoff conditions, *Rev. Sci. Instrum.* **68**: 3061–3067.

# 7

# Comparison of Square and Near-Circular Multicusp Waveguides

## Introduction

In Chapter 4, we saw that a plasma with a density greater than the O-mode cutoff density could be produced and maintained in a near-circular multicusp (10 poles) waveguide, with a dimension below the cutoff value. Chapter 6 dealt with a rectangular waveguide with a nonmulticusp magnetic field and a dimension below the cutoff value. The plasma properties and the power absorption in the ordinary (O-mode) and extraordinary (X-mode) modes have been studied. The individual study was useful because both the modes can occur when waves are launched into a multicusp, depending upon the orientation of the electric and magnetic field vectors. Experimental results showed that the plasma density is about an order higher in the X-mode as compared to that in the O-mode. The density is above the cutoff density in the X-mode. However, the uniform region of the plasma is smaller than in the case of multicusp magnetic field. This is because the particle motions were mainly confined along the one-directional field lines, which reduced the randomness of the spatial velocity distribution of the particles, thereby adversely affecting uniformity.

The purpose of this study is to understand the effect of the shape of cross-section of the tube on the plasma parameters, when the tube's dimension is smaller than the cutoff value. Although Chapters 4 and 5 deal with tubes of different geometry (near-circular and rectangular), the desired information cannot be obtained because these tubes have different nature of magnetic fields (multicups and nonmulticusp), and their area of cross-sections are different. In the current experiment, tubes of a square and a near-circular cross-section have been designed where both have multicusp magnetic fields, are of the same lengths, and have almost the same cross-sectional areas, so that the microwave power density in the tubes can be kept the same. This allows us to compare the efficiency of wave absorption depending upon the inner geometry of the tube. Moreover, in a four-sided geometry, a square waveguide is helpful because in a square cross-section one can construct the same number of magnetic poles on each side. This way the symmetry and uniformity of the magnetic field is improved. The results of the comparative study are presented, and possible mechanisms of the plasma maintenance and the effect of the cross-sectional shapes of the tubes are discussed.

The next section describes waveguide design followed by a section on experimental results. Finally, the results are discussed at the end.

## Waveguide Design

Figure 7.1a and b are schematic cross-sections of the near-circular and square waveguides, respectively. Black and white regions represent the north and south poles of the magnets, as indicated by N and S. The waveguides were constructed in a similar manner as described in previous chapters. The circular waveguide has 10 poles and the square tube 12 poles. The circular waveguide is similar to that described in Chapter 4 with a minimum inner radius $a = 3.4$ cm, which is made smaller than the cutoff radius $R_c = 3.6$ cm. The square waveguide has a width and height 2b. Three magnets are on each side of the cross-section, making $2b = 6.0$ cm, which is smaller than the cutoff value for the rectangular mode ($TE_{10}$).

A planar Langmuir probe was inserted through a radial port into the waveguides to measure the plasma (ion) density $N_+$, the space potential $V_s$, the electron temperature $T_e$, and the electron energy distribution function (EEDF) $f(E)$. $f(E)$ was obtained by using the second derivative ($I_p''(V_p)$) of the probe characteristics by the method described in Chapter 2.

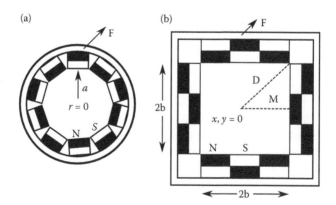

**FIGURE 7.1**
Schematic of the cross-sections for the (a) circular and (b) square tubes. The radius of the circular tube $a = 3.4$ cm and side 2b of the square tube $= 6$ cm. M and D are directions along the midline and the diagonal for the square tube, respectively. F is an aluminum frame. (Reprinted with permission from S. Bhattacharjee and H. Amemiya. 1999. Production of microwave plasma in narrow cross-sectional tubes; effect of the shape of cross section, *Rev. Sci. Instrum.* 70: 3332. Copyright 1999, American Institute of Physics.)

## Experimental Results

Figure 7.2 shows the variation of the plasma (ion) density $N_+$, with the feed power $P_f$ at two pressures for the cases of circular and the square cross-sections, denoted as "circular" and "square" in the figure. It is seen that $N_+$ is higher in the case of the circular tube. Moreover, $N_+$ is greater than the plasma cutoff density as denoted by $N_c$ for both the waveguides. At the lower pressure, for the circular case, $N_+$ is seen to rise monotonically with $P_f$, whereas for the square case $N_+$ increases up to 280–300 W, but tends to saturate at higher powers. It is interesting to note that $N_+$ is higher at the lower pressure in this pressure range. At this higher pressure, $N_+$ varies with $P_f$ stepwise. $N_+$ is higher than that obtained in an earlier experiment [1] in a circular waveguide with a larger radius (5.75 cm).

Figure 7.3 shows the variations of $V_s$ and the electron temperature $T_e$ with $P_f$ for the two cases of cross-sections. Both $V_s$ and $T_e$ are higher in the case of the circular tube. It is seen that $V_s$ and $T_e$ rise with $P_f$ in both the tubes, the rise being sharper in the case of the circular waveguide. $V_s$ is higher than the ionization potential of Ar (= 15.76 eV). With increase in $P_f$, the difference in the values of $V_s$ and $T_e$ for the two tubes becomes wider.

Figure 7.4 shows $f(E)$ for the circular tube, where $N_+ > N_c$. The circular tube has been chosen because of a higher value of the electron temperature. For this experimental condition, $N_+ = 1.9 \times 10^{11} \mathrm{cm}^{-3}$, and the average energy

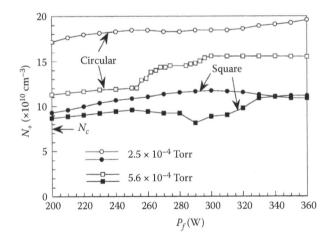

**FIGURE 7.2**
Variation of the plasma (ion) density $N_+$ with the feel power $P_f$ at pressures of $2.5 \times 10^{-4}$ and $5.6 \times 10^{-4}$ Torr for the circular and square tubes, respectively. (Reprinted with permission from S. Bhattacharjee and H. Amemiya. 1999. Production of microwave plasma in narrow cross-sectional tubes; effect of the shape of cross section, *Rev. Sci. Instrum.* **70**: 3332. Copyright 1999, American Institute of Physics.)

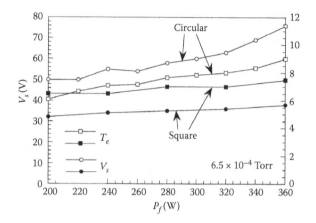

**FIGURE 7.3**

Variations of the space potential $V_s$ and electron temperature $T_e$ with the feed power $P_f$ at a pressure of $6.5 \times 10^{-4}$ Torr for the circular and square waveguides. (Reprinted with permission from S. Bhattacharjee and H. Amemiya. 1999. Production of microwave plasma in narrow cross-sectional tubes; effect of the shape of cross section, *Rev. Sci. Instrum.* **70**: 3332. Copyright 1999, American Institute of Physics.)

$E_{av} = 64$ eV. For a comparison, the Maxwellian and Druyvesteyn distributions with the same $E_{av}$ are drawn as $f_M$ and $f_D$, respectively. The population of the lower-energy electrons is strongly deficient as compared to $f_M$ and $f_D$, but appears in the medium energy range with the mean energy of 64 eV. A 10-times magnified view of the high-energy tail shows that the tail

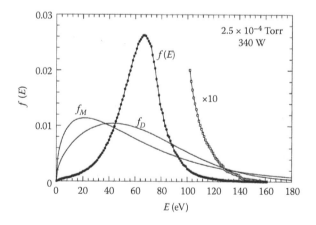

**FIGURE 7.4**

Electron energy distribution $f(E)$ for the circular tube at a pressure of $2.5 \times 10^{-4}$ Torr and a power of 340 W. $f_M$ and $f_D$ indicate the Maxwellian and Druyvesteyn distributions with the same average energy. (Reprinted with permission from S. Bhattacharjee and H. Amemiya. 1999. Production of microwave plasma in narrow cross-sectional tubes; effect of the shape of cross section, *Rev. Sci. Instrum.* **70**: 3332. Copyright 1999, American Institute of Physics.)

extends to at least 160 eV. The average energy is greater and the high-energy tail elongated as compared to those of an earlier experiment [2]. The fact that the high-energy tail as measured by the probe extends up to at least 160 eV indicates that direct ionization to $Ar^{8+}$ is possible.

## Discussion

### Comparison of the Effect of Cross-Section

The distribution of the magnetic field $B$ is different for the two types of tubes as shown in Figure 7.5, where the origins are taken as $r = 0$ and $x, y = 0$ for the circular and square cross-sections, respectively (Figure 7.1). In the case of circular cross-section, the magnetic field in the radial direction is denoted by "RC" and the magnet surface ($r = 3.4$ cm) is denoted by "WC." The ECR position ($B = 875$ G) lies at $r = 2.65$ cm. $B(r)$ can be fitted by $B(r) = k1r^{m/2-1}$, where $r$ is in cm and $B$ is in Gauss. $k1$ is a constant and $m$ is the number of poles. A good fitting was obtained with $k1 = 18$ and $m = 10$. In the case of the square cross-section, the field along the diagonal and midline shown with dotted lines D and M in Figure 7.1b is indicated as "DS" and "MS" in Figure 7.5. The field along the midline has a minimum-$B$ at $x = 0$. The field can be fitted by $B(x) = k2x^{m/2-1} + k3x^{n/2-1}$, where $k2$ and $k3$ are constants and $m$ and $n$

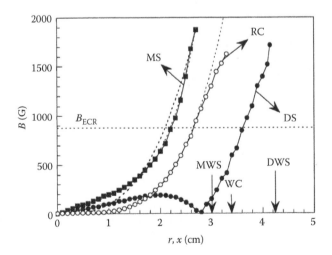

**FIGURE 7.5**
Distribution of the magnetic field $B$ in the cross-section of the circular and square tubes. (Reprinted with permission from S. Bhattacharjee and H. Amemiya. 1999. Production of microwave plasma in narrow cross-sectional tubes; effect of the shape of cross section, *Rev. Sci. Instrum.* **70**: 3332. Copyright 1999, American Institute of Physics.)

are multipolar contributions. A good fitting was obtained with the values $k2 = 10$, $k3 = 172$, $m = 12$, and $n = 4$.

This indicates that 12-polar and quadrupolar fields are superposed. The field along the diagonal has null points at $x = 2.8$ cm. The ECR condition is satisfied at about $x = 2.2$ cm along the midline and $x = 3.6$ cm along the diagonal. The wall along the midline is indicated as "MWS" at $x = 3$ cm and the corner of the square as "DWS" at $x = 4.24$ cm. A pair of ECR points near the corner, which are beyond the null points, may not be effective. This leads to a decrease in the number of effective ECR points to $24 - 8 = 16$. In the circular tube, on the other hand, all the ECR points, 20, can be effective.

If the microwave power density is the same in the tubes, we expect a nearly equal plasma density. The tubes have been designed to have a cross-section $\pi a^2 = (2b)^2 \sim 36$ cm$^2$ and equal lengths (37 cm). If the magnetic field $B = 0$ or is only a small perturbation along the tube periphery, the solution of the diffusion equation results in a cosine-type density distribution for the square cross-section and a Bessel-type radial density distribution for the circular cross-section. The densities $N_{cl}$ and $N_{sq}$ integrated over the respective cross-sections are given by

$$N_{cl} = \frac{N_o}{\pi a^2} \int_0^a J_0\left(\frac{2.4r}{a}\right) 2\pi r \, dr \tag{7.1}$$

$$N_{sq} = \frac{N_o}{b^2} \int_0^b \int_0^b \cos\left(\frac{\pi x}{2b}\right) \cos\left(\frac{\pi y}{2b}\right) dx \, dy \tag{7.2}$$

where $N_o$ is the density at the center. We get $N_{cl}/N_{sq} = 1.06$.

When we include the effects of $B$, the density distribution would be given by the modified diffusion coefficient $D_\perp$ in

$$D_\perp \frac{d^2N}{dr^2} + \frac{D_\perp}{r} \frac{dN}{dr} + \upsilon N = 0, \tag{7.3}$$

where $D_\perp = D_o/(1 + \omega_c^2 \tau^2)$ is the perpendicular diffusion coefficient, $D_o$ is the diffusion coefficient for the case when $B = 0$, $\omega_c = eB/m$ is the electron–cyclotron frequency, $\tau$ is the mean-free time, and $\upsilon$ is the ionization coefficient. For the circular waveguide, taking the radial variation of $B$ as $B = 18r^4(G)$, where $r$ is in cm and $\tau = 2.7 \times 10^{-7}$ s, as estimated from the experimental parameters of pressure $p = 2.5 \times 10^{-4}$ Torr, electron temperature $T_e = 10$ eV, $D_\perp$ can be written as $D_\perp = D_o/(1 + Cr^8)$, where $C$ (= $7.45 \times 10^3$) is a constant. Equation 7.3, normalized with respect to the tube's radius $a$, reads as

$$\frac{d^2N}{dX^2} + \frac{1}{X} \frac{dN}{dX} + \frac{a^2\upsilon}{D_o}\left(1 + Ca^8X^8\right)N = 0, \tag{7.4}$$

where $X = r/a$ and $a = 3.4$ cm. To find a solution to Equation 7.4, at first we consider two limiting cases. For, $Ca^8X^8 \ll 1$, Equation 7.4 can be written as

$$\frac{d^2N}{dX^2} + \frac{1}{X}\frac{dN}{dX} + \frac{a^2v}{D_o}N = 0, \tag{7.5}$$

which has a solution given by

$$N(X) = N_oJ_o\left(a\sqrt{\frac{v}{D_o}}X\right). \tag{7.6}$$

For $Ca^8X^8 \gg 1$, Equation 7.4 can be written as

$$\frac{d^2N}{dX^2} + \frac{1}{X}\frac{dN}{dX} + \frac{Cva^{10}}{D_o}X^8N = 0, \tag{7.7}$$

which reduces to a Bessel's equation with the substitution $y = X^5$, and has a solution given by

$$N(X) = N_oJ_o\left(\frac{1}{5}\sqrt{\frac{Cv}{D_o}}a^5X^5\right). \tag{7.8}$$

From the boundary condition, $N/N_o = 0$ at $X = 1$, we obtain $1/5\sqrt{Cv/D_o}\,a^5 = 2.4$. Hence, Equation 7.8 can be written as

$$N(X) = N_oJ_o(2.4X^2). \tag{7.9}$$

The solutions for the cases including $B$ and where $B = 0$ are shown in Figure 7.6.

In the case of the square tube, the magnetic field is complicated because of the two-dimensional geometry. Moreover, the field along the midline and the diagonal are nonuniform. To simplify the situation, we use a simpler representation: $B = 19x^3(G)$, where $x$ is in cm, which is a reasonably good fit of the midline field as shown by a thick-dotted curve in Figure 7.5. The normalized diffusion equation with respect to the tube's half-width $b$ can be written as

$$\frac{d^2N}{dX^2} + \frac{b^2v}{D_o}(1 + Cb^6X^6)N = 0, \tag{7.10}$$

where $X = x/b$, $b = 3$ cm, and $C$ ($= 8.3 \times 10^3$) is a constant.

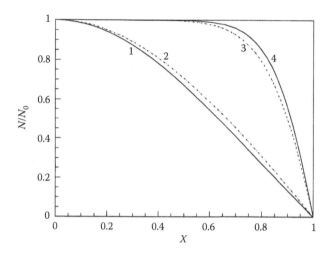

**FIGURE 7.6**
Distribution of the normalized plasma density in the circular and square tubes. Dotted lines 2 and 3 represent the square tube, and solid lines 1 and 4 represent the circular tube. Lines 1 and 2 are the $B = 0$ case. (Reprinted with permission from S. Bhattacharjee and H. Amemiya. 1999. Production of microwave plasma in narrow cross-sectional tubes; effect of the shape of cross section, *Rev. Sci. Instrum.* **70**: 3332. Copyright 1999, American Institute of Physics.)

For $Cb^6X^6 \ll 1$, Equation 7.10 can be written as

$$\frac{d^2N}{dX^2} + \frac{b^2v}{D_o}N = 0,$$ (7.11)

which has a solution given by

$$N(X) = N_o \cos\left(b\sqrt{\frac{v}{D_o}}x\right).$$ (7.12)

For $Cb^6X^6 \gg 1$, Equation 7.10 can be written as

$$\frac{d^2N}{dX^2} + gX^6N = 0,$$ (7.13)

where $g = CvB^8/D_o$. On substituting $N = n/\sqrt{X}$ and $y = X^4$ in Equation 7.13, it reduces to a modified Bessel's equation of the form

$$\frac{d^2n}{dy^2} + \frac{1}{y}\frac{dn}{dy} + \left(\frac{g}{16} - \frac{(1/8)^2}{y^2}\right)n = 0,$$ (7.14)

which has an analytical solution given by

$$N(X) = N_o \sqrt{X} J_{1/8} \left( \frac{1}{4} \sqrt{g} X^4 \right) \tag{7.15}$$

On applying the boundary condition to Equation 7.15, $N/N_o = 0$ at $X = 1$, we evaluate the constants, $\sqrt{g}/4 = 2.595$, which gives $b^2 v/D_o = 1.78 \times 10^{-5}$. Therefore, Equation 7.15 can be rewritten as

$$N(X) = N_o \sqrt{X} J_{1/8}(2.595 X^4). \tag{7.16}$$

To a good degree of approximation, the solution given by Equation 7.16 is valid for $X \geq 0.98$ and that of Equation 7.12 for $X \leq 0.21$. For $0.21 \leq X \leq 0.98$, we numerically solved Equation 7.10.

The density distribution for $B = 0$ and that including $B$ for the square tube are shown in Figure 7.6, where a comparison with the results of the circular tube is made.

The ratio of the average densities $N_{cl}$ and $N_{sq}$ integrated over the respective cross-sections for the case including $B$ gives the result $N_{cl}/N_{sq} = 1.08$, which is slightly larger than for $B = 0$. However, according to the experimental results of Figure 7.2, the plasma density for the case of the circular tube as compared to the square tube is higher by 25–50%. The circular cross-section supports a higher plasma density, although fewer number of magnetic poles are used.

The above discrepancy between the calculated and the experimental results is possibly due to two reasons: (a) efficiency of the ECR action— as discussed earlier, in a square tube only 16 ECR points are effective, although there are 12 poles of the magnets. A pair of ECR points near each corner, which are beyond the null points, may not be effective. In the near-circular tube, on the other hand, all 20 ECR points can be effectively operating and helping in the plasma production; and (b) simplified form of the field—the above treatment favors the square tube because the effect of confinement has been represented by the field distribution along the midline. In the diagonal direction, there are null points and the magnetic bottle is weaker; this will further suppress the average density from the value obtained above.

## Mechanisms of Plasma Maintenance

### *Ionization Effects of the Space Potential and the Electron Temperature*

The ionization rate $N_e \langle \sigma_i(\kappa T_e) v_e \rangle$, by considering the electron temperature $T_e$, and $N_{ew} \sigma_i(eV_s)(2eV_s/m)^{1/2}$, by considering the space potential energy $eV_s$, were calculated for the parameters in Figure 7.3. Here $N_e$ is the electron density, $N_{ew}$

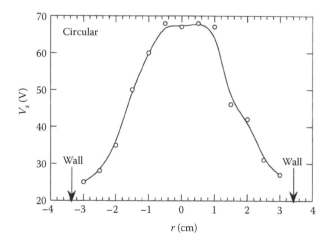

**FIGURE 7.7**
Radial variation of the space potential ($V_s$) in the circular waveguide. Note that $V_s$ is peaked at the center.

is the electron density near the wall ($N_{ew}/N_e \sim 1/10$), $\sigma_i$ is the ionization cross-section, $v_e$ is the electron thermal velocity, and $m$ is the electron mass. For the potential ionization effect, it was assumed that the electrons near the wall were dc-accelerated to the core plasma with an energy $eV_s$ and cause ionization.

Figure 7.7 shows a typical example of the radial variation of $V_s$ in the case of the near-circular tube. It is seen that $V_s$ is peaked at the waveguide center. A similar tendency was seen in the square waveguide. Calculations showed that at a lower power $P_f$ the ionization effects of $V_s$ and $T_e$ were comparable, whereas at a higher $P_f$ the effect of $T_e$ dominates over that of $V_s$. This tendency is pronounced in the case of the circular waveguide. The calculated results are shown graphically in Figure 7.8. Thus, the ionization effects of both $V_s$ and $T_e$ help in the production and maintenance of the overdense plasma, but the latter effect is higher as the microwave power is increased, particularly in the circular case.

### Reduction of the Wavelength λ due to UHR

Once the plasma is created, the medium inside the tube becomes inhomogeneous and anisotropic. In such a situation, the type of wave mode inside the tube can no longer be well-defined. However, depending upon the density distribution and the magnetic field, it can be expected that the plasma production and maintenance are supported by the UHR given by $\alpha^2 + \beta^2 = 1$, where $\alpha = \omega_p/\omega$, $\beta = \omega_c/\omega$, $\omega_p$, $\omega_c$ and $\omega$ are the angular plasma, electron–cyclotron, and wave frequencies, respectively. Due to the change in the refractive index in the magnetized regions, it can be explained that the waves can pass through the periphery of the central overdense plasma (near the UHR regions)

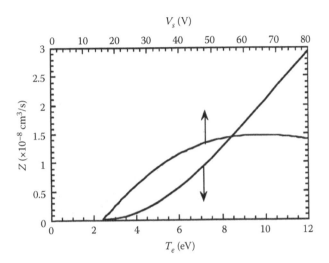

**FIGURE 7.8**
Comparison of the ionization efficiency of the electron temperature $T_e$ and the space potential $V_s$.

with a short wavelength, much smaller than the wavelength of microwaves in free space, the radial plasma extent, and the scale of magnetic nonuniformity $B/(dB/dr) = 0.08$ cm. The wavelength of the waves can be estimated from [3]

$$\left(\frac{k}{k_o}\right)^2 = \left(\frac{\lambda_o}{\lambda}\right)^2 = 1 - \frac{\alpha^2\left\{(1 - j\delta) - \alpha^2\right\}}{(1 - j\delta)^2 - \beta^2 - \alpha^2(1 - j\delta)} - \left(\frac{k_c}{k_o}\right)^2, \qquad (7.17)$$

where $k$, $k_o$, and $k_c$ are the wave numbers in the magnetized plasma in a waveguide, that in free space, and the cutoff wave number in the waveguide without plasma, respectively. $\delta = v_e/\omega$, where $v_e$ is the electron neutral collision frequency and $j^2 = -1$. As a typical example, the wavelength near the UHR region has been estimated theoretically and calculated using the experimental parameters $k_o = 0.5133$ cm$^{-1}$, $k_c = 0.5417$ cm$^{-1}$ (near circular tube), $\lambda_o = 12.24$ cm, and $\delta \leq 10^{-3}$. For $B = 500$ G, $N_+ = 5.0 \times 10^{10}$ cm$^{-3}$, we obtain $\lambda \cong 0.02$ cm. Under this condition, microwaves can propagate according to the one-dimensional wave theory for uniform plasma and the magnetic field. Such short wavelength waves would pass through the tube in the peripheral region and help to sustain the discharge.

### Existence of Pass Band for Wave Propagation

If the waves propagate with a small wavelength through the periphery of the central overdense plasma, the propagation characteristics do not deviate much from those in free space. Assuming a cold and collisionless plasma, the propagating waves in the magnetized region could be considered as a superposition of the ordinary ($E \parallel B$) and extraordinary ($E \perp B$) waves as discussed earlier. The X-wave has a resonance at UHR, which depends on $\alpha$ (plasma density) and $\beta$ (magnetic field). It is supposed that most of the power absorption occurs here. Moreover, for the X-wave as $E \perp B$, some of the power is also absorbed in ECR heating. The O-wave, on the other hand, does not have any resonance and is cut off once the density approaches its critical value $N_c = 7.4 \times 10^{10}$ cm$^{-3}$, irrespective of the value of $B$.

On analytically solving Equation 7.17, we find, depending upon the value of the magnetic field and the plasma density, the existence of finite regions of wave propagation (pass band) and nonpropagation (rejection band) for the extraordinary wave. Additionally, the ordinary wave is found to be damped inside the waveguide. Neglecting the conductivity of the medium, the propagation constant $\gamma$ can be written as $\gamma = jk$, then Equation 7.17 can be rewritten as

$$\left(\frac{\gamma}{k_o}\right)^2 = \left(\frac{k_c}{k_o}\right)^2 - \left\{1 - \frac{\alpha^2(1-\alpha^2)}{1-\beta^2-\alpha^2}\right\} \tag{7.18}$$

and a similar equation can be written for the case of the ordinary wave as

$$\left(\frac{\gamma}{k_o}\right)^2 = \left(\frac{k_c}{k_o}\right)^2 - (1-\alpha^2). \tag{7.19}$$

To further simplify Equations 7.18 and 7.19, we substitute $\Gamma = \gamma/k_o$, $K = (k_c/k_o)^2 - 1$, $X = \alpha^2$, and $Y = \beta^2$. The constant $K$ depends upon the cutoff wave number of the waveguide. For the case of the square and the near-circular waveguides, the calculated values of $k_c$ are 0.5236 and 0.5417 cm$^{-1}$, respectively. With $k_o$ being 0.5133 cm$^{-1}$, the value of $K$ becomes 0.04 and 0.11 for the square and the circular waveguides, respectively. The parameters $X$ and $Y$ depend upon the plasma density and the magnetic field in the waveguide.

With the above substitutions, Equations 7.18 and 7.19 can be written as

$$\Gamma^2 = K + \frac{X(1-X)}{1-Y-X} \tag{7.20}$$

and

$$\Gamma^2 = K + X, \tag{7.21}$$

respectively. The condition for wave propagation along the waveguide requires that the propagation constant $\gamma^2 \leq 0$, implying $\Gamma^2 \leq 0$. Therefore,

$$K + \frac{X(X-1)}{1-Y-X} \leq 0 \tag{7.22}$$

and

$$K + X \leq 0. \tag{7.23}$$

Considering Equation 7.23 first, it is seen that for the O-wave, the inequality is not satisfied because (a) $X$ is always $\leq 0$ and (b) $K > 0$ for the present cases. Therefore, the O-mode does not propagate. Coming to the case of X-waves, Equation 7.22 can be simplified as

$$y = \frac{X(X-1)}{1-Y-X} \geq K. \tag{7.24}$$

Several results could be immediately realized from the above inequality by considering the density and magnetic field distribution in the waveguide.

a. $X = 1$: There is no propagation.
b. $Y = 1$: The pass band lies in the range $0 \leq X \leq 1 - K$.
c. $Y = 0$: There is no propagation.
d. $X = 1 - Y$: A condition for UHR. There is propagation near the resonance region.
e. $X < 1 - Y$: The condition for propagation is $Y < 1$. The pass band lies in the range

$$\left[ \frac{-(K-1) + \sqrt{(K+1)^2 - 4KY}}{2} \right] \leq X \leq 1 - Y. \tag{7.25}$$

f. $X > 1 - Y$: The condition for propagation is $Y > 1$. The pass band lies in the range

$$\left[ \frac{-(K-1) - \sqrt{(K+1)^2 - 4KY}}{2} \right] \leq X \leq \left[ \frac{-(K-1) + \sqrt{(K+1)^2 - 4KY}}{2} \right]. \tag{7.26}$$

The limiting value of $Y = (K+1)^2 / 4K$, above which there is no pass band.

Additionally, Equation 7.24 has been plotted in Figure 7.9 ($y$ versus $X$) for three values of $\beta$ ($\beta < 1$, $\beta = 1$, and $\beta > 1$), which correspond to fields below ECR, at ECR, and above ECR, respectively.

The constant $K$ for the circular waveguide is shown by a thin-dotted line. The parameter $X$ has been varied from $X = 0 - 2$, which corresponds to a

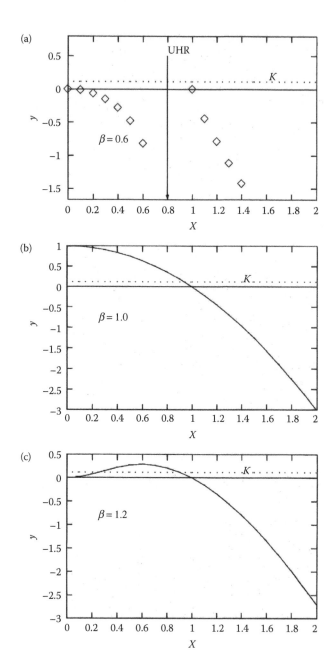

**FIGURE 7.9**
Graphical solution of Equation 7.24 for the circular waveguide for three different values of (a) $\beta < 1$, (b) $\beta = 1$, and (c) $\beta > 1$. Regions of $y \geq K$ represent pass band.

variation in plasma density from $\omega_p = 0 - \sqrt{2}$. The plasma density equals the cutoff density at $\omega_p = 1$. When $0 < \beta < 1$ (Figure 7.9a), UHR occurs at $X = 1 - \beta^2$, and the region of pass band is $1 - \beta^2 \leq X \leq 1$. For $\beta = 1$ (Figure 7.9b), the region of pass band is from $0 \leq X \leq 1$. Finally, for $\beta > 1$ (Figure 7.9c), the pass band has two limiting densities, lower and higher. The pass band becomes narrower as $\beta$ is increased and the limiting value of $\beta$ beyond which there is no pass band has been deduced as above ($Y = (K + 1)^2/4\,K$).

Thus, depending upon the magnetic field and the plasma density in the waveguide, propagation and rejection bands could exist even though the geometrical size of the waveguide is below cutoff.

## Conclusions

A plasma could be produced in a square waveguide with a dimension smaller than the cutoff value. Overdense plasmas with a density of $(0.8 - 2.0) \times 10^{11}$ cm$^{-3}$ are obtained in the range of $10^{-4}$ Torr for powers of 200–360 W. The electron temperature is 6–12 eV. Under the same experimental conditions, the plasma properties are compared with a circular waveguide with almost the same cross-sectional area. The plasma density and the electron temperature are round to be higher for the circular cross-section. The electron energy distribution of the overdense plasma in the near-circular waveguide considerably differs from the Maxwellian and the Druyvesteyn distributions in that the low- and very high-energy electrons are deficient in the distribution.

## Summary of Experimental Results in Continuous-Mode Microwaves

The experiments using continuous-mode microwaves have shown that a plasma can be produced in a waveguide even if the cross-section is smaller than the geometrical cutoff value. Additionally, overdense plasma can be maintained in the waveguide at the given wave frequency. Thus the objective of obtaining a high-density, narrow, and cross-sectional plasma has been realized using microwaves in the *cw* mode.

In the near-circular multicusp waveguide of 10 poles, a plasma density in the range of $10^{11}$ cm$^{-3}$ could be obtained at 200–360 W of power and $10^{-4}$ Torr range of pressure. The radial variation of the plasma density has shown that the plasma was overdense ($N_+ > N_c$) at least over a diameter of 4 cm and decreased toward the wall. The axial variation of the plasma density demonstrated uniform overdense plasma along the entire (37 cm) length of the waveguide. The radial measurement of the magnetic field inside the waveguide clearly demonstrated the minimum-B (bucket) field structure, with a distance of about 1 cm from the center being almost field free. The ECR condition was realized all along the waveguide at a distance of 0.75 cm from the inner wall. The electron temperature was 5–11 eV with slightly higher values near the ECR region, and the space potential was found to have a peak at the center of the waveguide with a value of about 70 V. Special characteristics such as an optimum pressure at $2.5$–$3.5 \times 10^{-4}$ Torr, where the density became a maximum, the presence of hysteresis in the ion current, and density jumps were observed depending upon power and pressure. The optimum pressure effect could be explained by a particle balance equation. The mechanisms of overdense plasma production were explained by field penetration, minimum-B confinement, potential ionization, and diffusion.

In the rectangular nonmulticusp waveguide, a linear magnetic field was formed by arranging three magnets on two sides of the waveguide. The field was almost uniform at the center of the waveguide, with values lying between 1125 and 1250 G and dropped at the side walls. The field was higher at the proximity of the magnets. At the join of the pipes and near the side walls, the field was weaker due to fringe effects. ECR condition was realized near the wall at about $x = 2.45$ cm from the center (Figure 6.4). A plasma density of about $2 \times 10^{10}$ cm$^{-3}$ was obtained in the $E \parallel B$ case and more than an order higher ($\sim 2.5 \times 10^{11}$ cm$^{-3}$) in the $E \perp B$ case. In the $E \parallel B$ mode, the plasma density, electron temperature, and space potential were maximum at an optimum pressure (0.2 mTorr). However, in the $E \perp B$ case, the plasma parameters increased with increase in power and pressure, indicating efficient ionization at all experimental conditions. The electron temperature was 3–11 eV and the space potential fell in the range of 20–70 V in both the modes. The particle trajectories were studied from single particle theory, which indicated that in the ordinary mode the particles describe a cyclotron motion, with the guiding center drifting in the direction of the magnetic field. In the extraordinary mode, the component of the velocity of the particles along the field is a constant.

In the square multicusp waveguide of 12 poles, the measured plasma properties were compared to those of the near-circular waveguide (10 poles) with almost the same cross-sectional area. The absolute values of the plasma parameters were smaller than that of the near-circular waveguide. At 200 W, the plasma density in the square waveguide was about $9.5 \times 10^{10}$ cm$^{-3}$ and the electron temperature was 5–8 eV. The experimental results of the comparative study were explained on the basis of the difference in the geometrical density distribution and the effect of the magnetic field distribution in the two waveguides. In the case of the square cross-section, the field along the midline (Figures 7.1 and 7.5) has a minimum-B at the center, and it is understood that the field distribution is a superposition of 12-polar and quadrupolar fields. The field along the diagonal has an additional null point at $x = 2.8$ cm. The ECR condition is satisfied at $x = 2.2$ cm along the midline and $x = 3.6$ cm along the diagonal. The mechanism of the overdense plasma production was explained by a comparison of the ionization rate due to the electron temperature and the space potential. Wave penetration through the peripheral region of the overdense plasma and UHR effects were also considered to be important. The wave propagation aspects were also studied from pass band theory, which showed the existence of pass bands or rejection bands depending upon the magnetic field and the plasma density in the waveguide. The properties of the plasma obtained in all the three waveguides are summarized in Table 7.1.

## Summary of the Plasma Production Mechanisms

### Plasma Ignition at the Waveguide Entrance

As the radial dimension of the waveguide is smaller than cutoff, it is difficult for the waves to propagate in vacuum. Despite this restriction, ignition can occur at the waveguide's entrance due to field penetration of the spatially decaying wave (evanescent wave). Once the plasma is created at this location, electrons can ionize the gas while making bounce motion in the arched field lines and help the plasma to spread along the waveguide. Thereafter, the wavelength of the fundamental mode is determined more by the magnetized plasma properties like the refractive index, rather than the geometry of the waveguide.

### Potential Ionization and Confinement

In the central region of the waveguide, the plasma is overdense and it is difficult for the waves to propagate through the plasma because of the plasma density cutoff. Due to the high value of the space potential $V_s$ in the center, electrons can be accelerated from the peripheral regions toward the center. Ionization by the dc field can occur and raise $N_+$. The high $V_s$ also helps to confine electrons in both radial and axial directions to maintain the high-density plasma.

### Confinement by the Minimum-$B$ Magnetic Field

Owing to the structure of the minimum-$B$ confinement by the magnetic bottle, charged particles tend to move toward the bottle's center. Moreover, the outer radial diffusion from the center across the magnetic field is suppressed except at the cusp regions. Both these effects help to accumulate the plasma in the center of the waveguide.

**TABLE 7.1**

A Summary of the Important Results Obtained in the Three Waveguides

| Cross-Section | Type of Field | No. of Poles | Continuous-Mode Microwaves | | | | |
| | | | Type of Plasma | Max. Density ($cm^{-3}$) | Optimum Pressure (Torr) | Electron Temp. (eV) | Space Potential (V) |
|---|---|---|---|---|---|---|---|
| A. Near-circular | Multicusp | 10 | Uniform (overdense) | $1.7 \times 10^{11}$ | $\sim 3 \times 10^{-4}$ | 5–11 | 50 |
| B. Rectangular | Nonmulticusp | 6 | | | | | |
| 1. $E \perp B$ | (Linear) | (3 magnets each on two sides) | Filamentary (underdense) | $2.5 \times 10^{11}$ | Nil | 3–11 | 60 |
| 2. $E \| B$ | (Linear) | (3 magnets each on two sides) | Filamentary (underdense) | $2.0 \times 10^{10}$ | $\sim 2 \times 10^{-4}$ | 4–10 | 35 |
| C. Square | Multicusp | 12 | Uniform (overdense) | $9.5 \times 10^{10}$ | $\sim 2 \times 10^{-4}$ | 5–10 | 30 |

*Note:* A comparison of the three waveguides and the plasma produced in the maximum plasma density at 200 W, the space potential at 200 W and 0.2 mTorr.

## Plasma Diffusion into the Waveguide

By the provision of the multicusp, the radial diffusion of the particles is suppressed and the plasma can stream into the waveguide in the axial direction. The axial variation of the plasma density $N_+(z)$ would follow, $N_+(z) = N_{+0} \exp\left[-(2.4/R)\sqrt{D_r/D_z}\,z\right]$ (section on Discussion), where $R$ is the radius of the waveguide, $D_r$ and $D_z$ are the radial and axial diffusion coefficients, and $N_{+0}$ is the plasma density at the entrance (microwave input side) of the waveguide ($z = 0$).

## Waveguide's Reduced Dimension

Owing to the smaller dimension of the waveguide, the microwave power density increases and can favor resonance heating such as UHR and electron–cyclotron resonance in the region outside the central overdense plasma (region II, Figure 5.5). This would enhance ionization at the resonance regions and help in the production and maintenance of an overdense plasma.

## Wave Propagation with a Reduced Wavelength and Existence of Pass Bands

The wavelength of the waves can be calculated from the dispersion relation (Equation 7.17), as described in this chapter. At or near UHR, the wavelength of the waves becomes much smaller than the free space wavelength, the scale length of magnetic nonuniformity, and the radial plasma extent. Hence, the uniform infinite plasma theory may be applicable and the waves can propagate through the periphery of the overdense plasma with a small wavelength. It has also been seen that depending upon the magnetic field and the plasma density inside the waveguide propagation bands could exist even though the geometrical size is below cutoff.

---

# References

1. M. Maeda and H. Amemiya. 1994. Electron cyclotron resonance plasma in multicusp magnets with axial magnetic plugging, *Rev. Sci. Instrum.* **65**: 3751–3755.
2. H. Amemiya and M. Maeda. 1996. Multicusp type machine for electron cyclotron resonance plasma with reduced dimensions, *Rev. Sci. Instrum.* **67**: 769–774.
3. W.P. Allis, S.J. Buchsbaum, and A. Bers. 1963. *Waves in Anisotropic Plasmas* (MIT Press, Cambridge, MA).
4. S. Bhattacharjee and H. Amemiya. 1999. Production of microwave plasma in narrow cross-sectional tubes; effect of the shape of cross section, *Rev. Sci. Instrum.* **70**: 3332.

# 8

## Plasma Production in Pulsed-Mode Microwaves: General Experiment

### Introduction

In this chapter, the experimental set-up for plasma production and its properties will be described using pulsed-mode microwaves. There are two parts of the experiment, as described in this chapter and Chapter 9.

The first part is of a general nature, where a waveguide with a dimension much larger than the cutoff value is used to study the general characteristics of a plasma produced by high-power, short-pulse microwaves. The waveguide was placed in a large vacuum chamber referred to as the "space chamber," as the chamber was primarily meant for space physics experiments. The chamber is evacuated using a turbo-molecular pump. Three different gases possessing varied properties, such as atomic, molecular, molecular mixture, electropositive, electronegative, possessing metastable states, and so on, are used in the research, and the behavior of the pulsed discharge in the power-off phase is studied. The values of the electron temperature and the plasma density are obtained. The study is conducted by oscilloscope measurements of the current drawn from a Langmuir probe.

With some understanding of the plasma characteristics of a high-power, short-pulse microwave plasma from the above experiments, the second experiment reverts to the cutoff problem using a circular waveguide with a dimension below the cutoff value. Plasma diagnostics and measurements are done by using a Boxcar integrator, digital oscilloscopes, a computer for storing data, an x–y recorder, and a probe for electric field measurements. A tiny leak valve L (cf. Figure 5.3) is connected from the chamber to the diffusion pump so that the experiment can be performed under gas flow conditions, and the plasma is kept clean. The chamber used here is the same as in experiments with continuous-mode microwaves. The temporal evolution of the plasma parameters is measured both from the probe characteristics and the oscilloscope data. In addition, the axial variation of the electric field and the plasma are measured along the waveguide.

In the next section, "Apparatus and Methods in the General Experiment," the experimental apparatus used in the first experiment is described. In the following section, "Apparatus and Methods of the Experiment with a Circular Waveguide with a Dimension below Cutoff," the experimental apparatus used for investigating the plasma production in the narrow circular waveguide is described. Some additional details on the experimental apparatus and measurement techniques are also included.

In Chapters 5 through 7, we studied the application of continuous-mode microwaves for producing a plasma in magnet-containing waveguides with a dimension smaller than the cutoff value. The results suggested that cw mode microwaves can be used to produce a high-density, narrow, and cross-sectional plasma at lower pressures (~$10^{-4}$ Torr) and moderate values of microwave power (200–360 W). At this pressure range, the electron- and ion-neutral mean-free paths, $\lambda_e$ and $\lambda_i$, are much larger than the system size, and the plasma can be considered effectively collisionless. Moreover, as the electron–cyclotron radius $r_e$ is much smaller than $\lambda_e$, ECR action can occur efficiently, favoring the absorption of microwaves and production of high-density plasma.

However, depending upon the application, if one wants to operate the source at higher pressures (~ a few Torr), where $\lambda_e$ and $\lambda_i$ become smaller than the system size, and $r_e$ is comparable or larger than $\lambda_e$, in such a situation the plasma is collisional and resonance effects may not be important for particle production in the discharge. Moreover, diffusion effects start playing an important role. As an alternative, it then becomes necessary to increase the power of the microwaves to obtain a reasonably high density. Pulsed microwaves are then a useful alternative because higher powers can be conveyed to the plasma in short-duration pulses. Furthermore, if high power is applied to the plasma in short pulses, steady-state conditions in the parent plasma can be avoided and an active plasma with a higher electron temperature can be produced in the power-off phase of the discharge. Moreover, due to the highly nonequilibrium state of the parent plasma, it is likely that the plasma build-up would be extended beyond the end of the microwave pulse. Pulsed microwave plasmas can be useful for the production of species with high internal energy, like multicharged ions, radicals, and metastables. The plasma has a greater controllability by varying the pulse width, pulse repetition frequency, and duty cycle—an option that does not exist in cw mode microwaves.

In the early studies on high-power, short-pulse microwave discharges were mainly oriented toward the determination of breakdown thresholds and microwave propagation through a gaseous medium [1–6]. However, characteristics of the plasma generated between the pulses of high-power waves, which would be quite different from the plasma during the pulse, have not been investigated fully. Further, these plasmas would be considerably influenced by the transient parent plasma if the pulse width is short.

In recent reports on a pulsed microwave discharge for negative ion production [7] in $Cl_2$ and damage-free etching [8], the 2.45 GHz microwave power was modulated with equal lengths of pulse duration ($t_w$) and inter-pulse time ($t$) of about 100 μs and a pulse height ($h$) ≈ 1 kW. Rousseau et al. used a pulsed microwave discharge for H production [9], where $t_w$ = 10 ms, $t$ = 50 ms, and $h$ about 1 kW. In an ECR-pulsed discharge, achievement of higher currents has been described [10] using pulses of about $t_w$ = 10–50 ms, $t$ = 20–100 ms, and peak power $h$ ≈ 1–1.6 kW. These, however, belong to the usual afterglow because the pulse duration is long, and the peak power is low. Moreover, here the afterglow is mainly governed by recombination and diffusion whereby the electrons turn into thermal equilibrium with the gas, thus the electron temperature $T_e$ has a lower value (<1 eV).

In this work, we study an afterglow of a different nature. The difference mainly arises from the pulse characterized by a large $h$, a short $t_w$, and a long $t$. If a high power (60–100 kW) is supplied to the plasma by narrow-width pulses ($t_w$ = 0.5–1 μs), the active parent plasma is followed by a comparatively long-time plasma, which is like a quasi-active state with electrons having still a high-average energy. We define this as an *"interpulse plasma"* to distinguish it from conventional afterglows. Recently, such short-pulse, high-power microwave discharges have been explored for environmental applications [11], with pulses of $t_w$ = $1.5 \times 10^{-8}$–$5 \times 10^{-7}$ s, $t$ = 0.1–3.0 ms, and $h$ = 10–50 GW, where a wide range of applicability of the interpulse phase can be expected. Moreover, interpulse plasmas could also be proposed for basic plasma stud-ies in the laboratory, such as plasma production in waveguides, plasma heat-ing in nuclear fusion experiments, or even for ionospheric studies, especially the D-layer to E-layer transition, where an argon interpulse plasma without negative ions can represent the E-layer plasma and the discharge in stan-dard air could represent the negative-ion-rich D-layer. The prime advantage of such an interpulse phase of the discharge is that the still-high $T_e$ in the power-off phase favors the creation of ions and radicals, as the plasma is electric-field-free with lower noise levels.

The experiment in this chapter comprises of studying the nature of the interpulse phase of a high-power, short-pulse microwave discharge. In the section "Experimental Results," the experimental results in a monoatomic gas (Ar), a molecular gas ($N_2$), and a mixture of gases ($O_2$ + $N_2$, 1:4) containing an electronegative gas ($O_2$) will be presented. Modeling and discussion will be presented in the "Discussion" section.

### Apparatus and Methods in the General Experiment

Figure 8.1 shows a schematic of the experimental apparatus. It consists of a large vacuum chamber (space chamber) in which a multicusp-type plasma bucket MC of 0.20 m $\phi$ and 0.30 m length was placed. The bucket, together with the chamber, serves as the grounded reference electrode. Pulsed microwaves of 1 μs pulse width, 60 kW peak power from a 3 GHz mag-netron oscillator PM, with a pulse repetition frequency of about 500 Hz were introduced into the bucket through an attenuator ATT. The waves

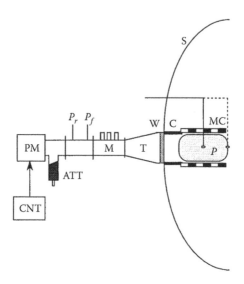

**FIGURE 8.1**

Schematic of the experimental apparatus. PM, magnetron oscillator; CNT, controller; ATT, attenuator; $P_f$ and $P_r$, power monitors for the feed and reflected wave powers; M, matching element; T, wave mode transformer; W, quartz window; C, cylindrical waveguide; MC, multicusp device; S, space chamber; and P, Langmuir's probe. (Reprinted with permission from S. Bhattacharjee and H. Amemiya. 1998. Interpulse plasma of a high-power narrow-bandwidth pulsed microwave discharge, *J. Appl. Phys.* **84**: 115. Copyright 1998, American Institute of Physics.)

were then passed through a matching element M, a mode transformer T for changing the rectangular $TE_{10}$ mode into the cylindrical $TE_{11}$ mode and then through a quartz window W into MC. Power monitors $P_f$ and $P_r$ indicated the feed and reflected powers. $P_r$ was controlled as little as possible by adjusting the matching M. C is a cylindrical waveguide to guide the microwaves into the multicusp MC. The chamber was pumped down to 1 mTorr by a rotary pump, thereafter the evacuation was backed by a turbomolecular pump until a base pressure of about $1 \times 10^{-7}$ Torr was obtained. Ar, $N_2$, or a mixture of $N_2$ and $O_2$ with a volume ratio of 4:1 was then introduced into the chamber through a needle valve. The working pressure was set at 0.2–1 Torr.

The bucket MC has been constructed in a similar fashion as described in Chapters 5 through 7 and consists of a multicusp having 16 poles. The confinement of the discharge into some volume was necessary for measurement purposes.

Measurements of plasma parameters and the temporal decay were made with a Langmuir probe P placed at some positions of MC. The probe was a disk of 6 mm diameter made of stainless steel of thickness 0.1 mm. Further construction details are described in Chapter 3. The probe was positioned on the axis of MC, where the magnetic field $B$ was minimal, almost zero.

The time-varying probe current ($I_p$) at a fixed probe voltage ($V_p$) is obtained using a digital oscilloscope. By varying $V_p$ in small steps in the range −40 to +30 V, a series of transient $I_p - t$ data is obtained, and from the data the probe characteristics are reconstructed.

For the measurement of the plasma density, the saturation region of the ion current of the plane probe characteristics is used, and the electron temperature is obtained from a slope of the logarithmic plot of the probe current in the range of probe voltage above the floating potential, as described previously.

## Apparatus and Methods of the Experiment with a Circular Waveguide with a Dimension below Cutoff

Figure 8.2 shows a schematic of the experimental apparatus with an inset showing the characteristics of the pulse modulation. Microwaves of 3 GHz with a power (60–100 kW) produced by a magnetron PM were pulse-modulated with a short pulse (0.05–1.0 µs) and a repetition frequency of 10–500 Hz. In the previous general experiment, the wave modulation was not varied. The waves were guided through a coaxial cable Cx, a power monitor PM, a matching M, a rectangular bend H, and a quartz window W into a circular tube CW (37 cm in length, 5.74 cm $\phi$) located inside a vacuum chamber C. The diameter of CW is less than the cutoff value. The cutoff diameter ($2R_c$) for the circular TE$_{11}$ mode = 5.87 cm (3 GHz). CW was inserted into a multicusp MC (10 poles).

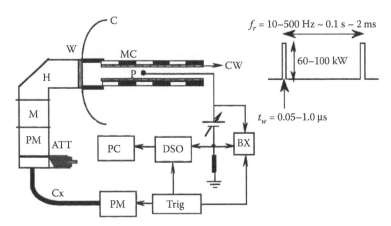

**FIGURE 8.2**
Schematic of the experimental apparatus. The inset shows the characteristics of the pulse modulation. $f_r$ = pulse repetition frequency, $t_w$ = pulse width. (Reprinted with permission from S. Bhattacharjee and H. Amemiya. 2000. Production of pulsed microwave plasma in a tube with a radius below the cutoff value, *J. Phys. D: Appl. Phys.* **33**: 1104–1116. Copyright 2000, Institute of Physics.)

A trigger signal, Trig, was used for pulse initiation and as a reference signal for diagnostics. A personal computer (PC) was used for storing data. The current decay profiles and probe characteristics were measured by a Langmuir probe P (plane, 5 mm $\phi$) using a digital oscilloscope DSO with time averaging (64 sots) and a Boxcar integrator BX. For measuring the decay profiles of ion ($I_+$) and electron currents ($I_e$), the probe was biased at –70 and 20 V, respectively.

For obtaining the probe characteristics using the Boxcar integrator, the gate aperture was set at 0.1 μs and the decay to 30–50 μs, measured in steps of ~1 μs by a slow scanning of the probe voltage. From the probe characteristics, the temporal variation of $T_e$ and $N_e$ was deduced. The electric field $E$ was measured using a needle-type probe with a 1.5 mm length aligned parallel to the electric field in the circular mode. The electric field impulse was rectified by a diode responsive to the GHz range microwave field. The axial variations of $I_+$ and $E$ were measured in DSO by moving the respective probes along the tube and recording the peak current and the field signal. The finer details of the plasma build-up within the pulse were measured with a two-channel liquid crystal display digital oscilloscope.

---

## Some Additional Details of the Experimental Set-Up

### Microwave Source

A magnetron is used as a source for pulsed microwaves. It is basically a marine radar system (RTR-018), which produces high-power (S-band, 60–100 kW) microwaves using a cavity oscillator in a vacuum tube located in a strong magnetic field. The microwaves have a frequency of 3 GHz, where the pulse width can be varied in the range 0.05–1.0 μs, the peak power could be varied in the range 0–100 kW, and the pulse repetition frequency was varied in the range 0–500 Hz. The pulse on time or pulse width can be selected during the experiment.

Figure 8.3 is a schematic diagram to show the power supply requirements for pulse width selection (power supply A (PLA) and power supply B (PLB)) and trigger signal (TX-SIG). The connection of the trigger pulse (TX-TRIG) is also shown. The tap connections numbered 18, 13, 25, 4, and 19 belong to the modulator trigger board.

The voltage requirements and the maximum frequency for safe operation for a particular pulse width are shown in Table 8.1.

### Boxcar Integrator

A Boxcar integrator has been used to make time-resolved measurements of the discharge after the end of the microwave pulse. It is useful to trace the probe characteristics at different temporal positions in the power-off phase

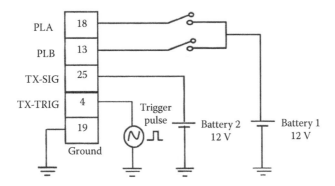

**FIGURE 8.3**
A schematic to show the power supply and trigger pulse connections that control the magnetron operation. The tap connections 18, 13, 25, 4, and 19 are part of the modulator trigger board.

of the discharge, and also for measuring the axial variation of the ion current at a particular temporal position. The gate aperture is set at 0.1 μs, and the ion-current decay set to 30–50 μs and measured. Beyond this time, the ion current is very small, and its detection is difficult. Measurement is begun about 2 μs after the end of the pulse because prior to that time the plasma has a greater disturbance. The magnetron and the Boxcar integrator are synchronized by the trigger pulse generator, which is a square wave of 12 V dc peak voltage and a frequency equal to the pulse repetition frequency.

## Digital Oscilloscopes

A fast response 4-channel (100 MHz, DL 1200, Yokogawa) digital oscilloscope and a 2-channel liquid crystal display (100 MHz, TDS 220, Tektronix) real-time digital oscilloscope are used to study the time-varying probe current signals. Since a Boxcar integrator uses time integration, rapid changes in the probe current signals cannot be observed. Therefore, the current build-up and decay time constants are obtained using the digital oscilloscopes. The Tektronix oscilloscope is useful to study the phenomena at very small time scales (<250 μs), for instance, during the plasma build-up within the pulse.

**TABLE 8.1**

Voltage Requirements for Selecting the Pulse Width and the Maximum Operational Pulse Repetition Frequency

| Pulse Width (μs) | PLA (V) | PLB (V) | Frequency Limit (Hz) |
|---|---|---|---|
| 1.2 | +12 | +12 | 600 |
| 0.6 | 0 | +12 | 1200 |
| 0.2 | +12 | 0 | 2400 |
| 0.08 | 0 | 0 | 2400 |

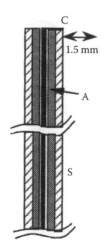

**FIGURE 8.4**
A schematic diagram to show the probe for measuring the electric field.

## Probe for Measuring the Electric Field

Figure 8.4 shows a schematic view of the probe used for measuring the axial variation of the electric field in the waveguide. A copper wire C with a diameter of 1 mm is covered with a ceramic insulator A, with an outer surface covered with a stainless-steel pipe S. The naked 1.5 mm length of the wire serves as the antenna and is curved at a right angle to the rest of the body of the probe. The antenna is aligned parallel to the central peak field in the $TE_{11}$ circular mode (cf. Figure 8.4) and moved axially along the tube. The electric field impulse is rectified by a diode responsive to the GHz range microwave field.

## Experimental Results

In this chapter, the experimental results pertaining to the "Apparatus and Methods in the General Experiment" are described. The experimental results of a below cutoff dimensional waveguide are described in a subsequent chapter. Probe measurements were done at two positions of MC. One position was at the center while the other was at the exit of MC (shown by dotted lines in Figure 8.1), both the positions situated along the axis. In what follows, the result of measurement at the center will be explained. The density decay at the exit was similar, but the absolute density was smaller. The data shown have been time-averaged 16 times, and by this acquisition procedure the obtained data were reproducible. From the probe characteristics, the space potential $V_s$ was seen to be around 20 V in Ar, and both the ion and electron currents were well saturated. $V_s$ decreased with time. $V_s$ in $N_2$ and

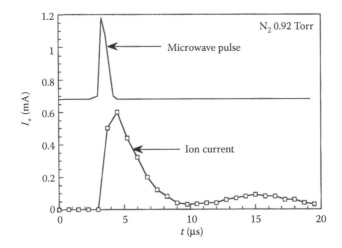

**FIGURE 8.5**
Example of an ion current pulse in $N_2$ at 0.92 Torr. The microwave pulse is also shown in the top of the figure, with pulse height in arbitrary units.

air were in the range 15–20 V. The plasma density and the electron temperature were about $10^{10}$ cm$^{-3}$ and 6–7 eV, respectively, with $T_e$ being rather high (~10 eV) during the initial stages of the decay.

Figure 8.5 shows an example of the data for the pulse shape of the ion current decay as obtained in $N_2$ at 0.92 Torr. We replotted this curve from the end of the microwave pulse to represent the interpulse decay, and compared it with the numerical modeling as detailed in the "Discussion" section. Experiments in Ar and standard air have been represented in the same way. The microwave pulse is also shown at the top of Figure 8.5 and the pulse duration is <1 µs. Although the trigger pulse is rectangular, the microwave pulse form is a little distorted.

Figure 8.6 shows the decay of the normalized ion current ($I/I_o$), where $I_o$ is the peak current at the end of the pulse in Ar at $V_P = -40$ V for pressure $p = 0.24$ and 0.92 Torr. The ion current resembled a monotonic exponential decay. The solid curve represents the fitting done with the solution of the rate equation as will be explained in the "Discussion" section.

Figure 8.7 shows the decay of the $I/I_o$ in $N_2$ at $V_P = -16$ V for $p = 0.24$ Torr. The characteristics are markedly different from that of Ar, in that an additional bump occurs after about 6 µs superposed on the monotonic exponential-like decay as observed in Ar. The peak height of the bump was about 10–20% of the initial ($I/I_o$) peak, depending on the pressure. The solid curve represents the fitting done with the solution of the rate equations, as will be explained in the "Discussion" section.

Figure 8.8 shows the case of discharge in standard air at 0.24 and 0.58 Torr. The decay characteristics could be broadly divided into three parts. The first

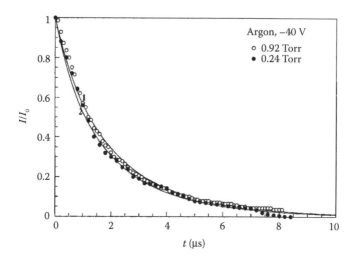

**FIGURE 8.6**

Normalized ion current $(I/I_0)$ at probe voltage $V_P = -40$ V, pressure $p = 0.24$, and 0.92 Torr versus time $t$ in Argon. Solid lines 1 and 2 show fitting to the experimental data as done with Equation 8.4. (Reprinted with permission from S. Bhattacharjee and H. Amemiya. 1998. Interpulse plasma of a high-power narrow-bandwidth pulsed microwave discharge, *J. Appl. Phys.* **84**: 115. Copyright 1998, American Institute of Physics.)

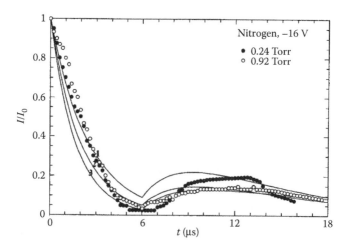

**FIGURE 8.7**

Normalized ion current $(I/I_0)$ at probe voltage $V_p = -16$ V, pressure $p = 0.24$, and 0.92 Torr versus time $t$ in nitrogen. Solid lines 1, 2, and 3 show fitting to the experimental data done with Equations 8.3 and 8.4. (Reprinted with permission from S. Bhattacharjee and H. Amemiya. 1998. Interpulse plasma of a high-power narrow-bandwidth pulsed microwave discharge, *J. Appl. Phys.* **84**: 115. Copyright 1998, American Institute of Physics.)

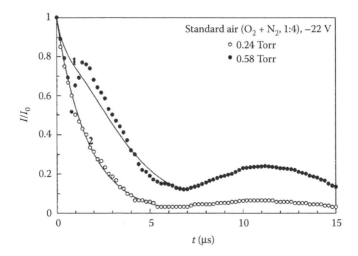

**FIGURE 8.8**

Normalized ion current ($I/I_0$) at $V_p = -22$ V, pressure $p = 0.24$, and 0.58 Torr is standard air. Solid lines 1 and 2 show the fitting done with Equations 8.5 and 8.6. (Reprinted with permission from S. Bhattacharjee and H. Amemiya. 1998. Interpulse plasma of a high-power narrow-bandwidth pulsed microwave discharge, *J. Appl. Phys.* **84**: 115. Copyright 1998, American Institute of Physics.)

part, about $t = (0-1)$ μs, occurs on the fastest time scale, while the second part, $t = (1-6)$ μs, decays at a relatively slower rate. The third part is a bump similar to that observed in $N_2$, but at the same pressure (for, e.g., 0.24 Torr), the bump is smaller as compared to that of the $N_2$ discharge. The demarcation of the first and second part of the decay is seen to be pronounced at 0.58 Torr, appearing as a splitting as shown in Figure 8.8. The splitting, however, disappeared as $p$ was raised beyond 0.6 Torr, where it was difficult to get a stable discharge. Here, also, the solid curve represents fitting to the experimental data, as will be explained in the following section.

## Discussion

The interpulse plasma seems to be more active than conventional afterglows because of a high $T_e$, even during the later stages of the decay. The high $T_e$ could be an after effect of the plasma during the pulse, where free electrons are accelerated to high energies. Moreover, as the decay time scales are faster (~20 μs), electrons are lost before they can come to thermal equilibrium with the gas.

The qualitative explanation of the decay process may be made as follows.

## Argon

The temporally decaying plasma may be represented by a set of two-coupled equations,

$$\frac{\partial N_e}{\partial t} = -a_r N_e^2 - (b_d - b_i)N_e + gN_a N_e + fN_a^2 \tag{8.1}$$

and

$$\frac{\partial N_+}{\partial t} = kN_e - hN_a N_e - gN_a N_e - qN_a - 2fN_a^2, \tag{8.2}$$

where $N_e$ and $N_a$ are the density of the electrons and metastables; $a_r$ is the recombination coefficient; $b_d$ and $b_i$ are the diffusion and the ionization coefficients; $g$ is the rate coefficient for collision of metastables, with electrons leading to ionization and loss of metastables; $f$ is the rate coefficient for metastable–metastable collisions leading to the production of an electron and loss of two metastables; $k$ is the rate coefficient for the electronic excitation of the gas molecules, resulting in the generation of metastables; $h$ is the rate coefficient for the loss of metastables through the combined effects of super-elastic collisions and quenching to resonant states; and $q$ is the rate coefficient for the combined effects of diffusion of the metastables and quenching by other two-body and three-body processes. By assuming charge neutrality ($N_+ = N_e$) at each moment of time, the equation of positive ions may be dropped.

As the ion current decay was found to decrease monotonically with time, this suggests that no appreciable effects of ionization due to metastables were occurring. Thus, if the effects of metastables are negligible and only volume recombination and diffusion dominate, the plasma could be modeled by a simplified form of Equation 8.1 as

$$\frac{\partial N_e}{\partial t} = -a_r N_e^2 - (b_d - b_i)N_e. \tag{8.3}$$

Taking $N_0$ as the peak electron density at the initial time $t = 0$ (at the end of a pulse) and $t_0$, a characteristic time (here 1 μs), the above equations reduce to similar forms but dimensionless variables by the following transformations:

$$N_e/N_0 \rightarrow N_e, \quad N_a/N_0 \rightarrow N_a, \quad t/t_0 \rightarrow t, \quad a_r N_0 t_0 \rightarrow a_r,$$

$$b_d t_0 \rightarrow b_d \quad b_i t_0 \rightarrow b_i, \quad gN_0 t_0 \rightarrow g, \quad fN_0 t_0 \rightarrow f, \quad k t_0 \rightarrow k,$$

$$hN_0 t_0 \rightarrow h, \quad q t_0 \rightarrow q.$$

Analytical solution of Equation 8.3 is given by

$$\frac{I(t)}{I_0} = \frac{e^{-b}d^t}{1 + N_0(a/b_d)(1 - e^{-b}d^t)},$$ (8.4)

where $I_0$ is the initial ion current $t = 0$ and $N_0(t = 0) = 10^{10}$ cm$^{-3}$.

Here it is assumed that the spatial profile of the plasma is constant and given by a Bessel function of zeroth order with the boundary condition that $N_e = 0$ at $r = r_c$, where $r_c$ is the radius of the multicusp MC. The diffusion term has been approximated as $\nabla^2(n_e) = n_e/\Lambda^2$, where $\Lambda$ is the characteristic diffusion length and depends upon the system geometry, and $b_d$ is the ambipolar diffusion coefficient.

The experimental data in Ar could be fitted well with Equation 8.4 and are shown by labels 1 and 2 in Figure 8.6 for 0.92 and 0.24 Torr, respectively. From the fitting, we obtained the recombination coefficient, $a = (0.5–0.9) \times 10^{-6}$ cm$^3$/s and $b_d - b_i = (0.4–0.5) \times 10^6$ s$^{-1}$. It should be noted that the purpose of the current experiment is not for the determination of the recombination and diffusion coefficients, because the rather high $T_e$ could bring about additional ionization during the decay process.

The experimental results could be explained rather well with the above solution, including diffusion and recombination. The recombination rate is dependent on $N_e^2$, therefore, its effect is more pronounced during the early stages of the decay when the plasma density is high. The diffusion rate is proportional to $N_e$, and therefore its effect is more in the later phase of the discharge, when the recombination effect is small. At a fixed input power, the density decay is faster at lower $p$, which could be because $b_d$ is inversely proportional to $p$.

The concentration of the metastable atoms in Ar is unlikely to be large. It has been estimated that their concentration does not exceed $10^{-5}$ of the concentration of ground-state atoms [12]. Since the two metastable states in Ar (11.5 and 11.7 eV) are positioned high in the energy level diagram, close to the ionization level, their excitation rate is not so high. Therefore, we have neglected the effects of metastables.

## Nitrogen

In N$_2$, we could expect additional effects due to the long-lived metastables, aside from the effects of recombination and diffusion. The cumulative ionization at the later stage of the decay could be due to the effects of metastable ionization.

Therefore, we consider the simultaneous solution of Equations 8.1 and 8.2, including metastables. Owing to the nonlinearity between the two-coupled equations, we have solved them numerically by time-integrating in time steps of $dt = 10^{-3}$ μs. Tentative values of the reaction rates for each process estimated from known data [12–15] were substituted in the numerical analysis as a first approximation. These values were later improved according to the fitting to our experimental data. Especially, good reproduction of the experimental results were obtained by changing the values of $f$, $g$, and $k$ at a later stage (>6 μs). This could be because a comparatively lower $T_e$ favors ionization by electron–metastable and metastable–metastable collisions. On the other hand, the ionization cross-section for the formation of the metastables $k$ would be considerably reduced. In the simulation, the initial values for the plasma density, normalized electron density, and normalized metastable density were $N_0(t = 0) = 10^{10}$ cm$^{-3}$, $N_e(t = 0) = 1$, and $N_a(t = 0) = 0$.

During the pulse, the electron temperature is very high (>10 eV), and it can excite the ground state $N_2$ molecules to a number of excited electronic states [15]. However, assuming that the important metastable states [13–16] are $A^3 \Sigma_u^+$(6.17 eV), $W^3\Delta_u$(7.36 eV), and $a'^{\,1}\Sigma_u^-$(8.40 eV) we have modeled the decaying plasma by considering ionization due to collisions of the type where one is in the $a'$ state and the other is in the same or in the $A$ state or $^3\Delta_u$ state. This is represented by $f$ in Equations 8.1 and 8.2, where $g$ represents ionization due to collisions of electrons with metastables. From our numerical simulation, we found that the influence of $f$ on the height of the bump was more important than that of $g$, implying that the metastable–metastable collisions play an important role in the ionization process. Further, one of the reasons for the time lag in the appearance of the bump could be because of a finite time required for the metastable density to build up in the discharge, which is around 6 μs from our simulation. Solid lines in Figure 8.7 show the result of numerical simulation for three different cases, where the rate coefficients $b_n = b_d - b_i$ and $k$ have been varied as $b_{n1} < b_{n2} < b_{n3}$ and $k_1 < k_2 < k_3$, and the numerals 1, 2, and 3 refer to the simulation labels.

The dimensionless parameters used for fitting the experimental data by the curves 1, 2, and 3 are as follows. Suffixes I and II refer to parameter values before and after $t = 6$ μs. For optimum fitting: $a_r = (5 - 6) \times 10^{-3}$, $b_n = 0.4 - 0.6$, $k_I = 20$, $k_{II} = 6 - 10$, $f_I = 10^{-8}$, $f_{II} = (1.5 - 2) \times 10^{-4}$, $g_I = 2 \times 10^{-10}$, and $g_{II} = (8 - 10) \times 10^{-3}$. The values of $a_r$ and $b_n$ were chosen to be the same as in the case of Ar. Owing to the plural number of fitting parameters, there are many possibilities of obtaining nearly the same fitted curve. The above values are not unique. Moreover, as the parameters are normalized with $N_o$ and $t_o$, they cannot always be accurate cross-sectional data relevant to atomic collisions.

## Standard Air

The decay characteristic in standard air has been described in three parts in the section "Experimental results." The third part has a striking resemblance with that of the bump observed in $N_2$. In order to understand the first and second parts of the decay, we have considered the effect of the presence of negative ions. The negative ions can be formed because of the electronegative gas $O_2$.

The decaying plasma containing negative ions could be represented by another set of coupled equations for electrons and negative ions,

$$\frac{\partial n_e}{\partial t} = -an_e(n_n + n_e) - (b_d - b_i)n_e - pn_e + dn_en_n, \tag{8.5}$$

$$\frac{\partial n_n}{\partial t} = pn_e - sn_n(n_n + n_e) - rn_e - dn_en_n, \tag{8.6}$$

where $n_e$ and $n_n$ are the electron and negative ion density, $p$ and $d$ are the rate coefficients for electron attachment and detachment, and $r$ and $s$ are the rate coefficients for diffusion and recombination of negative ions, respectively.

We assume that charge neutrality is maintained by the equation $n_i = n_n + n_e$. Here, also, the equations could be reduced to dimensionless forms by the transformation:

$$n_n/n_o \rightarrow n_n, \quad pt_o \rightarrow p, \quad rt_o \rightarrow r, \quad sn_ot_o \rightarrow s \quad \text{and} \quad dn_ot_o \rightarrow d.$$

During the first part of the decay (0–1 µs), electron–ion recombination leads to a rapid decrease in the ion density. Moreover, in this phase, production of negative ions can occur through dissociative attachment [17], which has a cross-section with threshold at 4.2 eV and a peak around 6.7 eV. The rather high $T_e$ in this phase could assist $n_n$ formation by this process, resulting in a sudden decrease in electron density.

The second phase of decay (1–6 µs) is slower and is mainly attributed to the negative ion–positive ion recombination. Negative ions can be destroyed by electron impact detachment [18], which has a threshold at 1.46 eV and peak at about 30 eV. On the other hand, the dissociative attachment cross-section has a threshold at 4.2 eV and peak at 6.7 eV. Therefore, when $T_e < 4.2$ eV, the detachment process could dominate over the dissociative attachment process and release electrons from the negative ions. The released electrons cannot take part in further attachment by three-body process unless their temperature is sufficiently low. Thus, the rate of decay of $n_e$ is expected to be smaller during the second phase (1–6 µs) of the interpulse decay.

In $O_2$, several types of metastable states are possible [17,18]. However, we assume that the major metastable states [19,20] are $a^1\Delta_g$(0.977 eV) and $b^1\Sigma_g^+$(1.63 eV). Although having long lifetimes, these states are very low-lying compared to the ionization energy level (12.07 eV). Hence, ionization from the metastable states of $O_2$ is very unlikely and the bump at the later stage of the discharge is more probably due to the $N_2$ metastables.

In the simulation, to begin with, we have assumed tentative values of the reaction rates obtained from known data [17,18]. These were later improved from the fitting to the experimental curves. The effect of the negative ions could be best modeled by changing the values for $p$ and $d$ after 1.0 µs, by considering the change of $T_e$. Solid lines in Figure 8.8 labeled as 1, 2 show the result of numerical simulation of the first and second parts of the decay at 0.24 and 0.58 Torr, respectively. The decay at the lower $p$ could be fitted rather well. The higher $p$ exhibited a splitting with a rise in the ion density. The reason for the splitting is not known, and one possibility could be the generation of $NO_x$ ions (e.g., $NO^+$, $NO_2^+$, etc.), which have not been taken into consideration in our model.

The experimental data fitted with curves 1 and 2 represent the following dimensionless parameter values. Suffixes I and II refer to the values before and after $t = 1$ μs, which were used to obtain an optimum fitting. For curve 1: $a = 0.06$, $p_I = 1.0$, $p_{II} = 0.2$, $d_I = 1.0$, $d_{II} = 0.5$, $b_I = 0.6$, $b_{II} = 0.5$, $q = 0.6$, and $s = 0.1$, while for curve 2: $a_r = 0.07$, $b_n = 0.6$, $p = 1.9$, $d = 1.0$, $s = 0.03$, and $q = 0.55$.

## Conclusions

This study has qualitatively clarified the temporal behavior of the interpulse phase of a high-power, short-pulse microwave discharge. The experimental results obtained from Langmuir's probes have shown a rather clear difference in the interpulse regime of the three gases, which are attributed not only to the effects of recombination and diffusion but also those of metastables and negative ions. The plasma density at the end of each microwave pulse was of the order $10^{10}$ cm$^{-3}$ and the electron temperature was 6–8 eV, much higher than that of the usual afterglow. Numerical simulation of the decay characteristics showed a fair agreement with the experimental results. Using the physical nature and process of the interpulse plasma, new areas of application will be realized.

## References

1. D.J. Rose and S.C. Brown. 1957. Microwaves gas discharge breakdown in air, nitrogen, and oxygen, *J. Appl. Phys.* **28**: 561–563.
2. A.D. MacDonald. 1996. *Microwave Breakdown in Gases* (Wiley, New York).
3. S.J. Tetenbaum, A.D. MacDonald, and H.W. Bandel. 1971. Pulsed microwave breakdown of air from 1 to 1000 Torr, *J. Appl. Phys.* **42**: 5871–5872.
4. W.M. Bollen, C.L. Yee, A.W. Ali, M.J. Nagurney, and M.E. Read. 1983. High-power microwave energy coupling to nitrogen during breakdown, *J. Appl. Phys.* **54**: 101–106.
5. C.A. Sullivan, W.W. Destler, J. Rodgers, and Z. Segalov. 1988. Short-pulse high-power microwave propagation in the atmosphere, *J. Appl. Phys.* **63**: 5228–5232.

6. S.P. Kuo, Y.S. Zhang, and P. Kossey. 1990. Propagation of high-power microwave pulses in air breakdown environment, *J. Appl. Phys.* **67**: 2762–2766.

7. T. Mieno and S. Samukawa. 1997. Generation and extinction characteristics of negative ions in pulse time modulated electron cyclotron resonance chlorine plasma, *Plasma Sources Sci. Technol.* **6**: 398–404.

8. H. Ohtake and S. Samukawa. 1996. Charge-free etching process using positive and negative ions in pulse time modulated electron cyclotron resonance plasma with low frequency bias, *Appl. Phys. Lett.* **68**: 2416–2417.

9. A. Rousseau, L. Tomasini, G. Gousset, C. Boisse-Laporte, and P. Leprince. 1994. Pulsed microwave discharge: A very efficient H atom source, *J. Phys. D: Appl. Phys.* **27**: 2439–2441.

10. P. Sortias. 1992. Pulsed ECR ion source using the afterglow mode, *Rev. Sci. Intrum.* **63**: 2801–2805.

11. W.D. Getty and J.B. Geddes. 1994. Size-scalable, 2.45-GHz electron cyclotron resonance plasma source using permanent magnets and waveguide coupling, *J. Vac. Sci Technol.* **12**: 408–414.

12. I.Yu. Baranov, N.B. Kolokolov, and N.P. Penkin. 1985. Investigation of step-wise excitation processes in the plasma of an argon afterglow, *Opt. Spektrosk.* **58**: 268.

13. H. Brunet and J. Rocca Serra. 1985. Model for a glow discharge in flowing nitrogen, *J. Appl. Phys.* **57**: 1574–1581.

14. H. Brunet, P. Vincent, and J. Rocca Serra. 1983. Ionization mechanism in a nitrogen glow discharge, *J. Appl. Phys.* **54**: 4951–4957.

15. Y. Itikawa, M. Hayashi, A. Ichimura, K. Onda, K. Sakimoto, K. Takayanagi, M. Nakamura, H. Nishimura, and T. Takayanagi. 1986. Cross-sections for collisions of electrons and photons with nitrogen molecules, *J. Phys. Chem. Ref. Data* **15**: 985–1010.

16. R.E. Lund and H.J. Oskam. 1969. The production and loss of $N_{2+}$ ions in the nitrogen afterglow, *Z. Phys.* **219**: 131–146.

17. Y. Itikawa, A. Ichimura, K. Onda, K. Sakimoto, K. Takayanagi, Y. Hatano, M. Hayashi, H. Nishimura, and S. Tsurubuchi. 1989. Cross sections for collisions of electrons and photons with oxygen molecules, *J. Phys. Chem. Ref. Data* **18**: 23–42.

18. V. Vahedi and M. Surendra. 1995. A Monte-Carlo collision model for the particle-in-method: Applications to argon and oxygen discharges, *Comp. Phys. Commun.* **87**: 179–198.

19. C. Yamabe and A.V. Phelps. 1983. Excitation of the $O_2$ $(a^1\Delta_g)$ state by low energy electrons in $O_2$–$N_2$ mixtures, *J. Chem. Phys.* **78**: 2984–2989.

20. S.A. Lawton and A.V. Phelps. 1978. Excitation of the $b^1\Sigma_g^+$ state of $O_2$ by low energy electrons, *J. Chem. Phys.* **69**: 1055–1068.

21. S. Bhattacharjee and H. Amemiya. 1998. Interpulse plasma of a high-power narrow-bandwidth pulsed microwave discharge, *J. Appl. Phys.* **84**: 115.

22. S. Bhattacharjee and H. Amemiya. 2000. Production of pulsed microwave plasma in a tube with a radius below the cutoff value, *J. Phys. D: Appl. Phys.* **33**: 1104–1116.

# 9

## Plasma Production in Pulsed-Mode Microwaves: Circular Multicusp Waveguide with a Dimension below Cutoff

### Introduction

The plasma produced by high-power, short-pulse microwaves is of interest [1,2]. In the previous chapter, we saw that the plasma state between the pulses is a useful phase of the discharge [3]. The active plasma in the inter-pulse regime is favorable for the production of radicals, metastables, multi-charged ions [4], negative ions [3], or for various plasma–chemical reactions. Other important applications include laser guiding through plasma in narrow tubes for particle acceleration [5] and inner-surface processing of small-diameter tubes [6] and so on.

The high power during the pulse generates a strong electric field that raises the electron temperature $T_e$. The high $T_e$ can be favorable for ionization in the interpulse regime. An important aspect of high-power, short-pulse microwave plasma is the plasma buildup process that depends upon the wave electric field, the electron temperature, and the seed plasma produced during the power-on time. Our earlier experiment in a narrow waveguide has indicated that the plasma buildup is delayed and reaches the peak density even tens of microseconds after the end of the microwave pulse. Other important aspects are anisotropic diffusion of the seed plasma and characteristic decay time.

In regard to the plasma production in a waveguide under the geometrical cutoff limitation, two simultaneous effects are useful: (a) production of the seed plasma at the plasma waveguide (tube) entrance during the power-on time, and (b) spatiotemporal diffusion of the seed plasma into the tube during the power-off phase. The plasma density cutoff limitation can be similarly understood because it is related to the density of the seed plasma produced at the tube's entrance during the power-on time—a higher density would favor axial diffusion into the tube.

Thus, for the production and maintenance of a plasma in a narrow tube, we should first suppress the radial diffusion. This has been achieved as in earlier experiments, in continuous-mode microwaves by an arrangement of magnets around the tube. Second, the use of high-power microwaves helps to increase the seed plasma density and the electron temperature, both are very important for the plasma buildup process. Third, short-pulse microwaves are helpful to obtain higher powers, avoid steady-state effects in the power-on time (parent) plasma, and have a longer interpulse plasma. After having these attributes, the next important thing is to control the power-off time (interpulse) plasma, the plasma decay, and the plasma buildup using pulse repetition frequency and/or pulse width (duration). With high-power microwaves, effects such as particles driven by potential and ponderomotive forces, nonlinear dispersion, and, additionally, the anisotropic diffusion of the plasma brought about by the arrangement of the magnets are expected.

Based upon the above viewpoints, as discussed in Chapters 5 through 7 where continuous mode microwaves were used to produce and maintain a high-density plasma in narrow conducting waveguides with a transverse dimension smaller than the cutoff value [7–9], this chapter is related to these possibilities in the case of pulsed microwaves. Additionally, the general properties of a high-power, short-pulse microwave plasma are studied in greater detail. The plasma production and properties are discussed at a relatively higher-pressure regime.

## Experimental Setup and Procedures

Figure 9.1 shows a schematic of the experimental apparatus. The pulsed microwaves produced by a magnetron oscillator OSC have a frequency of 3 GHz, output peak power 60–100 kW, and a pulse duration $t_w$ 0.05–1.0 μs in full-width at half-maximum (FWHM). The pulse repetition frequency $f_r$ is varied in the range 10–500 Hz. The waves are guided through an isolator ISO, an attenuator ATT, a coaxial cable Cx, a power monitor PM, a matching element M, a rectangular bend H, and a quartz window W, and launched into a circular conducting tube WG (37 cm in length, 5.74 cm inner diameter) located inside a vacuum chamber C. A trigger signal Trig is used for pulse initiation and as a reference signal for diagnostics. A personal computer (PC) is used for storing data obtained through a four-channel digital oscilloscope DSO.

The cutoff radius of the $TE_{11}$ mode is given by $R_c = \rho'_{11}\lambda_o/2\pi$, where $\rho'_{11} = 1.841$ is the first root of the derivative of the Bessel function $J_1$, and $\lambda_o$ is the vacuum wavelength of microwaves. For microwaves of 3 GHz, $\lambda_o = 10$ cm and $R_c = 2.93$ cm. The inner diameter of WG is therefore smaller than the cutoff value. WG is inserted into a multipolar magnet structure (MC), henceforth referred to as a "multicusp," constructed with permanent magnets

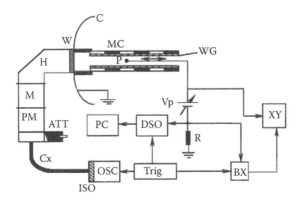

**FIGURE 9.1**
Schematic of the experimental apparatus; magnetron oscillator (OSC), isolator (ISO), coaxial cable (Cx), attenuator (ATT), power monitor (PM), matching element (M), rectangular bend (H), window (W), chamber (C), multicusp (MC), waveguide (WG), Langmuir probe (P), computer (PC), digital oscilloscope (DSO), trigger signal (Trig), probe voltage ($V_p$), Boxcar's integrator (BX), X–Y recorder (XY), and resistor (R). (Reprinted from *Vacuum* **58**, S. Bhattacharjee and H. Amemiya, Pulsed microwave plasma production in a conducting tube with a radius below cutoff, 222, Copyright 2000, with permission from Elsevier.)

(10 poles) to provide a minimum-$B$ field for particle confinement. The configuration results in a field that is mainly transverse to the axis ($z$) of the tube. The waves are launched perpendicular to the transverse field. The designs of the multicusp and the field distribution have been described earlier.

The temporally varying profiles of ion ($I_+$) and electron currents ($I_e$) and probe characteristics were measured by a Langmuir probe (plane, 5 mm diameter) denoted by P in Figure 9.1, using the DSO with time averaging (64 shots) and a Boxcar's integrator BX, respectively. The probe was inserted through an axial port and could be moved along WG. For measuring $I_+$ and $I_e$, the probe was biased at fixed negative and positive voltages against ground, respectively, whereas the voltage was swept for obtaining the probe characteristics. The gate aperture of BX was set at 0.1 μs and the probe characteristics were measured to a delay of 30–50 μs from the end of the microwave pulse in steps of ~1.0 μs, by a slow scanning of the probe voltage. From the probe characteristics, the temporal variation of the electron temperature $T_e$ and the plasma (electron) density $N_e$ were deduced.

During the probe measurements, the probe surface was set parallel to the wave vector. WG together with MC served as the reference electrode, which was connected with the grounded chamber. The ratio of the inner surface of WG where the plasma was confined to the probe area was ~1700, which may be considered adequate for obtaining reliable probe characteristics. Measurements were mostly made along the axis of the tube (cf. Figure 9.1), where the magnetic field is almost zero. In this position, the ion and electron currents to the probe are very little influenced by the field.

The ion current $I_+$ was measured at a deep negative bias so that its dependence on the probe voltage $V_p$ was small, and $I_+$ is proportional to $N_+$. On the other hand, the electron current $I_e$ was measured in the retarding potential region, that is

$$I_e = N_+ eS \sqrt{\frac{\kappa T_e}{2\pi m}} \exp\left\{\frac{-e(V_s - V_p)}{\kappa T_e}\right\}. \tag{9.1}$$

From the ratio of $I_+$ and $I_e$, the space potential $V_s$ can be estimated from the following equation:

$$V_s = \frac{\kappa T_e}{e} \ln\left(\frac{I_+}{I_e}\right) + \xi, \tag{9.2}$$

where $\xi$ is a constant to be determined by considering the boundary condition that $V_s \to 0$ at $t \to \infty$.

The microwave electric field $\tilde{E}$ inside the tube was measured using a needle-type probe with a 1.5 mm length aligned parallel to the central electric field in the circular mode. The electric field impulse was rectified by a crystal diode responsive to the GHz range microwave field. The axial variations of $I_+$ and $E$ were measured in DSO by moving the respective probes along the waveguide and recording the peak current and the field signal, respectively. The variation of $N_+$ was also measured for checking reproducibility.

---

## Experimental Results

Figure 9.2a and b shows typical spatiotemporal profiles of $I_+$ and $I_e$ at a pressure $p = 2.0$ Torr with the axial distance along the waveguide $z$ (cm) as a parameter. The values of the pulse repetition frequency and the pulse width are shown in Figure 9.2a and b. $z = 0$ corresponds to the entrance of the waveguide at the microwave input side. Time $t = 0$ corresponds to the beginning of the pulse whose width $t_W$ is ~1 μs. It may be noted from Figure 9.2a and b that the plasma is produced inside the narrow waveguide at all axial positions, and that the peak current spatially falls along the waveguide, but rises after ~$z = 18$ cm to a second maximum around ~$z = 26$–$28$ cm, and finally falls beyond 30 cm. Thus, two regions with higher plasma density have been identified, the peak current of the latter being usually ~50% of the first in the case of $I_+$ and smaller in the case of $I_e$. The electron current is seen to have a decay time constant more than four to five times longer than that of the ion current.

Figure 9.3a through c shows the form of $I_e$ buildup and decay at $z = 2$, 10, and 20 cm with regard to its dependence on $p$. It is clearly seen that the

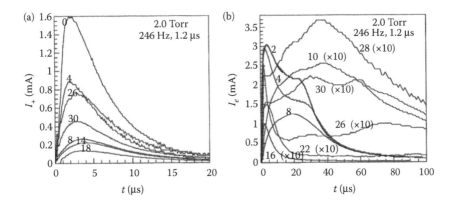

**FIGURE 9.2**
(a) Spatiotemporal ion current profiles. (b) Spatiotemporal electron current profiles.

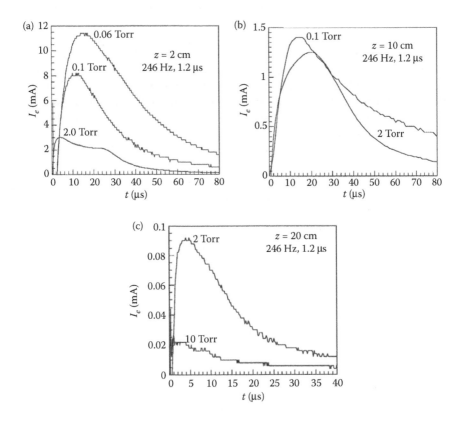

**FIGURE 9.3**
(a) Electron current buildup at $z = 2$ cm. (b) Electron current buildup at $z = 10$ cm. (c) Electron current buildup at $z = 20$ cm.

buildup occurs after the end of the pulse in a timescale of 2–20 µs. The buildup is faster at a higher $p$ at a position near the entrance, $z = 2$ cm. At locations farther away from the entrance, $z = 10$–15 cm (e.g., $z = 10$ cm, Figure 9.3b), the buildup was delayed at a higher $p$. However, again at $z = z_m = 20$–30 cm (e.g., $z = 20$ cm, Figure 9.3c, i.e., the region of second current peak in Figure 9.2a and b), the buildup followed that of $z = 2$ cm. Therefore, the buildup process is different in the area between the two regions, with a higher plasma density. The same feature has also been found for the ion current ($I_+$) buildup.

Figure 9.4 shows the temporal variation of $T_e$ in the interpulse regime at $z = 20$ cm. $T_e$ shows a value of about 6–10 eV and is higher just after the end of the pulse. At a higher $p$ (0.1 and 0.3 Torr, shown with dark symbols), $T_e$ is seen to decay faster temporally. The $T_e$ decay timescales are longer than the particle decay times, as may be seen by a comparison with Figure 9.5.

Figure 9.5 shows the temporal variation of the electron density $N_e$ deduced from the time-resolved probe characteristics measurements using the Boxcar's integrator. $N_e$ is seen to have a peak value over $10^{10}$ cm$^{-3}$ depending upon the pressure and a decay time constant of 20–30 µs.

Figure 9.6a and b shows the effect of varying the pulse repetition frequency $f_r$ (10–500 Hz) on the peak electron current $I_m$ and the decay time constant $\tau_d$ of the electrons. The measurements were made at an axial distance $z = 4$ cm along the tube. It is seen that $I_m$ and $\tau_d$ weakly depend on $f_r$. This suggests that any charge accumulation by pulse repetition was small. The slight rise of the peak current at lower frequencies is due to a longer decay time constant at lower frequencies (Figure 9.6b).

Figure 9.7a and b shows the effect of varying the pulse width $t_w$ (0.08–1.2 µs) on the peak electron current $I_m$ and the decay time constant $\tau_d$ at $z = 4$ cm. An increase in $t_w$ leads to an increase in $I_m$ and $\tau$. This effect is more pronounced at a higher $p$ (e.g., 0.5 and 1.0 Torr), which is possible due to a greater energy

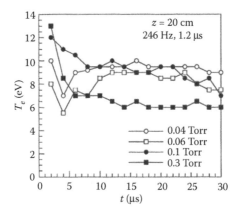

**FIGURE 9.4**
Variation of the electron temperature.

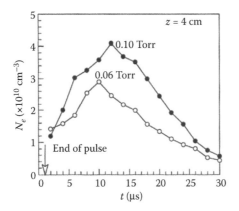

**FIGURE 9.5**
Temporal variation of the electron density.

input to the plasma and, consequently, an increase in the ionization rate with an increase in $p$.

Figure 9.8 shows the axial variation of the ion density $N_+$ along the waveguide, with $p$ measured at $t = 10$ μs after the end of the pulse. Plasma density is greater than $10^{10}$ cm$^{-3}$. It is seen that $N_+$ is peaked near $z = 0$ cm and a second peak appears at $z_m = 20$–30 cm, as has been described earlier with regard to Figure 9.2a and b.

Between the two regions, $N_+$ decreases to a smaller value ($z = 10$–15 cm). These measurements were made with the Boxcar's integrator and correspond rather well to the spatiotemporal current profiles shown in Figure 9.2a and b measured with the DSO. Although $N_+$ at $z = 7$–15 cm appears to be very small on the linear scale, the actual value lies in the range of $10^9$ cm$^{-3}$ when plotted in the semilog scale.

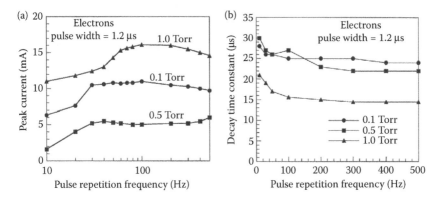

**FIGURE 9.6**
(a) Variation of the peak electron current pulse repetition frequency. (b) Variation of the electron with current decay constant and pulse repetition frequency.

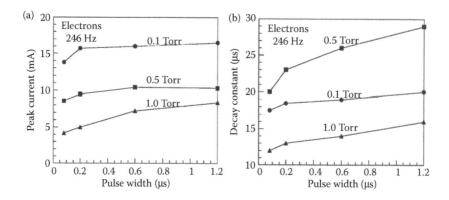

**FIGURE 9.7**
(a) Variation of the peak electron current with pulse width. (b) Variation of the electron current decay constant with pulse width.

Figure 9.9a through c shows the axial variation of the electric field $E$ at some $p$ and in vacuum. The axial variation of $E$ at a higher $p$ (1.0 and 5.0 Torr) shows a maximum near the entrance and a second maxima around 25–30 cm near the waveguide end away from the microwave input side. In between the two maximas, a few peaks are noticed for which magnitudes show an increase toward the second maxima and are approximately spaced at an interval of ~5 cm.

As $p$ is lowered (Figure 9.9b, 0.2 and 0.06 Torr), the peaks become diffused and broader. However, the presence of a higher field near the entrance and the farther end of the waveguide is clearly understood. Measurements in vacuum condition show that the field is higher near the entrance and decreases along the tube, with the appearance of sharp peaks spaced at almost the same distance as described earlier (during higher $p$), except that their magnitude

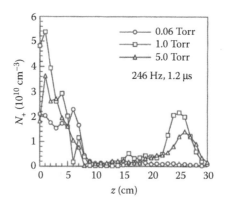

**FIGURE 9.8**
Axial variation of the ion density.

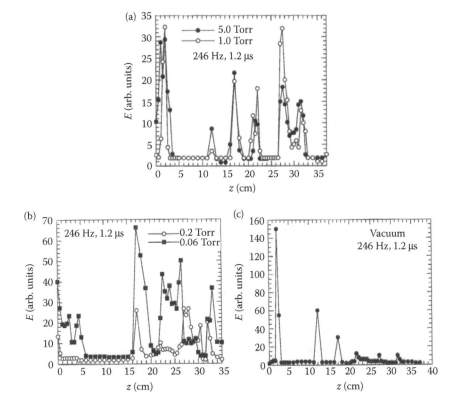

**FIGURE 9.9**
(a) Variation of the electric field $E$ with axial distance $z$ at $p = 5.0$ and 1.0 Torr. (b) Variation of the electric field $E$ with axial distance $z$ at 0.2 and 0.06 Torr. (c) Variation of the electric field $E$ with the axial distance $z$ in a vacuum.

decreases with $z$. The axial variation of $E$ during the presence of the plasma can thus be said to broadly resemble that of $N_+$ in Figure 9.7, suggesting a possible correlation between the two.

Figure 9.10 shows a typical temporal profile of the ion current $I_+$ at pressure $p = 0.7$ Torr measured at an axial distance $z = 22$ cm inside the tube. Unless otherwise stated, the peak power, the pulse repetition frequency, and the pulse duration were set at 60 kW, 246 Hz, and 1.2 μs, respectively. Distance $z = 0$ corresponds to the entrance of the waveguide at the microwave input side. Time $t = 0$ corresponds to the beginning of the pulse. It is noted that the plasma is produced inside the narrow tube, and the buildup is delayed from the end of the microwave pulse marked as "end of pulse": the plasma buildup time is about 2.0 μs. From the time delay of buildup $\tau$ at $z$, the ion drive velocity $v$ was determined by $v = z/\tau$.

Figure 9.11 shows the dependence of the drive velocity $v$ on the pressure $p$. $v$ was determined by shifting the probe along $z$ near the tube entrance and

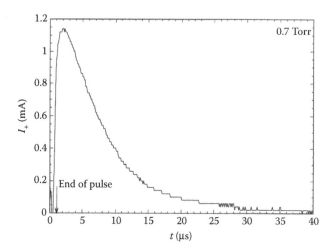

**FIGURE 9.10**

Temporal variation of the ion current at a pressure $p = 0.7$ Torr and axial position $z = 22$ cm. (Reprinted from *Vacuum* **58**, S. Bhattacharjee and H. Amemiya, Pulsed microwave plasma production in a conducting tube with a radius below cutoff, 222, Copyright 2000, with permission from Elsevier.)

recording the temporal shift of the peak ion $I_{m+}$ and electron $I_{me}$ currents. It is seen that $v$ is of the order of ~$10^6$ cm/s and slightly increases with pressure. A comparison of $v$ with the ion acoustic velocity $C_s$ (=$0.3 - 0.5 \times 10^6$ cm/s) shows that they are of the same order.

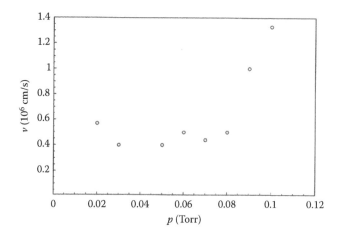

**FIGURE 9.11**

Dependence of the drive velocity $v$ with pressure $p$. $v$ is found to be comparable to the ion acoustic velocity. (Reprinted from *Vacuum* **58**, S. Bhattacharjee and H. Amemiya, Pulsed microwave plasma production in a conducting tube with a radius below cutoff, 222, Copyright 2000, with permission from Elsevier.)

Figure 9.12 shows the temporal variation of the space potential $V_s$ as determined by Equation 9.2 at some pressures measured at $z = 2$ cm. It is noted that $V_s$ has a higher value during the initial stages and gradually decreases with time. At lower pressures (e.g., 0.06 and 0.1 Torr), the slope $dV_s/dt$ is larger and the plasma seems to reach the potential equilibrium more quickly.

The temporal variation of the electron temperature $T_e$ determined from the probe characteristics shows a value of about 6–10 eV for at least a few tens of microseconds after the end of the microwave pulse. At a time close to the end of the pulse, $T_e$ was higher. At a higher $p$, $T_e$ decays faster temporally, and with increase in time, the dependence on $p$ becomes small. The electron density $N_e$ showed a peak value over $10^{10}$ cm$^{-3}$. In evaluating the plasma density, an assumption of a Maxwellian plasma was made, which will introduce a certain error in the density if the electron energy distribution is non-Maxwellian. The error was estimated to be in the range 5–9%.

Figure 9.13a shows the axial variation of the plasma (ion) density $N_+$ with $p$ measured at $t = 10$ µs after the end of the pulse. Measurements at a time close to the end of the pulse were difficult because the ion current had a disturbance. The maximum value of $N_+$ is $>10^{10}$ cm$^{-3}$. It may be noted that $N_+$ is peaked near $z \sim 0$ and a second peak appears at $z = z_m \sim 20$–30 cm, finally falling beyond 30 cm. The value of $N_+$ in the intermediate region lies in the range of $10^9$ cm$^{-3}$. Thus, two regions with higher plasma density have been identified, the peak current of the latter being usually 50% of the first.

Figure 9.13b shows the axial variation of the microwave electric field $\tilde{E}$ for some pressures $p$. The axial variation of $\tilde{E}$ shows a maximum near the tube

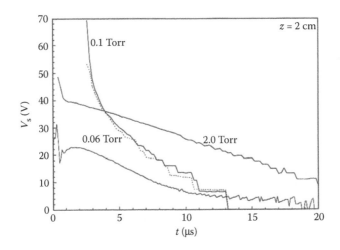

**FIGURE 9.12**
Temporal variation of the space potential $V_s$ measured at an axial position $z = 2$ cm for some pressures $p$. (Reprinted from *Vacuum* **58**, S. Bhattacharjee and H. Amemiya, Pulsed microwave plasma production in a conducting tube with a radius below cutoff, 222, Copyright 2000, with permission from Elsevier.)

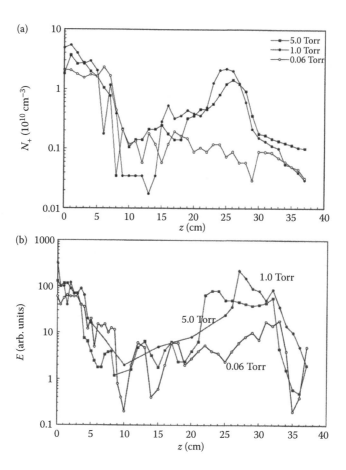

**FIGURE 9.13**

Axial variation of (a) plasma (ion) density $N_+$ measured at a time $t = 10\ \mu s$ from the end of the microwave pulse and (b) the microwave electric field $\tilde{E}$, at pressures $p = 5.0$, 1.0, and 0.06 Torr, respectively. (Reprinted from *Vacuum* **58**, S. Bhattacharjee and H. Amemiya, Pulsed microwave plasma production in a conducting tube with a radius below cutoff, 222, Copyright 2000, with permission from Elsevier.)

entrance and another maximum around $z = 20$–30 cm near the exit. Between the two maxima, a few crests and troughs are seen at an interval of ~5 cm. As $p$ is lowered (0.06 Torr), the peaks become diffused and broader. However, the presence of higher fields near the entrance and exit of the waveguide are still noticeable. Measurements showed that the field is higher near the entrance and decreases exponentially along the tube. The field pattern is distinctly different than that during the presence of the plasma, depending upon the plasma density and its gradient at the entrance of the tube. The axial variations of $\tilde{E}$ and $N_+$ have a spatial correlation.

Figure 9.14 shows the effect of varying the pulse repetition frequency $f_r$ (10–500 Hz) on the axial profile of the peak ion current $I_{m+}$ at pressure $p = 1.0$

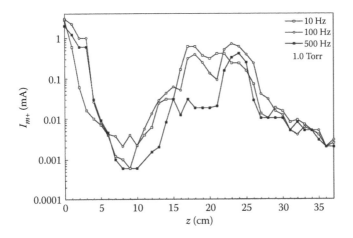

**FIGURE 9.14**
Effect of varying the pulse repetition frequency $f_r$ on the axial profile of the peak ion current $I_{m+}$ at a pressure $p = 1.0$ Torr. (Reprinted from *Vacuum* **58**, S. Bhattacharjee and H. Amemiya, Pulsed microwave plasma production in a conducting tube with a radius below cutoff, 222, Copyright 2000, with permission from Elsevier.)

Torr. Two prominent current peaks are observed but the magnitude of the current and the spatial profile only weakly depends on $f_r$. At $p \leq 0.1$ Torr, the axial current gradually decreases along the tube and the second peak is not observed, indicating that the plasma uniformity is better at lower pressures.

Figure 9.15 shows the effect of varying the pulse duration $t_w$ on the peak electron current $I_{me}$ at pressures $p = 0.1$, 0.5, and 1.0 Torr, respectively,

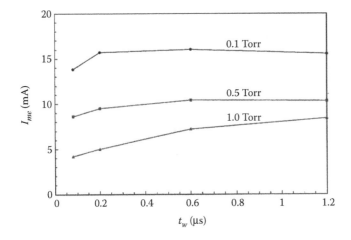

**FIGURE 9.15**
Effect of varying the pulse duration $t_w$ on the peak electron current $I_{me}$. (Reprinted from *Vacuum* **58**, S. Bhattacharjee and H. Amemiya, Pulsed microwave plasma production in a conducting tube with a radius below cutoff, 222, Copyright 2000, with permission from Elsevier.)

measured at $z = 20$ cm. An increase in $t_w$ leads to an increase in $I_{me}$ for pressures above 1 Torr.

## Discussion

### On the Mechanisms of Plasma Production

The quicker buildup with the pressure of $I_e$ (Figure 9.2a and c) at $z \sim 0$ and $z_m$ can be explained by the higher ionization efficiency with $p$ since the electric field was stronger at these positions (cf. Figure 9.8a and b), while the slower buildup with pressure at the intermediate region ($z = 10$–15 cm) can be due to the diffusion. The coincidence between the axial variations of $N_+$ and $E$ (cf. Figures 9.7 and 9.8a and b) supports this view. Similar interpretation can be made for the buildup of $I_+$. The facts that $I_+$ and $I_e$ continue to build up after the end of the microwave pulse (cf. Figures 9.1 and 9.2) can be due to the continuation of high $T_e$ (Figure 9.3) even after the end of the microwave pulse. This helps to maintain the interpulse plasma. Although two peaks were observed in the axial density distribution (Figure 9.7), the distribution became more uniform at a lower $p$ (e.g., $p = 0.06$ Torr). However, the result at higher $p$ helped us to investigate the mechanisms of plasma production in the narrow waveguide.

The production of plasma in the narrow waveguide may be described by the ionization due to the penetrating $E$ and the high $T_e$ as well as the diffusion. Although the waveguide's dimension is below cutoff, the spatially decaying $E$ field of the high-power wave creates a dense plasma near the entrance of the tube, which diffuses into the tube. A convex shape of the plasma at the entrance diverges the wave, which would be reflected by the wall of the tube and converged in the second region ($z = z_m$). This plasma lens effect is plausible because under vacuum conditions we do not observe a second maxima at $z_m$. Once the plasma is created at $z_m$, the waveguide would behave as a cavity and give rise to the standing waves whose nodes are observed as sharp peaks in the intermediate region (cf. Figure 9.8a). At a lower $p$, the plasma density is lower, and due to the weak diverging effect of the plasma at the entrance, the field is spread over a larger region, as seen in Figure 9.8b. The results in vacuum (Figure 9.8c) have shown that despite the dimension of the waveguide being below cutoff, the evanescent wave can actually penetrate into the waveguide. This is considered to be due to the finite length effect of the waveguide.

### Modeling of the Spatiotemporal Current Profiles

In this model, it is assumed that during the power-on time, the microwave pulse generates a plasma (of a finite width) at the entrance of the waveguide. The experimental results also conform to this view (cf. Figure 9.7). The plasma generated at the entrance diffuses in space and time along the wave-guide. The radial diffusion is suppressed by the magnetic field. The temporal behavior and the axial variation of the plasma are considered to be governed by the particle balance between the production and loss, obtained by assuming a diffusion model at a higher pressure $p$. The second peak in the plasma density at $z_m$ is considered to be the effect of the focused electric field at that position by the combined effects of wave refraction and reflection. This has been taken into account in the model by considering additional ionization at $z_m$. Starting from the continuity equation

$$\frac{\partial N_+}{\partial t} + \vec{\nabla} \cdot \vec{J} = \alpha N_+, \tag{9.3}$$

where $N_+$ is the charged particle density, $J$ is the current density, and $\alpha$ is the ionization rate. The right-hand side of Equation 9.3 is the source term representing particle generation. Considering a cylindrical coordinate system, the current density $J$ can be decomposed into radial and axial components given by

$$J_z = N_+ v - D_z \frac{\partial N_+}{\partial z} \tag{9.4}$$

and

$$J_r = -D_r \frac{\partial N_+}{\partial r}, \tag{9.5}$$

where it is assumed that the azimuthal current is zero. $v$ is the drive velocity in the axial direction, which depends upon the charged particle mobility and the electric field gradient along $z$ near the entrance of the waveguide. It is considered that the electric field drives the particles through the inlet of the waveguide with a velocity $v$.

Substituting Equations 9.4 and 9.5 into Equation 9.3 and simplifying, we get

$$\frac{\partial N_+}{\partial t} + v \frac{\partial N_+}{\partial z} - D_z \frac{\partial^2 N_+}{\partial z^2} - D_r \frac{1}{r} \frac{\partial}{\partial r} \left( r \frac{\partial N_+}{\partial r} \right) - \alpha N_+ = 0. \tag{9.6}$$

Using the method of separation of variables, the plasma density can be divided into radial and axial components given by

$$N_+(r,z,t) = N_R(r) N_Z(z,t), \tag{9.7}$$

where it has been assumed that the radial particle density distribution is time independent.

On the substitution of Equation 9.7 in Equation 9.6, we can get two independent equations for the radial and the axial density. The equation of the radial part can be written as

$$\frac{\partial^2 N_R(r)}{\partial r^2} + \frac{1}{r}\frac{\partial N_R(r)}{\partial r} + \frac{1}{D_r/(\alpha + \beta)}N_R(r) = 0, \tag{9.8}$$

which is a Bessel's equation (of order zero) and has a solution given by

$$N_R(r) = N_{+0}J_0\left(\frac{r}{[D_r/(\alpha + \beta)]^{1/2}}\right), \tag{9.9}$$

where $N_{+0}$ is the plasma density at the center of the tube ($r = 0$). To satisfy the boundary condition $N_R(r) = 0$ at $r = R$, where $R$ is the radius of the waveguide and for a finite $N_{+0}$, we must set $a/[D_r/(\alpha + \beta)]^{1/2}$ equal to a constant $c$, where $c = 2.4$ for the first zero of $J_0$. $\beta$ is then given by

$$\beta = (c/R)^2 D_r - \alpha, \tag{9.10}$$

$\beta$ can be considered a decay constant given by the radial diffusion rate $D_r$ and the ionization rate $\alpha$.

The equation for the axial part can be written as

$$\frac{\partial N_Z(z,t)}{\partial t} + v\frac{\partial N_Z(z,t)}{\partial z} - D_z\frac{\partial^2 N_Z(z,t)}{\partial z^2} + \beta N_Z(z,t) = 0, \tag{9.11}$$

which has a solution given by

$$N_Z(z,t) = \frac{2}{\sqrt{\pi}}\exp(-\beta t)\frac{1}{\sqrt{4D_zt}}\exp\left(-\frac{(z - vt)^2}{4D_zt}\right), \tag{9.12}$$

where the axial diffusion coefficient $D_z$ is in cm²/s, the drive velocity $v$ is in cm/s, and the decay constant $\beta$ is in cm²/s. Equation 9.12 can be considered the basic equation to explain the spatiotemporal evolution of the plasma density along the tube. It may be noted that the density $N_+(z,t)$ is given in cm⁻¹ and is a line density because it describes only one component of the total density ($N = N_r + N_\theta + N_z$) in three dimensions.

Considering the finite pulse width of duration $t_w$ and a finite pulse on time plasma width $\Delta z$, the integrated plasma density at a particular axial location $z$ and at a decay time $t$ after the end of the pulse can be written as

$$N_+(z,t) = \frac{2}{\sqrt{\pi}} \sum_{j=1}^{2} N_j \int_{z_i}^{\Delta z} \int_0^{t_w} \frac{\exp\{-\beta(t+t')\}}{\sqrt{4D_z(t+t')}} \exp\left\{-\frac{(z-z'+z-v_j(t+t'))^2}{4D_z(t+t')}\right\} dt'dz',$$

(9.13)

where $t'$ is the moving time coordinate in the pulse of duration $t_w$ with respect to the decay time $t$, $z'$ is the moving position coordinate in the plasma of width $\Delta z$ with respect to the position $z$, $z_i$ is the initial position of the plasma width $\Delta z$ as seen from the location $z$, $N_j$ represents the pulse on time plasma density at $z \sim 0$, $z_m$ (corresponding to the two peaks in $N_+(z)$), and $v_j$ is the drive velocity at $z = 0$, $z_m$. The subscript $j = 1, 2$ indicates a label for the positions $z = 0$, $z_m$. As discussed earlier, the stronger fields at $z = 0$ and $z_m$ give rise to additional ionization at these locations.

As a remark to Equation 9.13, if the tube has a diameter above the cutoff value, the wave propagates freely into it, then we may put $D_z \rightarrow \infty$ into Equation 9.13 and the spatial integration becomes unity and drops out. The temporal integral reduces to $N_+ \sim \exp(-\beta t)$, which was previously obtained for the case of a recombination free decay [3].

At a pressure of 1 Torr and an electron temperature of 6 eV, the mean free path is ~0.025 cm. The electron and the ion cyclotron radii at a position near the wall where $B = 1000$ G are ~0.01 cm and ~1.03 cm, respectively. Thus, the radial diffusion of the ions may not be influenced much; however, electron diffusion would be suppressed as they move radially outward to regions of higher magnetic field. Thus, it is understood that the radial diffusion coefficient $D_r = (\kappa T_e v_e)/(m\omega_c^2)$ would be greater for ions than electrons. Therefore, the decay constant for the ions is greater than the electrons ($\beta_+ > \beta_-$) (cf. Equation 9.8), which has also been experimentally observed (cf. Figure 9.1a and b).

Considering Equation 9.4, the drive velocity $v$ can be written as a product of the mobility of the charged particles $\mu_m$ and the electric field. $E$ can be considered a static electric field due to a denser plasma at the entrance, which has a higher plasma potential $V_s$. As $\mu_m = q/mv_e$ and $D_z = \kappa T_E/mv_e$, where $m$ is the particle mass, $\mu_{m-} > \mu_{m+}$ and $D_{z-} > D_{z+}$, where the plus and minus signs denote ions and electrons, respectively. However, the axial motion of the electrons will be retarded and those of the ions accelerated because of the high $V_s$ at the entrance, although the magnitude of the electron velocity will be larger because of the greater mobility. The following conditions have been taken into account in the numerical model for obtaining the spatiotemporal profiles of electrons and ions (a) $\beta_+ > \beta_-$, (b) $D_{z-} > D_{z+}$, and (c) $v_- > v_+$.

Figure 9.16a and b shows the spatiotemporal ion ($I_+$) and electron ($I_e$) current profiles plotted using Equation 9.13 with the above considerations. The values of the parameters are displayed in Figure 9.16a and b. It can be seen that the decay profiles resemble the experimentally obtained results (Figure 9.1a and b) rather well. Figure 9.9c shows the axial variation of the ion current with time. The development of the current peaks near the entrance and at

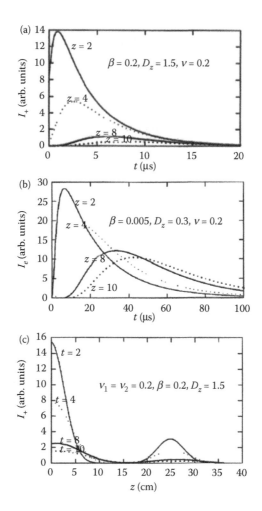

**FIGURE 9.16**
Numerical simulation of the spatiotemporal (a) ion ($I_+$), (b) electron ($I_e$) current profiles, and (c) the axial variation of $I_+$ with time.

$z = z_m$ is clearly reproduced as in the experimental result of Figure 9.7. With increase in time, the density profiles become flatter.

Before performing the numerical simulation, the effect of the parameters $\beta$, $D_z$, and $v$ on the decay time constant $\tau_d$, the peak current height $I_m$, and the temporal position of $I_m$, $t_m$ have been studied. An increase in the parameters produces the following changes in the spatiotemporal profiles, as shown in Tables 9.1 and 9.2.

The results have been interpreted as follows: The radial diffusion of the electrons is suppressed due to the magnetic field near the waveguide wall. Although electrons have a higher mobility, $V_s$ retards their axial diffusion along the waveguide and accelerates those of the ions. Thus the combined

**TABLE 9.1**

Effect of Increasing the Parameter $\beta$, the Axial Diffusion Coefficient $D_z$, and the Drive Velocity $v$ on the Decay Time Constant $\tau_d$, the Peak Height $I_m$, and the Temporal Position of the Peak Height $t_m$ of the Current Profile

|   |        | $\tau_d$          | $I_m$     | $t_m$     |
|---|--------|-------------------|-----------|-----------|
| 1 | $\beta$   | Decreases         | Decreases | Decreases |
| 2 | $D_z$     | Slightly decreases | Increases | Decreases |
| 3 | $v$       | Slightly decreases | Increases | Decreases |

**TABLE 9.2**

Range of Parameters Obtained for the Best Fit of the Ion and Electron Spatiotemporal Profiles as Obtained from the Numerical Simulation

|   |        | Ion Current ($I_+$) | Electron Current ($I_e$) |
|---|--------|----------------------|--------------------------|
| 1 | $\beta$   | 0.1–0.2              | 0.001–0.005              |
| 2 | $D_z$     | 1.5–3.0              | 0.3–0.5                  |
| 3 | $v$       | 0.08–0.3             | 0.08–0.2                 |

*Note:* The numbers represent only relative values.

result could cause them to have almost similar $v$ as those of ions. The axial diffusion of ions is found to be slightly greater than those of the electrons, although theoretically it should be smaller, as discussed earlier. This discrepancy is thought to be once again due to the retarding effect of the $V_s$ on the electron axial diffusion.

Although initially it was thought that the electric field cannot penetrate the tube, experimental results have shown the appearance of the field inside the waveguide. This is a new and useful result that helps in the plasma production and maintenance in the narrow waveguide.

## Discussion

The results indicate that higher-density plasmas are produced at the tube entrance ($z \sim 0$) and at $z = z_m \sim 20$–30 cm near the tube exit. A high electron temperature helps plasma production and maintenance in the interpulse regime. From regions of higher density, the plasma can diffuse axially toward the tube center where the density is lower. Additionally, a shift in the axial position of the peak current indicates that the charged particles are driven through the tube inlet. With regard to this, the effect of the electrostatic and ponderomotive forces may be considered.

## Electric Field Penetration

Although the radius of the tube is smaller than the cutoff value, experimental results indicate that even under vacuum conditions the field can penetrate into the tube. This is considered to be due to an effect of the evanescent high-power wave, the finite length effect of the tube, and the power-loss effect due to a finite conductivity of the wall. These effects will cause discrepancies from an infinitely long-conducting wall tube.

The strong field near the tube entrance will help in the production of a dense plasma at that location. After the plasma is created at the entrance, the refractive index $\eta$ is modified from the vacuum. Calculations using the cold plasma dispersion, including collisions, for waves launched perpendicular to $B$ show that $\eta(r)$ is constant (0.875) from the tube center until $r = 1.4$ cm, but increases sharply toward unity near the wall. The profile of $\eta(r)$ suggests that the plasma at the first maximum $N_+$ behaves similarly to a concave lens.

As a possible mechanism of the second peak in the microwave electric field, it is considered that the waves are refracted by the plasma lens and diverge toward the inner wall of the tube. After they are reflected by the wall, phase mixing will occur and a second point of maximum amplitude will occur. This corresponds to the region $z_m$ as found experimentally.

## Plasma Buildup

The high power in the pulse generates strong electric fields of short intervals that can accelerate electrons to high energies. The high-energy (hot) electrons gradually lose energy by collision with neutrals, in the event the ionization increases due to convolution of collision cross section with electron energy distribution function (peak ~50 eV for singly charged Ar ions), before eventually decreasing at lower energies. This slowing-down process could be a reason for the observed delay in the plasma buildup.

The higher plasma densities at $z \sim 0$ and $z_m$ can be explained by the higher ionization efficiency since the electric field was stronger at these positions. The correlation between the axial variations of $N_+$ and $\tilde{E}$ supports this view.

Figure 9.12 shows that at the entrance of the tube ($z = 2$ cm), the space potential $V_s$ drops rapidly in about 1 μs. Such a potential drop would repel electrons and force them to move in the axial directions. Ions would eventually follow electrons by the space charge field. We define here $v_s$ as an electrostatic drive velocity caused by this potential effect.

Besides this, high-power microwaves during the pulse can exert a ponderomotive force on the electrons. The force is expressed as $F_{pm} = -(\omega_p^2 / \omega^2)\nabla(\epsilon_o \tilde{E}^2 / 2)$, where $\epsilon_o$ and $\tilde{E}$ are the dielectric constant in vacuum and the microwave electric field, respectively [10,11]. $F_{pm}$ can be written as $F_{pm} = -\nabla(P/2cA)$, where $P$, $c$, and $A$ are the power, the light velocity, and the cross section of the wave, respectively. If we assume that $\omega_p \cong \omega$ and the wave cross section is equal to that of the waveguide ($A = \pi d^2/4 = 25.9 \times 10^{-4}$ m$^2$), we have $\epsilon_o \tilde{E}^2/2 = (3-4) \times 10^{-2}$ (J m$^{-3}$) for $P = 60$ kW. On the other hand, the force $F_{pl}$ due to the pressure gradient is given by $F_{pl} = \nabla(N_e \kappa T_e)$. For typical experimental conditions of $N_e = 10^{16}$–$10^{17}$ m$^3$ and $T_e \cong 10$ eV, $N_e \kappa T_e = (2-16) \times 10^{-2}$ (J m$^{-3}$). Therefore, under the present experimental situation, $F_{pm}$ and $F_{pl}$ become of the same order. We define $v_p$ as a drive velocity due to the ponderomotive force. The measured drive velocity $v$ shown in Figure 9.11 is considered to be the sum of two kinds of driven velocity, $v = v_s + v_p$, which drives the particles through the tube inlet.

The minimum-$B$ structure of the multicusp magnetic field suppresses radial diffusion. At $p = 1$ Torr and $T_e = 6$ eV, the electron gyroradius at a position near the wall ($B = 1000$ G) is ~0.01 cm. Therefore, the plasma would be transported mostly by the axial diffusion. The delayed buildup, high electron temperature, and particle transport in the power-off phase are new properties as compared to the conventionally known afterglows.

## Pulse Repetition Frequency and Pulse Duration

It was seen that the pulse repetition frequency $f_r$ does not significantly change the ion current and the spatial profile. This suggests that any charge accumulation by pulse repetition is small. The phenomena at each pulse can therefore be considered as independent.

However, an increase in the pulse duration increases the energy input to the plasma. This leads to higher values of peak current and decay time constant, thereby allowing additional control of the plasma in the tube.

## Conclusions

A pulsed microwave plasma has been produced in a circular conducting tube with a radius smaller than the cutoff value. The principal mechanisms of plasma production and maintenance are gaseous breakdown by electric field penetration and wave propagation by nonlinear dispersion. Charged particles are axially driven into the tube by the plasma electrostatic and ponderomotive force of the high-power wave. The anisotropic diffusion of the charged particles is favorable for axial plasma uniformity in the tube. The interpulse plasmas, characterized by a high electron temperature and a delayed buildup, are new features of a plasma without an energy source.

Main results indicate that the electron temperature $T_e$ is about 6–14 eV and continues to be high even after the pulse ends. The plasma density builds up after the end of the pulse, with a peak value above $10^{10}$ cm$^{-3}$. Although the above experimental results are concerned with higher pressures, the plasma uniformity increases at a lower pressure. The results will be useful for certain applications requiring a pulsed plasma in a narrow cross-sectional waveguide.

## References

1. C.A. Sullivan, W.W. Destler, J. Rodgers, and Z. Segalov. 1988. Short-pulse high-power microwave propagation in the atmosphere, *J. Appl. Phys.* **63**: 5228–5232.
2. G.A. Askaryan, G.M. Batanov, A.E. Barkhudarov, S.I. Gritsinin, E.C. Korchagina, I.A. Kossyi, V.P. Silakov, and N.M. Tarasova. 1984. A freely localized microwave discharge for removal of chlorofluorocarbon contamination from the atmosphere, *J. Phys. D: Appl. Phys.* **27**: 1311–1318.
3. S. Bhattacharjee and H. Amemiya. 1998. Interpulse plasma of a high power narrow bandwidth pulsed microwave discharge, *J. Appl. Phys.* **84**: 115–120.
4. S. Biri, T. Nakagawa, M. Kidera, Y. Miyazawa, M. Hemmi, T. Chiba, N. Inabe et al. 1999. Production of highly charged ions in the RIKEN 18 GHz ECR ion source using an electrode in two modes, *Nucl. Instrum. Meth. B* **152**: 386–396.
5. F. Dorchies, J.R. Marques, B. Cros, G. Matthieussent, C. Courtois, T. Velikoroussov, P. Audebert et al. 1999. Monomode guiding of $10^{16}$ W/cm$^2$ laser pulses over 100 Rayleigh lengths in hollow capillary dielectric tubes, *Phys. Rev. Lett.* **82**: 4655–4658.
6. H. Fujiyama, Y. Tokitu, Y. Uchikawa, K. Kuwahara, K. Miyake, K. Kuwahara, and A. Doi. 1998. Ceramics inner coating of narrow tubes by a coaxial magnetron pulsed plasma, *Surf. Coatings Technol.* **98**: 1467–1472.
7. S. Bhattacharjee and H. Amemiya. 1997. Transversely magnetized microwave plasma in a rectangular waveguide under cutoff conditions, *Rev. Sci. Instrum.* **68**: 3061–3067.
8. S. Bhattacharjee and H. Amemiya. 1998. Microwave plasma in a multicusp circular waveguide with a dimension below cutoff, *Jpn. J. Appl. Phys. (Part 1)* **37**: 5742–5745.
9. S. Bhattacharjee and H. Amemiya. 1999. Production of microwave plasma in narrow cross sectional tubes: Effect of the shape of cross section, *Rev. Sci. Instrum.* **70**: 3332–3337.
10. F.F. Chen. 1984. *Introduction to Plasma Physics and Controlled Fusion* (Vol. 2, Plenum Press, New York) p. 305 (Chapter 8).
11. H. Ito, Y. Nishida, and N. Yugami. 1996. Formation of duct and self-focusing in plasma by high-power microwave, *Phys. Rev. Lett.* **76**: 4540–4543.
12. S. Bhattacharjee and H. Amemiya. 2000. Pulsed microwave plasma production in a conducting tube with a radius below cutoff, *Vacuum* **58**: 222.

# 10

## Plasma Buildup by Short-Pulse, High-Power Microwaves

### Introduction

Short-pulse, high-power microwaves were earlier applied for several studies such as gaseous breakdown [1–6], pulse propagation [7], material surface processing [8], excimer discharges [9], and even for environmental amelioration [10,11]. These studies are concerned with the determination of the breakdown of electric fields [1–6], effect on pulse propagation through a plasma layer [7], surface modification by plasma formed near the sample [8], sustainment of high-pressure discharges for lighting applications [9], and environmental remediation using air plasma chemistry in the atmosphere [10,11]. However, little information can be obtained from them regarding the characteristics of the plasma produced by such short-pulse, high-power microwaves. Of interest is knowledge of the plasma parameters and characteristics of the temporally evolving plasma after the launching of the high-power waves in a short pulse.

Recently, we have investigated the plasmas produced by this scheme of pulsing [12–16], and found some novel properties particularly in the power-off phase. Notable among them are: the plasma buildup is extended beyond the pulse duration, with the peak density attained a few to tens of microseconds later in the power-off phase. A high value of electron temperature (~6–10 eV) continues for a time comparable to the buildup time. The steady-state phase is transient or absent. The plasma state between the pulses defined as an interpulse plasma [12] covers both buildup and decay.

Using the properties of the interpulse plasma, short-pulse, high-power microwaves have been applied for the production of a narrow cross-sectional, pulsed plasma in a conducting tube with a radius below the cutoff value [16]. The plasma properties in the narrow tube have been studied and the production mechanisms explained on the basis of electric field penetration and some nonlinear effects associated with the large amplitude of the field. The wave propagation was found to be modified by the dispersion of a magnetoplasma. Results indicated that the plasma buildup plays an important role in discharge production and maintenance, and further investigation would be required for an understanding of the process. Such a study will be useful from the viewpoint of applications of the pulsed plasma.

With this motivation, the present study addresses the buildup of inter-pulse plasmas over a pressure range of 10 mTorr–10 Torr. The temporal evo-lution of the electron current and the total optical intensity have been studied through parameters such as peak current, buildup time, growth, and decay constants. The results are explained by the growth of the electron tempera-ture during the pulse and the temporal evolution of the plasma following the end of the pulse. The effects of varying the repetition frequency and the pulse duration on the buildup are also investigated.

This chapter has been arranged as follows. In the section "Experiments," the experiment is described and the results are presented. Discussion on the important aspects of the buildup, including the experimental results, is presented in the "Discussion" section. Finally, this chapter ends with the "Conclusions" section.

## Experiments

Figure 10.1 shows a schematic of the experimental system. The experi-ment is conducted in a cylindrical tube waveguide (WG; length $L = 37$ cm and inner radius $R = 2.87$ cm) kept inside a vacuum chamber C. The pulse duration and the repetition frequency of the microwaves are varied in the range 0.05–1.2 µs and 10–500 Hz, respectively. Microwaves of 3 GHz are pro-duced by a magnetron oscillator OSC with a peak output power of 100 kW. The waves are guided through a coaxial cable Cx, a power attenuator ATT, a directional coupler DC with fast diodes for monitoring the forward and reflected powers, a stub tuner M for adjustment of impedance mismatch, a rectangular bend H for changing the wave direction, and finally through a quartz window W into WG. Permanent magnets are arranged around WG in the form of a multicusp [17] MC (10 poles) to provide a minimum-$B$ field for plasma confinement. The waves are launched with the wave vector perpen-dicular to the magnetic field. The circulator ISO protects the magnetron from the reflected power. A trigger signal TRIG is used for pulse initiation and as a reference signal for diagnostics. A PC is used for storing data obtained through a digital oscilloscope DOS.

Before the experiments, the chamber C was evacuated with a diffusion pump to pressures below $10^{-6}$ Torr. Argon served as the experimental gas and was maintained in the pressure range of 10 mTorr–10 Torr with a mass flow controller. The absolute pressure of the test gas was measured by a transducer gauge (Baratron) in the pressure range of 0.1–100 Torr. To avoid contamination of the discharge, the experiment was performed under gas flow conditions.

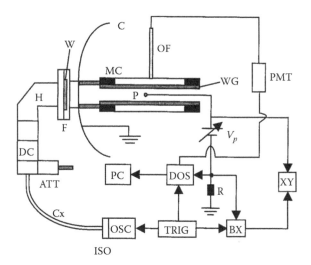

**FIGURE 10.1**
Schematic of the experimental apparatus; OSC: magnetron oscillator, ISO: isolator, Cx: coaxial cable, ATT: attenuator, PM: power monitor, M: matching element, H: rectangular bend, W: window, C: chamber, MC: multicusp, WG: waveguide, OF: optical fiber, P: Langmuir's probe, PC: computer, DOS: digital oscilloscope, BX: Boxcar's integrator, XY: X–Y chart recorder, TRIG: trigger signal, $V_p$: probe voltage, R: resistor. (Reprinted with permission from S. Bhattacharjee, H. Amemiya, and Y. Yano. 2001. Plasma build-up by short-pulse high-power microwaves, *J. Appl. Phys.* **89**: 3573. Copyright 2001, American Institute of Physics.)

The temporally varying profiles of the ion ($I_+$) and electron ($I_e$) currents are measured using an electrostatic probe P (plane, 5 mm diameter) at fixed negative and positive biases $V_p$ and recorded in the digital oscilloscope. Data averaging is applied in the DOS, usually averaging 64 shots, to obtain a steady pattern with repeatability. The planar probe P is inserted through an axial port and placed at the center of WG, where the magnetic field is almost zero. The probe surface was set parallel to the wave vector. WG together with MC served as the reference electrode, which was connected to the grounded chamber C. Time-resolved measurements of the probe ($I$–$V$) characteristics were made using a Boxcar integrator BX and plotted on an X–Y chart recorder XY. The measured electron temperature and the plasma density have been reported elsewhere [16]. In all measurements, the probe P was kept at the tube center.

The temporal variation of the total optical intensity $I_{op}$ in the visible range was measured by inserting an optical fiber OF through a radial port onto the top of a tiny aperture in WG, at $z = 18.5$ cm. The obtained light signal was guided into a photomultiplier tube PMT and the amplified signal observed in DOS. The voltage in PMT was set at 1 kV.

Figure 10.2 shows a typical profile of the microwave pulse and the trigger signal as recorded in the digital oscilloscope. The shape of the pulse is

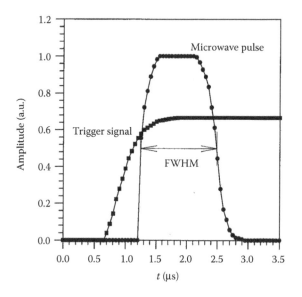

**FIGURE 10.2**
Temporal response of the microwave power and the trigger signal as recorded in the digital oscilloscope. FWHM: full width at half maximum 1.2 μs. (Reprinted with permission from S. Bhattacharjee, H. Amemiya, and Y. Yano. 2001. Plasma build-up by short-pulse high-power microwaves, *J. Appl. Phys.* **89**: 3573. Copyright 2001, American Institute of Physics.)

roughly rectangular with the edges smoothed by characteristic rise and fall times. Measurements at FWHM show that the pulse duration is about 1.2 μs. The temporal response indicates that the power delivery is delayed with respect to the trigger signal by ~0.75 μs, considered to be the effect of generator electronics and impedance conditions with the rest of the microwave circuit. The characteristic rise and fall times of the microwave pulse, defined as the time to attain or decrease to 90% of the peak amplitude, are about 150 and 175 ns, respectively.

Figure 10.3 shows typical profiles of the temporal variation of the electron current $I_e$ and the total optical intensity $I_{op}$ as replotted from the data obtained through the digital oscilloscope. The end of the pulse is marked as "end of pulse" in the figure. Unless otherwise stated, the pulse repetition frequency $f_r$ and the pulse duration $t_w$ have been set at 246 Hz and 1.2 μs, respectively. It is noted that the buildup of $I_e$ and $I_{op}$ begins during the pulse; however, it is extended well into the power-off phase of the discharge. The buildup time, defined as the time required from the end of the pulse to attain the maximum intensity, for both $I_e$ and $I_{op}$, is about 9 μs for this case. However, it varies in the range of 1–40 μs depending upon the pressure. The peak current corresponds to a density of over $10^{10}$ cm$^{-3}$.

The variation of the peak values of $I_e$ and $I_{op}$, including their buildup times, was studied as a function of pressure. Figure 10.4 shows the variation of

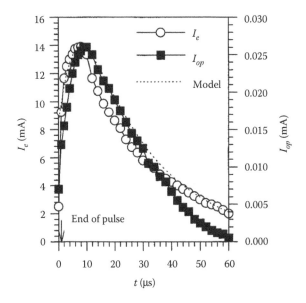

**FIGURE 10.3**
Temporal variation of the electron current $I_e$ and the optical intensity $I_{op}$ at a pressure $p = 5$ Torr. The dotted line shows the fitting of $I_e$ with Equation 10.22. (Reprinted with permission from S. Bhattacharjee, H. Amemiya, and Y. Yano. 2001. Plasma build-up by short-pulse high-power microwaves, *J. Appl. Phys.* **89**: 3573. Copyright 2001, American Institute of Physics.)

the peak electron current $I_{me}$ and the buildup time $t_{ep}$ with pressure $p$. With increase in $p$ from the lower-pressure side, $I_{me}$ and $t_{ep}$ decrease to a minimum at about 1 Torr for $I_{me}$ and 0.6 Torr for $t_{ep}$. As the pressure is further increased, $I_{me}$ and $t_{ep}$ increase, rather sharply until ~2 Torr. $I_{me}$ shows a saturation tendency; however, it again increases beyond 7 Torr. $I_{me}$ and $t_{ep}$ have a minimum value over the pressure range.

Figure 10.5 shows the variation of the peak of the total optical intensity $I_{mo}$ and the buildup time $t_{op}$ with $p$. The profiles are similar to those observed for $I_{me}$ and $t_{ep}$ (Figure 10.4), except that the decrease to the minimum is more gradual. Moreover, the minimum values of $I_{mo}$ and $t_{op}$ are observed to be slightly shifted to higher pressures (i.e., at 2 and 0.9 Torr, respectively). Beyond these pressures, the profiles show a sharp increase without any tendency for saturation within the pressure range.

The particle growth ($\zeta$) and decay ($\rho$) constants were studied as a function of pressure. $\zeta$ and $\rho$ were obtained from the temporal profiles of $I_e$ (e.g., Figure 10.3) by assuming an exponential growth and decay. Figure 10.6 shows the variation of $\zeta$ and $\rho$ with $p$. With increase in $p$ from the lower-pressure side, $\zeta$ and $\rho$ gradually increase to a maximum at ~0.6 Torr, and then decrease as $p$ is further increased. It may be noted that the pressure at which the peaks occur coincides with the pressure at which the minimum of $t_{ep}$ is observed (cf. Figure 10.4).

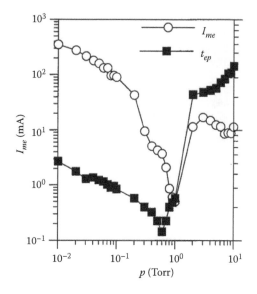

**FIGURE 10.4**
Variations of the peak electron current $I_{me}$ and its buildup time $t_{ep}$ with pressure $p$. (Reprinted with permission from S. Bhattacharjee, H. Amemiya, and Y. Yano. 2001. Plasma build-up by short-pulse high-power microwaves, *J. Appl. Phys.* **89**: 3573. Copyright 2001, American Institute of Physics.)

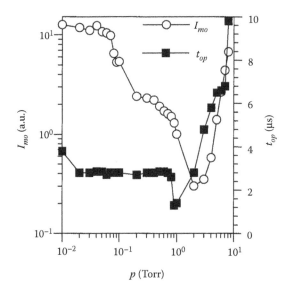

**FIGURE 10.5**
Variation of the peak of the total optical intensity $I_{mo}$ and buildup time $t_{op}$ with pressure $p$. (Reprinted with permission from S. Bhattacharjee, H. Amemiya, and Y. Yano. 2001. Plasma build-up by short-pulse high-power microwaves, *J. Appl. Phys.* **89**: 3573. Copyright 2001, American Institute of Physics.)

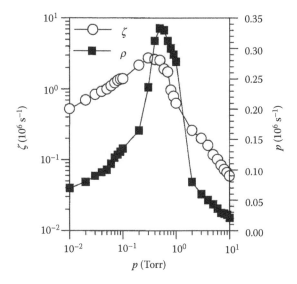

**FIGURE 10.6**
Variation of the growth constant $\zeta$ ($10^6$ s$^{-1}$) and the decay constant $\rho$ with pressure $p$. (Reprinted with permission from S. Bhattacharjee, H. Amemiya, and Y. Yano. 2001. Plasma build-up by short-pulse high-power microwaves, *J. Appl. Phys.* **89**: 3573. Copyright 2001, American Institute of Physics.)

The effect of the pulse repetition frequency $f_r$ on the buildup was investigated through its influence on $I_{me}$ and $t_{ep}$. Figure 10.7 shows the variation of $I_{me}$ and $t_{ep}$ with $f_r$ varied in the range 10–500 Hz. With increase in $f_r$, a gradual increase in $I_{me}$ is seen initially until ~30–50 Hz, thereafter the dependence upon $f_r$ is small. The buildup time $t_{ep}$ shows a small increase after 100 Hz; however, the overall trend suggests that both $I_{me}$ and $t_{ep}$ weakly depend upon $f_r$. Such a weak dependence is also shown by $\zeta$ and $\rho$ when plotted against $f_r$.

The effect of the pulse width $t_w$ on the buildup was also investigated. The variation of $I_{me}$ and $t_{ep}$ with pulse energy $U$ (2–100 mJ), defined as the product of the pulse duration ($t_w = 0.08, 0.2, 0.6,$ and 1.2 μs) and the peak power, indicated that $I_{me}$ increases with $U$, with the increase being sharper at a lower $p$ (e.g., at $p = 0.01$ Torr). The buildup time $t_{ep}$, however, decreases with increase in $U$.

## Discussion

The plasma buildup comprises of two important phenomena: the growth of electron temperature during the pulse and the temporal evolution of the plasma following the end of the pulse. Further study on these effects is made and thereafter the experimental results will be discussed.

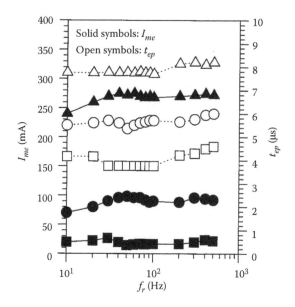

**FIGURE 10.7**
Variation of the peak electron current $I_{me}$ and the buildup time $t_{ep}$ with pulse repetition frequency $f_r$. The symbols are square: 1.0 Torr, circle: 0.1 Torr, and triangle: 0.01 Torr. (Reprinted with permission from S. Bhattacharjee, H. Amemiya, and Y. Yano. 2001. Plasma build-up by short-pulse high-power microwaves, *J. Appl. Phys.* **89**: 3573. Copyright 2001, American Institute of Physics.)

## Electron Temperature Growth during the Pulse

The electron temperature growth during the microwave pulse is determined by a balance between the heating and the loss rates, given by

$$\frac{dT_e}{dt} = \frac{v_c \left[ eE_p \sin \omega t \right]^2}{2m\left( \omega^2 + v_c^2 \right)} - L_c(T_e), \tag{10.1}$$

where $v_c$ is the elastic collision frequency, $\omega$ is the angular wave frequency, $e$ and $m$ are the charge and mass of an electron, and $E_p$ is the amplitude of the electric field in the tube. The first term on the right-hand side represents the heating rate (i.e., the average energy gained per second by the electrons between collisions), and the second term represents the loss rate (i.e., the average energy lost per second due to electron neutral collisions).

When $T_e$ is small (<8 eV), the loss rate due to collisions $L_c(T_e)$ is governed by elastic collisions [18]: $L_c(T_e) = v_c (3 m/M) T_e$, where $M$ is the ionic mass. Then, Equation 10.1 can be written as

$$\frac{dT_e}{dt} = c - \epsilon T_e, \tag{10.2}$$

where $c = v_c[eE_p \sin \omega t]^2 / 2m(\omega^2 + v_c^2)$ and $\epsilon = v_c(3 \, m/M)$. A solution to Equation 10.2 is given by

$$T_e = \frac{c}{\epsilon}\left[1 - \exp(-\epsilon t)\right]. \tag{10.3}$$

The time $(= t_a)$ when $T_e$ reaches 8 eV $(= T_1)$ is given by

$$t_a = \frac{1}{\epsilon}\ln\left[\frac{1}{1 - (\epsilon/c)T_1}\right] \cong \frac{T_1}{c}. \tag{10.4}$$

From Equation 10.3, $T_e \rightarrow c/\epsilon$ if the loss rate is always given by elastic collisions. However, after $T_e$ attains a value $T_1$ $(t_a \sim 10^{-10}$ s$)$, the loss due to inelastic collisions starts to become significant, thereby reducing the net energy gain of the electrons. Then, $L_c(T_e)$ is given by

$$L_c(T_e) = v_i E_i + v_{ex} E_{ex}, \tag{10.5}$$

where $v_i$ and $v_{ex}$ are the ionization and excitation collision frequencies, respectively, and $E_i$ and $E_{ex}$ are the threshold energies for the collisions. Equation 10.1 can then be written as

$$\frac{dT_e}{dt} = c - \left(v_i E_i + v_{ex} E_{ex}\right). \tag{10.6}$$

The collision frequencies have the general form $v = N_g K$, where $N_g$ is the neutral (gas) density and $K$ is the collision rate constant. The rate constants can be determined empirically from experimental cross sections fitted by the Arrhenius functions [19]. The ones used in the numerical calculation (in units of m$^3$ s$^{-1}$) are

$$K_{e1} = 3.3 \times 10^{-13} \exp(-3/T_e), \tag{10.7}$$

$$K_i = 5.0 \times 10^{-14} \exp(-E_i/T_e), \tag{10.8}$$

$$K_{ex} = 2.0 \times 10^{-14} \exp(-E_{ex}/T_e), \tag{10.9}$$

with $E_i = 15.76$ eV and $E_{ex} = 11.6$ eV. Equation 10.6 does not have a simple analytical solution. It was therefore solved numerically. For typical values: $v_c \approx (1 - 10) \times 10^{10}$ s$^{-1}$, $\omega = 1.9 \times 10^{10}$ s$^{-1}$, and $E_p \cong 1.21 \times 10^5$ V/m, we obtain $T_e \approx 7 - 150$ eV and the temperature was found to saturate within the pulse duration: the saturation time $T_s \cong 10^{-7}-10^{-9}$ s. At lower pressures,

$v_c \approx (0.1 - 1) \times 10^9 \text{ s}^{-1}$, higher temperatures were obtained (order of keV), and the temperature buildup was found to continue beyond the end of the pulse. The attainment of such high temperatures is not unusual because this model does not include the production of multicharged states: the ionization probability to higher charge states would increase with increasing energy and lead to further energy losses. This effect can be pronounced at a lower pressure.

## Plasma Evolution after the End of Pulse

The plasma buildup can be modeled by considering jointly the rate equations for the plasma (electron) density and the electron temperature $T_e$ as

$$\frac{\partial n_e}{\partial t} = \langle \sigma v \rangle n_e - D\nabla^2 n_e, \qquad (10.10)$$

$$\frac{\partial T_e}{\partial t} = -\eta T_e, \qquad (10.11)$$

where $n_e$ is the electron density, $\langle \sigma v \rangle$ is the ionization rate ($\sigma$: ionization cross section, $v$ is the electron velocity) of the background gas, $D$ is the diffusion rate of the induced plasma, and $\eta$ is the decay rate of the electron temperature. After the end of the microwave pulse, $T_e$ decreases rapidly to below 10 eV within the first few microseconds (typically ~5 μs), after that it could be fitted with a weak exponential decay (cf. Ref. [16]). We may neglect the recombination rate. Equation 10.10 can be rewritten as

$$\frac{\partial n_e}{\partial t} = \langle \sigma v \rangle n_e - \rho n_e, \qquad (10.12)$$

where $\rho = D/\Lambda^2$ is the decay rate due to diffusion and $\Lambda$ is the characteristic diffusion length. By including the effect of the transverse magnetic field, $\Lambda$ can be estimated from Ref. [3]

$$\frac{1}{\Lambda^2} = \left(\frac{2.4}{R}\right)^2 + \left(\frac{\pi}{L}\right)^2 \chi, \qquad (10.13)$$

where $\chi = v_c^2/(v_c^2 + \omega_b^2)$ and $\omega_b$ ($= 1.76 \times 10^7 B(G)Hz$) is the electron–cyclotron frequency. For $L = 37$ cm, $R = 2.87$ cm, and $B = 0$, we obtain $\Lambda = 1.2$ cm. When $B$ ($= 300$–$1600$ G) is included, where the upper value is the field near the tube wall, and for $p = 10$ mTorr–10 Torr, $\chi$ lies in the range $10^{-5}$–$1$. The correction to $\Lambda$ due to the magnetic field is therefore negligibly small. Hence, $\Lambda = 1.2$ cm can be used as the effective value.

Assuming a Maxwellian electron energy distribution $F(E)$, $\langle \sigma v \rangle$ is given by

$$\langle \sigma v \rangle = N_g \int_{E_i}^{\infty} \sqrt{\frac{2E}{m}} \sigma(E) F(E) dE. \tag{10.14}$$

By substituting

$$F(E) = \frac{2}{\sqrt{\pi}} \frac{\sqrt{E}}{T_e^{3/2}} \exp\left(-\frac{E}{T_e}\right) \tag{10.15}$$

and assuming that $\sigma$ is almost linear near the threshold as

$$\sigma\left(\frac{E}{T_e}\right) = A\left(\frac{E}{T_e} - \frac{E_i}{T_e}\right), \tag{10.16}$$

where $A$ (= $2.5 \times 10^{-17}$ cm$^2$/eV) is the slope at the threshold of the ionization cross section, we obtain from Equation 10.14

$$\langle \sigma v \rangle = C\sqrt{\xi}\left(2 + \frac{1}{\xi}\right)\exp\left(-\frac{1}{\xi}\right), \tag{10.17}$$

where

$$C = \frac{2}{\sqrt{\pi}} \sqrt{\frac{2E_i}{m}} N_g A \tag{10.18}$$

and $\xi = T_e/E_i$. Substituting the values of the constants, we obtain $C = 2.35 \times 10^8$ $p$ (s$^{-1}$), where $p$ is the pressure in Torr.

The ionization rate $\langle \sigma v \rangle$ given by Equation 10.17 is plotted against $\xi$ to study its dependence. For the region of interest, $0.1 < \xi < 0.8$, $\langle \sigma v \rangle$ may be well approximated by

$$\langle \sigma v \rangle = C1.3 \, \xi^2. \tag{10.19}$$

Using Equation 10.19 and the solution of Equation 10.11: $T_e = T_{eo} \exp(-\eta t)$, where $T_{eo}$ is $T_e$ at $t = 0$ (end of the pulse), Equation 10.10 can be written as

$$\frac{\partial n_e}{\partial t} = C1.3 \, \xi_0^2 \exp(-2\eta t) n_e - p n_e \tag{10.20}$$

or

$$\frac{\partial n_e}{\partial t} = \zeta n_e \exp(-2\eta t) - p n_e, \tag{10.21}$$

where $\zeta = C1.3\xi_0^2$ and $\xi_0 = T_{eo}/E_i$.

A solution of Equation 10.21 is given by

$$n_e = n_0 \exp\left\{\frac{\zeta}{2\eta}\left[1 - \exp(-2\eta t)\right] - \rho t\right\}, \tag{10.22}$$

where $n_0$ is the initial free electron density. At a small $t$, the density grows exponentially as

$$n_e = n_0 \exp\left[(\zeta - \rho)t\right] \tag{10.23}$$

and reaches a peak at $t_p$ given by

$$t_p = \frac{1}{2\eta}\ln\left(\frac{\zeta}{\rho}\right), \tag{10.24}$$

which is the buildup time, and decays in the final phase as

$$n_e = n_0 \exp(-\rho t). \tag{10.25}$$

In Figure 10.6, the variations of the growth ($\zeta$) and decay ($\rho$) constants of the electron current with pressure have been shown. They have been obtained as follows. In accordance with Equation 10.23, ($\zeta - \rho$) was obtained from the growth of $I_e$ after the end of the pulse (Figure 10.3). The decay constant $\rho$ was obtained from the decay region in accordance with Equation 10.25. Thereafter, $\rho$ was added to ($\zeta - \rho$) yield $\zeta$ as shown in Figure 10.6.

The dotted profile labeled "model" in Figure 10.3 shows a fitting to the experimentally measured $I_e$ by using Equation 10.22. A reasonably good fit is obtained with parameter values $\zeta = 0.2 \times 10^6$ s$^{-1}$, $\rho = 0.04 \times 10^6$ s$^{-1}$, and $\eta = 0.1 \times 10^6$ s$^{-1}$, which are in agreement with those obtained from the experiment (cf. Figure 10.6). The overall characteristics of the temporal evolution of the current are well reproduced. However, some discrepancy appears during the initial stages of the decay, which could be because of the noninclusion of the recombination loss in the model.

### On the Experimental Results

The minimum in the peak electron current $I_{me}$ (Figure 10.4) can be understood as follows. The increase in $I_{me}$ at a higher pressure may be due to efficient power transfer from the wave electric field to the plasma: the effective electric field is given by $E_e = E_p v_c / (v_c^2 + \omega^2)^{1/2}$. At a higher pressure, $(v_c^2/\omega^2 \gg 1)\, E_e = E_p$, therefore, the total power can be conveyed to the plasma, whereas at a lower pressure, $(\omega^2/v_c^2 \gg 1)E_e = E_p v_c / \omega$, the effective power is reduced. For typical experimental values of $T_e = 6$–$10$, $\omega = 1.9 \times 10^{10}$ Hz, and $v_c = v_e/\lambda_e$ Hz, where $v_e \, (= 5.9 \times 10^7 \sqrt{T_e(\mathrm{eV})}$ cm/s) is the electron thermal velocity and $\lambda_e$ is the electron mean-free path, the pressure at which the transition ($v_c/\omega = 1$) occurs lies in the range 1–3 Torr and is in close agreement with the experimental results.

The increase in $I_{me}$ at lower pressures is considered to be due to better efficiency of electron–cyclotron heating and resonance (ECR) action ($B_{ECR} = 1070$ G, at $r \cong 2.8$ cm) and improvement of magnetic confinement due to the increase of $\omega_b \tau$, where $\tau$ is the mean collision time: the lateral diffusion rate $D = D_0 / \omega_b^2 \tau^2$, where $D_0$ is the diffusion rate for the case $B = 0$. Particle loss through the cusp regions [20] is considerably reduced. Heating efficiencies tend to be better for pressures satisfying $r_e \le \lambda_e$, where $r_e$ is the electron gyroradius and $\lambda_e$ is the mean-free path. For $T_e = 6$–$10$ eV and a lower limit of $B = 300$ G ($r \cong 2$ cm), we obtain $p < 0.5$–$1$ Torr.

The profile of the peak of the total optical intensity $I_{mo}$ (Figure 10.5) resembles that of $I_{me}$ (Figure 10.4), except that the position of the minimum is shifted to a slightly higher pressure region (2 Torr). $I_{mo}$ depends upon the excitation rate, which is a function of the electron temperature and density, and can be written as $\Gamma = n_e \langle v \sigma_{ex} \rangle$, where $\sigma_{ex}$ is the excitation cross section. The electron temperature has a slight linearly decreasing tendency with pressure. If we assume that $I_{me}$ in Figure 10.4 represents the variation of the electron density $n_e$, and $\langle v \sigma_{ex} \rangle$ can be represented by an equation like Equation 10.19, then $\Gamma \propto n_e (T_e/E_{ex})^2$, where $E_{ex}$ is the threshold excitation energy. The profile of $I_{me}$ then closely resembles $I_{mo}$, and the shift of the minimum to the higher-pressure side can be explained.

Equation 10.24 suggests that the pressure dependence of $t_p$ will depend on the pressure dependence of the parameters $\zeta$, $\rho$, and $\eta$. This is decided mainly by the ratio of the dependences of $\zeta$ and $\rho$, if we assume that the dependence of $\eta$ on $p$ is small. From Figure 10.6, it is understood that the minimum value of the ratio $\zeta/\rho$ occurs at ~0.6 Torr. Hence, the experimentally observed minimum value of $t_p$ (i.e., $t_{ep}$) at 0.6 Torr (Figure 10.4) is consistent.

Owing to a small duty cycle ($10^{-3}$–$10^{-2}$%) of the pulsed microwaves, charge accumulation by repetitive pulsing can be considered small. As a result, $t_{ep}$ and $I_{me}$ (Figure 10.7) show only a weak dependence on $f_r$. Increase in pulse duration, however, leads to a greater amount of energy being input to the plasma, hence, $I_{me}$ increases. The decrease in $t_{ep}$ with increase in $U$ could be because of an increase in the ionization rate due to a possible rise in $T_e$.

The continuation of the plasma buildup into the power-off phase of the discharge can be explained by the high-average energy of the electrons generated during the pulse, which can bring about ionization during the power-off phase. Additionally, the ionization effect due to collisional relaxation of the hot electrons in the tail of the distribution has also been considered to be important [16].

## Conclusion

The buildup of a plasma produced by short-pulse, high-power microwaves has been studied. It has been found that the buildup continues beyond the end of the microwave pulse into the power-off phase, with the buildup time varying between a few to tens of microseconds depending upon the pressure. The continuation of the buildup is considered to be due to the generation of a high electron temperature during the pulse. The plasma density and the buildup time can be higher both at a lower and at a higher pressure, with the occurrence of a minimum. At a lower pressure, higher plasma densities are obtained with the help of efficient electron–cyclotron heating and improved plasma confinement. At a higher pressure, efficient transfer of energy from the wave electric field to the plasma helps to produce a high-density plasma. The buildup phenomena consist of two important processes that is, the temporal evolution of the electron temperature during the pulse and the following plasma evolution after the end of the pulse.

## References

1. L. Gould and L.W. Roberts. 1956. Breakdown of air at microwave frequencies, *J. Appl. Phys.* **27**: 1162–1170.
2. D.J. Rose and S.C. Brown. 1957. Microwave gas discharge breakdown in air, nitrogen, and oxygen, *J. Appl. Phys.* **28**: 561–563.
3. A.D. MacDonald. 1966. *Microwave Breakdown in Gases* (Wiley, New York).
4. S.J. Tetenbaum, A.D. MacDonald, and H.W. Bandel. 1971. Pulsed microwave breakdown of air from 1 to 1000 Torr, *J. Appl. Phys.* **42**: 5871–5872.
5. W.M. Bollen, C.L. Yee, A.W. Ali, M.J. Nagurney, and M.E. Read. 1983. High-power microwave energy coupling to nitrogen during breakdown, *J. Appl. Phys.* **54**: 101–106.
6. C.A. Sullivan, W.W. Destler, J. Rodgers, and Z. Segalov. 1988. Short-pulse high-power microwave propagation in the atmosphere, *J. Appl. Phys.* **63**: 5228–5232.
7. S.P. Kuo, Y.S. Zhang, and P. Kossey. 1990. Propagation of high power microwave pulses in air Breakdown environment, *J. Appl. Phys.* **67**(6): 2762–2766.
8. R.B. James, P.R. Bolton, R.A. Alvarez, W.H. Christie, and R.E. Valiga. 1988. Melting of silicon surfaces by high-power pulsed microwave radiation, *J. Appl. Phys.* **64**: 3243–3253.
9. V. Rousseau, S. Pasquiers, C. Boisse-Laporte, G. Callende, P. Leprince, J. Marec, and V. Puech. 1992. Efficient pulsed microwave excitation of a high-pressure excimer discharge, *J. Appl. Phys.* **71**: 5712–5714.
10. G.A. Askaryan, G.M. Batanov, A.E. Barkhudarov, S.I. Gritsinin, E.G. Korchagina, I.A. Kossyi, V.P. Silakov, and N.M. Tarasova. 1994. A freely localized microwave discharge for removal of chlorofluorocarbon contamination from the atmosphere, *J. Phys. D* **27**: 1311–1318.

11. A.V. Gurevich, N.D. Borisov, S. Montecinos Geisse, and P. Hartogs. 1995. Artificial ozone layer, *Phys. Lett. A* **207**: 281–288.
12. S. Bhattacharjee and H. Amemiya. 1988. Interpulse plasma of a high-power narrow-bandwidth pulsed microwave discharge, *J. Appl. Phys.* **84**: 115–120.
13. S. Bhattacharjee and H. Amemiya. 1999. *Proceedings of the XXIV International Conference on Phenomena in Ionized Gases*, ICPIG, Warsaw, Poland, p. 133.
14. S. Bhattacharjee and H. Amemiya. 2000. Pulsed microwave plasma production in a conducting tube with a radius below cutoff, *Vacuum* **58**: 222–232.
15. S. Bhattacharjee. 2000. *Proceedings of the IEEE International Conference on Plasma Science*, New Orleans, p. 203.
16. S. Bhattacharjee and H. Amemiya. 2000. Production of pulsed microwave plasma in a tube with a radius below the cutoff value, *J. Phys. D Appl. Phys.* **33**: 1104–1116.
17. S. Bhattacharjee and H. Amemiya. 1998. Microwave plasma in a multicusp circular waveguide with a dimension below cutoff, *Jpn. J. Appl. Phys. Part 1* **37**: 5742–5745.
18. A. Von Engel. 1983. *Electric Plasmas: Their Nature and Uses* (Taylor & Francis, London) p. 79.
19. M.A. Lieberman and A.J. Lichtenberg. 1994. *Principles of Plasma Discharges and Materials Processing* (Wiley, New York) p. 80.
20. S. Bhattacharjee and H. Amemiya. 1999. Production of microwave plasma in narrow cross-sectional tubes; effect of the shape of cross section, *Rev. Sci. Instrum.* **70**: 3332–3337.
21. S. Bhattacharjee, H. Amemiya, and Y. Yano. 2001. Plasma build-up by short-pulse high-power microwaves, *J. Appl. Phys.* **89**: 3573.

# 11

## Power Absorption and Intense Collimated Beam Production in a Pulsed High-Power Microwave Ion Source

### Introduction

ECR ion sources have been the subject of active research for many years [1,2]. These ion sources normally operate at lower pressures ($10^{-5}$–$10^{-8}$ Torr) and moderate values of microwave power (~1 kW), primarily because of the reduction in the efficiency of ECR action at higher pressures (>$10^{-4}$ Torr) and stringent cooling requirements. While operation at a lower pressure is favorable for obtaining higher charge states, the beam intensity is considerably limited (approximately electron microamperes) due to a lack of sufficient number density of neutrals. To overcome this problem, several methods such as gas mixing, wall coating, and installation of an electrode into the plasma have been applied [1,2]. However, the maximum increments in beam intensity for a particular charge state are about two to three times larger than what is obtained normally.

To address the above challenges, we report the development of a novel pulsed high-power microwave ion source [3,4], which does not require any resonance effects for its operation. This allows us to raise the neutral density so that higher beam intensities can be obtained. Pulsing the source eliminates any cooling requirements. The achievement of higher-power densities gives rise to nonlinear effects, which lead to higher ionization frequency [4] and enhancement of plasma density and electron temperature [5,6]. The delayed plasma buildup is an important consequence [5,6]. The high electron temperature both within the pulse and its maintenance after the end of the pulse will be favorable for the generation of multicharged ions [4,5]. The possibility of obtaining current densities [4] of ~100 mA/cm² at an elevated pressure (10 mTorr) indicates a high ionization efficiency. Moreover, the improved radial plasma confinement and the narrow cross section of the plasma chamber are favorable for the generation of intense collimated ion beams.

## Ion Source Design and Experimental Setup

The ion source employs high-power (60–100 kW), pulsed (0.05–1.2 μs) micro-waves of 9.45 GHz that are launched into a circular conducting tube, which forms the plasma chamber (for a figure and details on the setup, see Refs. [3] and [6]). Four different chambers have been experimented whose radii $R$ are slightly smaller than $n\lambda/4$ and length = 30 cm. The chamber is surrounded by permanent magnets (Nd–Fe–B, surface magnetic field $\simeq$1 T) in a multi-cusp arrangement with end-plugging. The absence of electromagnetic coils makes the device compact. The resulting field structure with a minimum-$B$ (zero) at the center of the chamber provides plasma confinement and assists in wave propagation through a narrow cross section [5]. The wave launch mode belongs to $k \perp B$, where $k$ is the wave vector and $B$ is the magnetic field. Argon is taken as the test gas. For $n \leq 4$, the constricted chamber cross section enhances the microwave power density (~100 kW/cm²), which is considerably higher than conventional moderate power, larger bore-size ion sources (e.g., for power ~500 W and bore radius ~3 cm, power density ~0.01 kW/cm²).

The current measurements have been made at the center of the chamber where the magnetic field is almost zero. A circular plane Langmuir probe of diameter 4 mm, made of stainless steel of thickness 0.1 mm, is used in the fixed voltage (20 V for electron current and –70 V for ion current) mode to study the temporal profiles of the particle (electron and ion) saturation currents and in a sweep voltage mode (–100 to 100 V) to obtain the probe ($I$–$V$) characteristics. Both sides of the planar probe were used for charge par-ticle collection, and the probe surface was kept parallel to the wave vector. A Boxcar's integrator is used to make time-resolved measurements of the probe characteristics and the optical data. For the optical studies, the light emitted by the pulsed argon plasma is obtained by an optical probe, which consists of an optical fiber and a lens attached at the tip. The probe is inserted through a radial port onto a tiny gap in the multicusp and the plasma chamber. The obtained light is passed onto a monochromator and the signal amplified by a photomultiplier tube. We have studied the wavelength of Ar II (404.3 nm) corresponding to the transition $3s^2 3p^4(^1D)4s \rightarrow 3s^2 3p^4(^1D)4p$, which is the most intense in the visible region of the spectrum of the pulsed plasma.

## Experimental Results

Figure 11.1 shows the variation of the peak electron current density $J_{me}$ with pressure $p$ and the radius of the plasma chamber $R$ as a parameter. $J_{me}$ has a peaked profile over the pressure range. The pressure at which the peak occurs depends upon $R$. With a decrease in $R$, the peak is shifted to

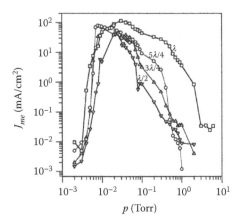

**FIGURE 11.1**

Variation of the peak electron current density $J_{me}$ with pressure $p$ and the chamber radius $R$ as a parameter. (Reprinted with permission from S. Bhattacharjee et al. 2002. Power absorption and intense collimated beam production in the pulsed high power microwave ion source at RIKEN, *Rev. Sci. Instrum.* **73**: 620. Copyright 2002, American Institute of Physics.)

higher-pressure regions and lies in the range of 7–30 mTorr. In general, $J_{me}$ decreases for smaller $R$; however, the profile at $R \simeq \lambda$ is of particular interest because $J_{me}$ has the highest value ($\simeq 100$ mA/cm²), which is an exception to the general trend. Moreover, higher currents are obtained over a wide pressure region (2 mTorr–10 Torr), particularly on the higher-pressure side.

Figure 11.2 shows the variation of the peak optical intensity $I_{op}$ (wavelength 404.3 nm) with pressure $p$ and the chamber radius $R$ as a parameter at full microwave power (100 kW). The profiles of $I_{op}$ also demonstrate the existence of a peak value over the pressure range, with the peaks being shifted to higher pressure regions for tubes of decreasing $R$ (similar to $J_{me}$ in Figure 11.1). However, a remarkable feature (as compared to $J_{me}$ in Figure 11.1) is that $I_{op}$ increases with a decrease in $R$. The profile at $R \simeq \lambda$ deviates from the normal trend and has a high intensity and performance with regard to output light intensity at the higher-pressure side.

The microwave power absorption by the plasma is investigated from observations of the variation of $J_{me}$ and $I_{op}$ with the normalized input power $P_i$. Figure 11.3 shows the dependence of $J_{me}$ on $P_i$ at $p = 8$ mTorr and $R$ as a parameter. The existence of a threshold power, $P_i = 0.25$ for the onset of the current, is noted. The largest chamber ($R \simeq \lambda/4$) shows a decrease in $J_{me}$ for $P_i > 0.65$, although the initial rate of increase, $dJ_{me}/dP_i$, is higher as compared to other waveguides. As $R$ is reduced, $dJ_{me}/dP_i$ decreases; however, beyond $P_i = 0.85$, $dJ_{me}/dP_i$ increases sharply, particularly for $R \simeq \lambda$, indicating good power response and absorption at higher power densities.

The optical intensity has a stronger dependence on the electron temperature $T_e$, which is an indication of microwave heating. Figure 11.4 shows the

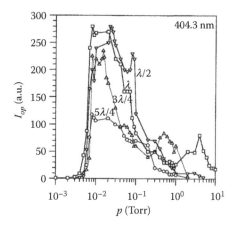

**FIGURE 11.2**

Variation of the peak optical intensity $I_{op}$ (wavelength 404.3 nm) with pressure $p$ and the chamber radius $R$ as a parameter. (Reprinted with permission from S. Bhattacharjee et al. 2002. Power absorption and intense collimated beam production in the pulsed high power microwave ion source at RIKEN, *Rev. Sci. Instrum.* **73**: 620. Copyright 2002, American Institute of Physics.)

variation of $I_{op}$ with $P_i$ for chambers of different $R$, at $p = 8$ mTorr. Waveguides with smaller radii ($R \simeq \lambda/2$ and $3\lambda/4$) show a rapid increase in $I_{op}$ with $P_i$ (especially for $P_i > 0.55$) with a saturation tendency at higher powers. It may be noted that the trend is different from Figure 11.3, where $J_{me}$ was found to decrease for smaller $R$. For $R \simeq 5\lambda/4$, $I_{op}$ shows a saturation tendency above

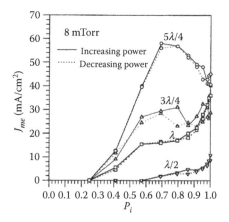

**FIGURE 11.3**

Variation of the peak electron current density $J_{me}$ with normalized input power $P_i$ at a pressure $p = 8$ mTorr. (Reprinted with permission from S. Bhattacharjee et al. 2002. Power absorption and intense collimated beam production in the pulsed high power microwave ion source at RIKEN, *Rev. Sci. Instrum.* **73**: 620. Copyright 2002, American Institute of Physics.)

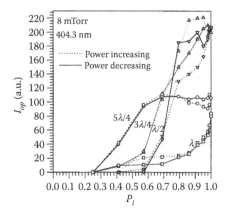

**FIGURE 11.4**
Variation of the peak optical intensity loop (wavelength 404.3 nm) with normalized input power $P_i$ at a pressure $p = 8$ mTorr. (Reprinted with permission from S. Bhattacharjee et al. 2002. Power absorption and intense collimated beam production in the pulsed high power microwave ion source at RIKEN, *Rev. Sci. Instrum.* **73**: 620. Copyright 2002, American Institute of Physics.)

$P_i > .0.7$, which corresponds to the decrease in $J_{me}$, as noted in Figure 11.3. The chamber with $R \simeq \lambda$ shows a constant increase in $I_{op}$, the increase is rather sharp at powers close to the maximum ($P_i = 1.0$) and no saturation is observed. The power absorption cannot be explained by linear theory.

## Discussion

The optical intensity $I_{op}$ depends upon the excitation rate, which is a function of the electron temperature $T_e$ and the electron density $N_{el}$ and can be written as $\Gamma = N_e \langle v\sigma_{ex} \rangle$, where $v$ is the electron velocity and $\sigma_{ex}$ is the excitation cross section. $N_e$ is related to $J_{me}$ by the relation $N_e = J_{me}/e(\kappa/2\pi m)\sqrt{T_e}$, where $e$ and $m$ are the charge and mass of an electron and $\kappa$ is the Boltzmann constant. $J_{me}$ is taken as the electron saturation current density measured by the probe. In a simplified form [6], $\langle v\sigma_{ex}\rangle$ can be expressed by $C(T_e/E_{ex})^2$, where $C$ is a constant, which depends upon the pressure, and $E_{ex}$ is the excitation threshold energy for the particular level. Substituting for $N_e$ and $\langle v\sigma_{ex}\rangle$, the optical excitation rate $\Gamma$ can be written as $\Gamma \propto J_{me}T_e^{3/2}$, which helps us to compare the variation of $T_e$ in the different plasma chambers. Using the result of $T_e$, one can also compare the variation of $N_e$. The results are shown in Figure 11.5 for $p = 25$ mTorr. It is noted that with a decrease in $R$, $T_e$ increases uniformly, being almost double for $R \simeq \lambda/2$, as compared to the largest chamber $R \simeq 5\lambda/4$. On the other hand, $N_e$ has an optimum $R$, at which it is maximum for the present experiment $R \simeq \lambda$, where $N_e \cong 2 N_e$ ($R \simeq 5\lambda/4$).

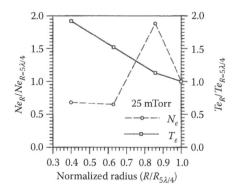

**FIGURE 11.5**

Variation of the ratio of the electron temperature $T_e$ and the electron density $N_e$ with respect to the chamber with $R = 5\lambda/4$, at a pressure $p = 25$ mTorr. (Reprinted with permission from S. Bhattacharjee et al. 2002. Power absorption and intense collimated beam production in the pulsed high power microwave ion source at RIKEN, *Rev. Sci. Instrum.* **73**: 620. Copyright 2002, American Institute of Physics.)

In the pressure dependence study (Figures 11.1 and 11.2), it is noted that $J_{me}$ and $I_{op}$ reach a maximum as the pressure is increased from $10^{-3}$ to 10 Torr. This may be understood qualitatively from the pressure dependence of the energy gain by the electrons from the oscillating electric field. At lower pressures, the energy gain efficiency is poor because electrons undergo many oscillations before undergoing a collision. There is an optimum pressure at which the energy gain is maximum. As the plasma becomes more and more collisional, the energy gain efficiency once again decreases. It is also remarked that $I_{op}$ drops off faster than $J_{me}$ as the pressure is increased.

From our experimental results (Figures 11.3 and 11.4), we notice that the power absorption is largely nonlinear except for the tube with the largest radius ($R = 5\lambda/2$), for which it may be considered to be linear until $P_i \simeq 0.6$. This behavior is considered to be because of the dependence of $J_{me}$ (and $I_{op}$) jointly on $T_e$ and $N_e$. A rapid enhancement of $T_e$ would be favorable for the generation of multicharged states, and $N_e$ for the increase of beam intensity.

## Conclusions

A pulsed high-power microwave ion source shows us the possibility of achieving higher current densities (~100 mA/cm²) at a relatively higher-pressure regime ($10^{-3}$–1 Torr). This is due to the enhancement of the ionization frequency, electron temperature, and the plasma density by the high-power wave absorption in the plasma, which shows a nonlinear behavior. The increase in $T_e$ brought about by narrowing the size of the plasma chamber will be useful for the production of multicharged states. Optimum operation may be achieved by selecting the chamber radius at $R$.

## Summary in Pulsed Microwaves

### Experimental Results

In the previous chapter, it was shown that a plasma can be produced in a waveguide with a dimension below the cutoff value also by using pulsed microwaves. The plasma production at a higher-pressure regime was studied by the use of high-power, short-pulse microwaves. Two sets of experiments were carried out: first, the general properties of the pulsed plasma were studied in a larger waveguide, and second, the plasma production in a waveguide with a dimension below cutoff was investigated. In the following, we summarize the results in two parts: (a) those belonging to the general plasma properties of high-power, short-pulse microwaves, and (b) those belonging to the properties of the plasma in the narrow waveguide.

### General Properties of the Plasma

The plasma created by high-power (60–100 kW), short-pulse (~1 µs) microwaves have demonstrated properties quite uncommon to conventional pulsed discharges of moderate power (~1 kW) and longer pulse duration (~10–100 µs). Duty cycle (defined as the pulse on time divided by the sum of pulse on and pulse off times) in the present experiment is very small (~$10^{-4}$) as compared to conventional pulsed DC or high-frequency discharges where it is usually of the order of 1. The electron temperature $T_e$ showed a value of about 6–10 eV in the interpulse regime, which continued for a timescale much larger than particle decay times. The plasma buildup is extended and reaches the peak density at a time 2–20 µs after the end of the pulse, depending upon the pressure. The pulse repetition frequency was found to have an almost negligible effect on the peak current $I_m$ and the decay time constant $\tau_d$ of the decay profile. On the other hand, $I_m$ and $\tau_d$ were seen to increase on increasing the pulse width. The peak plasma density was found to be over $10^{10}$ cm$^{-3}$.

The results of the study of the temporally decaying ion current profile in a rare gas (Ar), a molecular gas ($N_2$), and a molecular gaseous mixture ($O_2 + N_2$) showed different features for the three kinds of gases. In Ar, the ion current decay was found to decrease monotonically with time. This could be explained by the combined effects of volume recombination and diffusion. In $N_2$, the current decay characteristics had an additional bump after about 6 µs. This result could be explained by including additional effects of metastable ionization. It was found that ionization by metastable–metastable collisions is more important than electron–metastable collisions. In standard air ($O_2 + N_2$, volume mixing 1:4), the decay profile could be divided into three distinct regimes, a faster decay (0–1 µs), a slower decay (1–6 µs), and a bump similar to that observed in $N_2$ (>6 µs). Besides the effect of metastables, of which those of $N_2$ were considered to be important in the discharge, the effects of negative ions were also found in the decay characteristics. Thus, depending upon the gas, the interpulse plasma can be considered useful for the production of metastables, negative ions, radicals, multicharged ions, or for reactions with the gas phase.

## Plasma Characteristics in the Circular Waveguide

The plasma could be produced along the entire length (37 cm) of the waveguide using pulsed microwaves. The decay time constant of the ions were about 20–30 µs, and those of the electrons about four to five times longer. The axial variation of the plasma density indicated a peak near the entrance $z = 0$ cm and a second peak at $z_m = 20$–30 cm. The plasma density was above $10^{10}$ cm$^{-3}$ in the region of the peaks and over $10^9$ cm$^{-3}$ in the region between the peaks. At lower pressures, the plasma became more uniform along the waveguide.

The buildup of the plasma density along the tube with pressure as a parameter indicated that near the entrance and at $z_m$ the buildup was faster at a higher $p$. At the intermediate region, for example, $z = 10$–15 cm, the buildup was delayed at a higher $p$.

The electric field $E$ measurements along the waveguide showed a good correlation with the axial variation of the plasma density. $E$ was found to be higher at the entrance of the tube and at $z_m$. In between the two maximas, a few peaks resembling a standing wave pattern have been observed. The peaks were spaced at an interval of ~5 cm, which is roughly equal to half the free space wavelength ($\lambda_0/2$) of waves of 3 GHz. In vacuum, the electric field was exponentially damped inside the waveguide.

By assuming a diffusion model at a higher $p$, the spatiotemporal electron and ion current profiles could be modeled, and they represented the experimental results rather well. From the modeling, the difference in the decay constants of the electron and ion currents could be explained.

## Summary of the Plasma Production Mechanisms

### *Continuation of the High Electron Temperature*

The continuation of the high electron temperature (6–10 eV) after the end of the pulse is helpful in ionization and plasma production in the temporally longer interpulse regime.

### Penetration of the Electric Field

The presence of a dense plasma at the tube entrance refracts the waves, which are then reflected by the tube walls, thereby helping the field to be focused at a second region near the exit of the tube ($z_m$). The penetration of the electric field is important for ionization and plasma production.

### Particle Diffusion into the Waveguide

At a higher pressure, the plasma is collisional and the presence of density gradients is helpful for particle diffusion. The axial variation of the plasma density along the tube shows a higher-density plasma at the entrance, from where the plasma can diffuse into the tube. Moreover, the multicusp magnetic field suppresses the loss by radial diffusion and helps in axial plasma diffusion and confinement.

### Effect of Finite Length of the Waveguide

Under vacuum conditions, measurement of the electric field along the tube shows that the field actually penetrates into the waveguide even though its dimension is below cutoff. This is considered to be due to the effect of the finite length of the waveguide, under the conditions of which the wave is exponentially damped along the tube as $e^{-\gamma z}$, rather than being cut off at the entrance, which is true for an infinitely long waveguide ($z \rightarrow \infty$), where $\gamma$ is the propagation constant in the waveguide.

---

## References

1. R. Geller. 1996. *Electron Cyclotron Resonance Ion Sources and ECR Plasma* (IOP, Bristol), and references therein.
2. D. Hitz, G. Melin, and A. Girard. 2000. Fundamental aspects of electron cyclotron resonance ion sources: From classical to large superconducting devices (invited), *Rev. Sci. Instrum.* **71**: 839–845.
3. S. Bhattacharjee et al. 1999. Development of compact high current multicharged ion source for RIKEN RI Beam Factory, *RIKEN Accel. Prog. Rep.* **33**: 200.
4. S. Bhattacharjee et al. 2000. Plasma measurements on the pulsed high power microwave ion source at RIKEN, *RIKEN Accel. Prog. Rep.* **34**: 309.
5. S. Bhattacharjee and H. Amemiya. 2000. Production of pulsed microwave plasma in a tube with a radius below the cutoff value, *J. Phys. D* **33**: 1104–1116.
6. S. Bhattacharjee, H. Amemiya, and Y. Yano. 2001. Plasma build-up by short-pulse high-power microwaves, *J. Appl. Phys.* **89**: 3573–3579.
7. S. Bhattacharjee et al. 2002. Power absorption and intense collimated beam production in the pulsed high power microwave ion source at RIKEN, *Rev. Sci. Instrum.* **73**: 620.

# 12

## New Experiments in Continuous-Mode Microwaves: Formation of Standing Waves and Electron Trapping

### Overview of the New Experimental System

The experimental setup consists of a stainless-steel (grade: SS-304) vacuum chamber (VC) of length 50 cm and diameter 20 cm, with four cylindrical arms having numerous ports for pumping, gas inlet, vacuum gauges, and plasma diagnostics. The chamber can be evacuated to a base pressure of ~$5 \times 10^{-7}$ Torr using a Varian 301 Navigator turbomolecular pump (TMP) of pumping speed ~250 L/s, backed by a rotary pump with a speed of ~2 L/s, with respect to nitrogen. Argon (zero grade) is used as the experimental gas and is introduced into the chamber through the gas inlet (GI) with a stable flow maintained by an MKS-1179A (range: 50 sccm) mass flow controller. The neutral pressure inside the chamber is maintained by dynamic equilibrium between the inflow and outflow rates, and is measured by a Varian SenTorr system with a cold cathode ionization gauge (IG) for measurement from $10^{-3}$ to $10^{-7}$ Torr, and a ConvecTorr gauge (CG) for measurement from $10^{-3}$ to $10^{3}$ Torr. A controllable leak valve (LV) is used for venting the chamber. Wilson seals (WS) are used for introducing the plasma diagnostic probes from the radial (RP) and axial (AP) ports, as shown in Figure 12.1.

The microwave system consists of a Richardson Electronics microwave generator (MWG; model: Alter–TMA20), which produces microwaves of 2.45 GHz and is capable of supplying 180–1800 W in continuous mode and a peak power of 7.6 kW in the pulsed mode, which is controlled from the power supply and control module (PSC; model: Alter–PM740T). The pulsed microwaves are utilized during the experiments where pulsed waves are required. The microwaves are guided into the experimental chamber via standard WR340 waveguides manufactured by Gerling Applied Engineering, USA. The water-cooled isolator (ISO) protects the magnetron from any reflected power, a dual directional coupler (DC) is used for measuring the forward and reflected powers, and a triple stub tuner (TST) for impedance matching

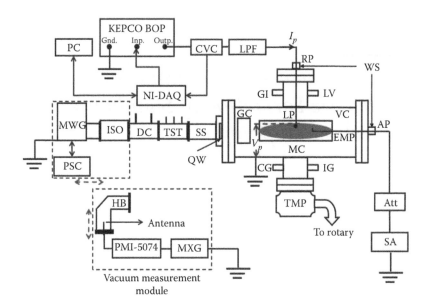

**FIGURE 12.1**
Schematic of the experimental setup: VC: vacuum chamber, GI: gas inlet, LV: leak valve, CG: ConvecTorr gauge, IG: ionization gauge, RP: radial port, AP: axial port, MC: multicusp, GC: guiding cylinder, TMP: turbomolecular pump, QW: quartz window, SS: straight section, TST: triple stub tuner, DC: directional coupler, ISO: isolator, MWG: microwave generator, PSC: power supply and control for the MWG, WS: Wilson seal, LP: Langmuir's probe, $V_p$: probe voltage, $I_p$: probe current, LPF: low-pass filter, CVC: current-to-voltage converter, KEPCO BOP: bipolar power supply, Gnd.: common ground, Inp.: input, Outp.: output, NI-DAQ: National Instruments Data Acquisition Device, PC: personal computer, EMP: electromagnetic probe, Att.: 30 dB attenuator, SA: Agilent N1996A spectrum analyzer, HB: H-Bend, PMI-5074: low-power isolator, MXG: Agilent N5183A microwave signal generator. (Reprinted with permission from I. Dey and S. Bhattacharjee. 2011. Penetration and screening of perpendicularly launched electromagnetic waves through bounded supercritical plasma confined in multicusp magnetic field, *Phys. Plasmas* **18**: 22101. Copyright 2011, American Institute of Physics.)

between the source and the load; the TST is tuned to keep the reflected power below 5% of the incident power in all experiments. The waves are allowed to pass through a straight waveguide section (SS) and then through a high-grade quartz window (QW) into the VC; QW maintains the vacuum integrity. Since the WR340 waveguides are rectangular, the microwave propagation mode in them is predominantly in the $TE_{01}$ mode. However, on entering the VC, there is a transition from a rectangular to a circular geometry ($TE_{11}$ mode) near the QW, which may lead to large-reflected power. Hence, a guiding cylinder (GC) of 104 mm inner diameter and 50 mm length is placed coaxially at the entrance, 10 mm away from the QW, to guide the microwaves into the multicusp (MC) with minimal reflection [1].

For measurement of the wave field intensity profiles in the vacuum for comparison with the plasma case, a separate vacuum measurement module having low power output (~100 mW) to prevent damage to the probes is employed. An Agilent N5183A microwave signal source (MXG) is used to launch microwaves at 2.45 GHz into an H-bend (HB) by a 3-cm-long quarter-wave antenna (Figure 12.1). The HB is connected to the DC for coupling the signal to the experimental chamber. A solid-state coaxial ISO PMI-5074 protects the MXG from any reflected signal.

Cylindrical, compact MC confinement geometries are employed for the experiments [1–5]. The NdFeB permanent magnets used are of grade N42H, having dimensions $20 \times 20 \times 50$ mm$^3$, with the magnetization direction perpendicular to the broader ($20 \times 50$ mm$^2$) side. The remanence and coercivity values are $B_r = 1.30$ T, $H_{cB} = 1040$ kA/m, and $H_{cJ} = 1430$ kA/m. The magnets are nickel-coated and have a surface field of ~0.5 T, with a curie temperature of ~200°C. A typical Poisson–Superfish calculation [6] for an 8-pole (octupole) is shown in Figure 12.2a, where the reference polar angle scheme is shown, as viewed from AP (Figure 12.1). The cusp field lines toward the boundary and the minimum-$B$ magnetostatic well at the center are clearly visible. Aluminum pipes with rectangular cross sections are used for holding the magnets, which are then fitted into slotted circular holding plates. Cylindrical shells of very small thickness (~1–2 mm) may be inserted into the structure for adjusting the diameter and also for providing continuous conducting boundary condition. MCs with six and eight poles have been fabricated with a bounding radius ($a$) in the range of 3.5–4.2 cm, around the cutoff radius ($r_c$ ~ 3.6 cm). A typical-fabricated structure is shown in Figure 12.2b.

Figure 12.3 shows the experimental plot of the total magnetostatic field ($|B_o(r, \theta)|$) as a function of the radius for the different fabricated MCs (a hexapole and an octupole of 3.5 cm radius and an octupole of 4.1 cm radius). The magnetic field data are measured between diametrically opposite poles (p–p) and the magnet gaps (g–g). The 4.1-cm octupole plots are fitted with the theoretical expression $B_o = B_p(r/a)^{2n-1}$ [7], where $B_p$ (~0.345 T for p–p) is the magnetic field at the pole at $r = a$ (~4.1 cm), and $n$ is the number of poles ($n = 8$). It can be seen that the octupole configuration provides a doubly broad well structure at the center, with ~5% steeper walls near the boundary (~3.0 cm)

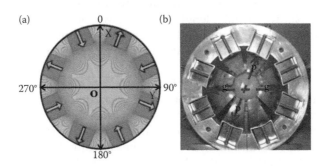

**FIGURE 12.2**
(a) Poisson's simulation of an octupole. (b) Picture of a 3.5-cm-radius octupole.

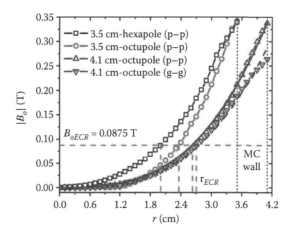

**FIGURE 12.3**
Experimental variation of the total magnetostatic field ($|B_o(r, \theta)|$) with the radius for three different multicusps (a hexapole and an octupole of 3.5 cm radius and an octupole of 4.1 cm radius).

compared to the hexapole configuration of the same radius (Figure 12.3). In the absence of an inner cylinder, the effective radius ($a_e$) of the MC is given by $a_e = (a/3)\left[1 + \sqrt{\{1 + (3n\tan(\pi/n))/\pi\}}\right]$ [1]. The radial points where the ECR condition (875.5 gauss for 2.45 GHz) is satisfied for the three MCs are also indicated (Figure 12.3).

## Introduction

The axial profiles of the plasma parameters ($N_i$ and $T_e$) discussed in Chapter 5 indicate that the plasma is uniform and overdense on average, along the minimum-$B$ axis over a major length (~80%) of MC. Such an observation provides the motivation for studying the axial profile of the wave field in order to understand the role of the input microwave power in maintaining the plasma uniformity.

In this chapter, the measurements of axial wave electric and magnetic field intensity profiles under various plasma conditions are reported. It is observed that electromagnetic standing waves (SWs) are sustained along the central overdense, minimally magnetized plasma column. The origin of SW is discussed here and analyzed in further detail in Chapters 13 and 15. The properties of the SWs in the plasma are investigated and compared with the vacuum (no plasma) case. Unlike damping of high-amplitude traveling waves in plasmas where the Landau damping [8] is the principal mechanism, it is found that in the weakly collisional regime of the present experiments, the dispersive nature of the plasma modifies the SWs, leading to effects such as elongation of the wavelength and decrease in wave amplitude. The effect of the plasma on the vacuum waveguide mode excited in the chamber indicates that the mode undergoes downshift in its wave number with amplitude damping. Plasma electron dynamics due to the SW pattern is discussed [9].

## Spatial Plasma Characteristics

The spatial (axial and radial) variation of the plasma parameters, as described in Chapter 5, at various combinations of input microwave power ($P_{in}$) and neutral pressure ($p$) [9] is reproduced in Figure 12.4a and b. Figure 12.4a shows the axial variation of $T_e$ and $N_i$ at certain combinations of power and pressure. Along the z direction, both $T_e$ and $N_i$ are uniform in the middle of the MC and fall off at entrance and exit (30 cm). The decrease is about 6% for $T_e$ over a length of 22.5 cm from the entrance and about 17% for $N_i$ over a length of 20.0 cm, extending from 2.5 to 22.5 cm.

Figure 12.4b shows the radial variation of $T_e$ and $N_i$ at $P_{in} = 180$ W and $p = 0.15, 0.35,$ and $0.45$ mTorr. The data are taken at the center (15 cm from the entrance). $T_e$ shows a peak around 2.25 cm from the center, which corresponds to the resonance zone, signifying that the electrons gain a large amount of energy in this region due to the resonance mechanisms (ECR and UHR) as indicated by the high values of $T_e$ in this region. With

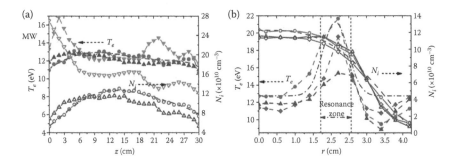

**FIGURE 12.4**

(a) Axial and (b) radial variation of electron temperature ($T_e$; dashed lines with solid symbols) and ion density ($N_i$) (solid lines with hollow symbols) at 0.15 mTorr (circle), 0.35 mTorr (diamond), and 0.45 mTorr (up triangle) with 180 W input power and at 450 W, 0.15 mTorr (inverted triangle). Axial and radial fitting on the 180 W, 0.15 mTorr $N_i$, and $T_e$ data are shown by a dashed and dashed-dot line, respectively. (Reprinted with permission from I. Dey and S. Bhattacharjee. 2008. Experimental investigation of standing wave interaction with a magnetized plasma in a minimum B field, *Phys. Plasmas* **15**: 123502. Copyright 2008, American Institute of Physics.)

increase in pressure, the height of the peak reduces and the FWHM broadens due to increase in collisions. In the central region ($r = 0$–2.5 cm), the wave propagation is hindered since the plasma is overdense (i.e., $>7.4 \times 10^{10}$ cm$^{-3}$ for $\omega_p/2\pi = \omega/2\pi = 2.45$ GHz), as shown by the $N_i$ data, and results in a lower $T_e$. Near the periphery (3.2–4.2 cm), $T_e$ shows an increase since the plasma density is much below cutoff and wave propagation can take place through this region [2]. The axial and radial variations of $T_e$ and $N_i$ show similar trends at other powers and pressures.

## Axial Wave Field Intensity Measurements

Axial wave field intensity measurements are carried out with the pickup antenna probes described in Chapter 3. The intensity peak corresponding to the input frequency of 2.45 GHz is recorded from the spectrum analyzer as a function of axial distance inside MC. Figure 12.5a shows the axial variation ($z$) of the wave electric field intensity $|E_{tot}|^2$ in the MC detected by the antenna probe, in the vacuum, and at three different combinations of power and pressure (180 W, 0.15 mTorr; 180 W, 0.45 mTorr; and 450 W, 0.15 mTorr) in the presence of plasma [9]. Figure 12.5b shows the variation of the axial component of the wave magnetic field $|B_z|^2$ under similar conditions as Figure 12.5a, measured by the B-dot probe with the plane of the loop perpendicular to the MC axis. The wave field intensity is reported in dBm scale, and has been upscaled by +30 dB to account for the 30 dB attenuator (Att) used during measurements (cf. Figure 12.1). Under vacuum conditions (~5 × 10$^{-7}$ Torr), prominent SWs are observed in the MC, with both the E-field and the B-dot probes. The detection of the SW pattern with the B-dot probe proves its electromagnetic character.

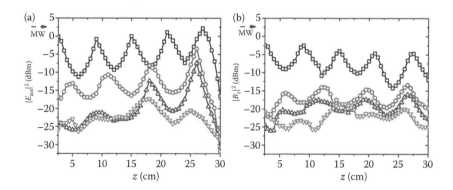

**FIGURE 12.5**

Axial variation of (a) wave electric field intensity ($|E_{tot}|^2$) and (b) $z$ component of the wave magnetic field intensity ($|B_z|^2$) in vacuum (squares) and in plasma at 0.15 mTorr (circles), 0.45 mTorr (triangles) with 180 W, and at 0.15 mTorr (inverted triangle) with 450 W.

In the presence of plasma (Figure 12.5a), the SW is observed to be damped along the plasma, the damping being nearer to the wave launch region (~3–15 cm). The typical e-folding damping lengths are of the order of a few centimeters, namely, for 180 W, 0.15 mTorr, it is 10.7 cm. It decreases with increases in power and pressure (8.0 cm for 180 W, 0.45 mTorr and 7.9 cm for 450 W, 0.15 mTorr), indicating more damping of the wave. There is also a subsequent increase of up to 15 cm in the wavelength compared to the vacuum wavelength of 12.24 cm. Figure 12.5b shows that in the presence of plasma, there is an overall reduction of $|B_z|^2$, since the electrons interact primarily with the wave electric field and are not much affected by the wave magnetic field, which is ~$10^{-8}$ times smaller in amplitude than the electric field.

In order to study the effect of the plasma on the SW characteristics, a Fourier transformation of the $|E_{tot}|^2$ versus $z$ data is performed, to obtain the corresponding variation of the normalized intensity amplitude ($|E_n|$) with the real part of the wave number ($\gamma$). The value of $\gamma$ will depend on the effective radius of the plasma chamber, and also on the medium inside it (vacuum or plasma). The detailed relations between them are provided in the section "Electron Trapping in SW Minima."

The resulting Fourier spectrum for $P_{in} = 180$ W, at two different pressures, and in the vacuum case is shown in Figure 12.6a, where the power has been normalized with the vacuum power for facilitating the comparison. In the vacuum case, we observe a primary peak at $\gamma_1 = 0.15242$ cm$^{-1}$, corresponding to a radius of 4.0 cm (radius to center of magnet face) of MC, and a much smaller secondary peak at $\gamma_2 = 0.29802$ cm$^{-1}$ corresponding to a radius of 4.4 cm (possibly radius to edge of magnet face; see the section

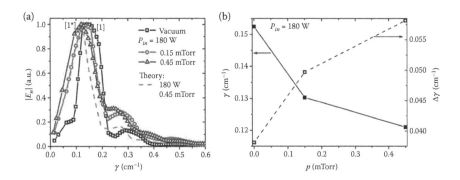

**FIGURE 12.6**

(a) The Fourier transform of the standing wave pattern showing variation of amplitude ($|E_n|$) normalized with respect to the maximum peak, with the real part of the wave number ($\gamma$) in vacuum (squares) and in plasma at 0.15 mTorr (circles), 0.45 mTorr (triangles) with 180 W. A theoretical variation is shown by dashed lines from the model calculations as 180 W, 0.45 mTorr. (b) Variation of $\gamma$ and full width at half maxima of the wave number ($\Delta\gamma$) with pressure ($p$). (Reprinted with permission from I. Dey and S. Bhattacharjee. 2008. Experimental investigation of standing wave interaction with a magnetized plasma in a minimum B field, *Phys. Plasmas* **15**: 123502. Copyright 2008, American Institute of Physics.)

"Summary" for details). These arise due to a near-circular boundary of the plasma chamber (cf. Figure 12.2b). The effective radius of such a near-circular geometry is calculated to be 4.2 cm [1]. In the presence of plasma, the peaks show a noticeable shift from their vacuum value due to dispersion. Figure 12.6b shows how $\gamma$ and its FWHM $\Delta\gamma$ change as we go from vacuum to plasma condition for the primary peak in Figure 12.6a. The wave number is downshifted by about 23%, as indicated in Figure 12.6a by [1] and [1*] for the 180 W, 0.45 mTorr case. The increase in $\Delta\gamma$ is attributed to the increase in collisionality with the increase in neutral pressure.

## Electron Trapping in SW Minima

Possible particle trapping in the potential troughs of the large-amplitude SW is investigated by performing an axial scan of the floating potential ($V_f$) using the Langmuir probe. The $V_f$ measurements are useful because they disturb the plasma minimally, since no net current is drawn. One typical result is shown in Figure 12.7, where the axial variation of the floating potential ($V_f$), the plasma potential ($V_s$), $T_e$, and $|E_{tot}|^2$ are compared at 180 W and 0.15 mTorr. It may be noted that $V_f$ is actually negative by convention, but is measured positive here. A high degree of correlation is noticed between the crests and troughs of $|E_{tot}|^2$ and $V_f$, especially near the exit of the MC, where the SW pattern is well pronounced. A lower value of $V_f$ in the troughs indicates some degree of localization (trapping) of electrons in those regions.

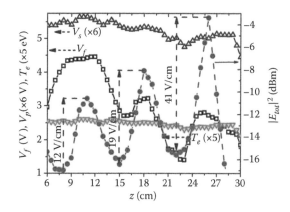

**FIGURE 12.7**

Axial variation of floating potential ($V_f$; squares), space potential ($V_s$; up triangle), average electron temperature ($T_e$; down triangle), and wave electric field intensity ($|E_{tot}|^2$; circles) at 180 W and 0.15 mTorr. (Reprinted with permission from I. Dey and S. Bhattacharjee. 2008. Experimental investigation of standing wave interaction with a magnetized plasma in a minimum B field, *Phys. Plasmas* **15**: 123502. Copyright 2008, American Institute of Physics.)

The electric field corresponding to the intensity difference (obtained by calibrating the E-field probe in vacuum) existing between the maxima and the minima of the SW is indicated in the plot. From the observations, it can be seen that the plasma electrons near the SW maximum would be influenced more, and would oscillate with larger amplitude compared to the electrons in the intensity minimum. The phenomenon is of interest and has been further investigated, where the effect of SW on the "hot" component of the electron temperature and the electron energy distribution is investigated in detail, which possibly leads to the undulations in the measurement of the space and floating potentials (Figure 12.7) since $V_s = V_f + K_0 T_e (K_0 \sim 4.7$ for argon) [10].

## Model Calculation

In wave–plasma interaction studies, there appear to be two approaches for analyzing wave propagation in plasmas [11]. In one, Maxwell's equations are solved simultaneously with the fluid equations describing the particle motions. In the second, Maxwell's equations are solved with the dielectric permittivity of free space ($\varepsilon_0$) replaced by $\varepsilon$ of the plasma. In the presence of an external magnetic field, $\varepsilon$ becomes a tensor. In this chapter, a preliminary effort has been made to explain the interaction of the SWs using the second approach.

Taking the microwave electric field to be of the form

$$\vec{E}(\vec{r},t) = E_0 \exp\left(j\omega t - \vec{k} \cdot \vec{r}\right)\hat{r}, \tag{12.1}$$

where $\vec{k}$ is the wave vector, $\omega$ is the wave frequency, and $j = \sqrt{(-1)}$.

Maxwell's equations for such a medium are

$$\vec{\nabla} \cdot \vec{D} = 0, \tag{12.2a}$$

$$\vec{\nabla} \cdot \vec{B} = 0, \tag{12.2b}$$

$$\vec{\nabla} \times \vec{E} = -\partial \vec{B}/\partial t = -j\omega\mu_0 \vec{H}, \tag{12.2c}$$

and

$$\vec{\nabla} \times \vec{H} = \partial \vec{D}/\partial t + \vec{J}_f = j\omega\left(\tilde{\varepsilon} - j\tilde{\sigma}/\omega\right)\vec{E}, \tag{12.2d}$$

where $\vec{D} = \tilde{\varepsilon}\vec{E}$ is the electric displacement; $\vec{B} = \mu\vec{H} \cong \mu_0\vec{H}$ is the magnetic field induction; $\vec{E}$ is the electric field; $\vec{H}$ is the magnetic field; $\vec{J}_f = \tilde{\sigma}\vec{E}$ is the free-current density; $\tilde{\varepsilon}$ and $\tilde{\sigma}$ are the complex dielectric constant and conductivity, respectively; and $\mu \approx \mu_0$ is the permeability of the medium. The inclusion of the free-current density takes into account the losses in the plasma medium due to flow of charges. We can write the effective dielectric permittivity of the plasma medium as $\tilde{\varepsilon}_{eff} = (\tilde{\varepsilon} - j\tilde{\sigma}/\omega)$.

In Equations 12.2a through 12.2d, since $\tilde{\varepsilon}$ and $\tilde{\varepsilon}_{eff}$ are functions of spatial coordinates $(\rho, z)$, taking curl on both sides of Equations 12.2c and 12.2d, we obtain

$$\nabla^2 \vec{E} - \vec{\nabla}(\vec{\nabla} \cdot \vec{E}) + \omega^2 \mu_0 \tilde{\varepsilon}_{eff} \vec{E} = 0, \tag{12.3a}$$

$$\vec{\nabla} \cdot (\tilde{\varepsilon}\vec{E}) = 0, \tag{12.3b}$$

and

$$\nabla^2 \vec{H} - \vec{\nabla}(\vec{\nabla} \cdot \vec{H}) + j\omega\vec{\nabla} \times (\tilde{\varepsilon}_{eff}\vec{E}) = \nabla^2 \vec{H} + j\omega\vec{\nabla} \times (\tilde{\varepsilon}_{eff}\vec{E}) = 0, \tag{12.3c}$$

since $\mu \cong \mu_0$ which is space independent.

The set of coupled Equations 12.3a through 12.3c has to be solved incorporating the spatial variation of $\tilde{\varepsilon}$ and $\tilde{\varepsilon}_{eff}$ subject to the boundary condition that the polar component of the electric field $(E_\theta)$ vanishes at the boundary of a grounded conducting cylindrical waveguide of radius $a$, and assuming that the dominant mode of propagation is the $TE_{11}$ mode $(E_z = 0)$.

The basic difficulty with the solution of Equations 12.3a through 12.3c arises due to anisotropy and inhomogeneity of the medium, which couple all the components of the electric and magnetic field vectors [12]. In such a situation, a numerical solution would be more appropriate, which will be expanded upon in detail in the next chapter. A qualitative understanding of the formation of the SWs may be obtained by analyzing the dielectric tensor.

For waves launched perpendicularly to the magnetic field $(\vec{k} \perp \vec{B})$, the three components of the tensor can be written as [11,13] $\tilde{\varepsilon}_{eff\perp}$, $\tilde{\varepsilon}_{eff\parallel}$ and $\tilde{\varepsilon}_{eff\times}$ which are the perpendicular, parallel, and cross (Hall) components of $\tilde{\varepsilon}_{eff}$, respectively. The boundary conditions of the plasma parameters and the magnetic field are taken into account to realize the bounded model of the dielectric tensor. Of these, $\tilde{\varepsilon}_{eff\times}$ is purely imaginary and hence causes attenuation, $\tilde{\varepsilon}_{eff\parallel}$ corresponds to the O-mode-like propagation, and $\tilde{\varepsilon}_{eff\perp}$ is the X-mode-like propagation in the bounded plasma. Since the plasma is overdense in a major part of the central region, the O-mode-like propagation is not of much interest, hence $\tilde{\varepsilon}_{eff\perp}$ is the main term to be considered, which interacts with the plasma.

Therefore, considering $\tilde{\varepsilon}_{eff\perp} = (\tilde{\varepsilon}_\perp - j\tilde{\sigma}_\perp/\omega)$, the variation of $\tilde{\varepsilon}_\perp$ and $\tilde{\sigma}_\perp$ with the plasma parameters is given by [11–13]

$$\tilde{\varepsilon}_{\perp}(r,z) = \tilde{\varepsilon}(r,z) = \varepsilon_0 \left[ 1 - \frac{\alpha_p^2 \left\{ (1 - j\delta) - \alpha_p^2 \right\}}{(1 - j\delta)^2 - \beta_c^2 - \alpha_p^2(1 - j\delta)} \right] \tag{12.4}$$

and

$$\tilde{\sigma}_{\perp}(r,z) = \tilde{\sigma}(r,z) = \varepsilon_0 \omega \left[ \frac{j\alpha_p^2 (1 - j\delta)^2}{\left\{ \beta_c^2 - (1 - j\delta)^2 \right\}} \right] \tag{12.5}$$

where $\alpha_p = \omega_p/\omega$, $\beta_c = \omega_{ce}/\omega$, $\delta = v_{coll}/\omega$ are functions of $r$ and $z$.

Here, $\omega_p(r,z) = \sqrt{n_e(r,z)e^2/m_e\varepsilon_0}\,(\text{rad/s})$ is the plasma frequency, $\omega_{ce}(r) = eB(r)/m_e(\text{rad/s})$ is the electron–cyclotron frequency, at any point in space in the MC at some power and pressure when plasma is present. $B_0$ is the static magnetic field and $v_{coll}(r,z) = 4\pi \times 10^{28}\,p\sqrt{T_e(r,z)}\sigma_{coll}$ is the electron–neutral collision cross section in m$^2$ obtained from experimental data available in the literature [14]. The plasma has been assumed to be azimuthally symmetric.

The nonuniformities in $N_i$, $T_e$, and $B_0$ are taken into account by applying appropriate fitting curves to the experimentally obtained data shown in Figures 12.2 and 12.3. The $B_0$ versus $r$ variation can be fitted with the equation $B_0(r) = C_0 r^{(b/2)-1}$, where $b$ is the number of poles ($b = 8$ and $C_0 = 0.048$ G/mm$^3$). The axial variation of $N_i$ and $T_e$ is fitted with fifth-order polynomials as shown in Figure 12.4a. The radial variation of $N_i$ is fitted with the equation of the form $N_i(r) = N_0 J_0(Cr^q)$ [3], where $N_0$ is the density at $r = 0$, $J_0(X)$ is the zeroth-order Bessel function, $C$ is a constant depending upon geometry and ionization rates, and $q$ is the exponent obtained on solution of the diffusion equation [3]. The radial variation of $T_e$ can be fitted approximately with the equation of the type $T_e(r) = T_0 + D_0 r^2 \exp\{-W_0(r - r_0)^2\}$, where $T_0$, $D_0$, $W_0$, and $r_0$ are fitting parameters. Using the above fitting equations and the dependence of the permittivity on them, the value of $\tilde{\varepsilon}_{eff} = \tilde{\varepsilon}_{eff\perp} = (\tilde{\varepsilon}_{\perp} - j\tilde{\sigma}_{\perp}\omega)$ can be extrapolated for any $r$ and $z$ in the MC at a given power and pressure.

The variation of the amplitude of the wave along the axial direction is proportional to $\exp(-j\tilde{\beta}z)$ in the cylindrical coordinate system $(r, \theta, z)$, which is evident from the experimental data (Figure 12.5a and b). Here, $\tilde{\beta} = \sqrt{(\tilde{k}^2 - k_c^2)}$ is the complex propagation constant within the waveguide, where $\tilde{k} = (\tilde{\varepsilon}_{eff}/\varepsilon_0)k_0$ [ $\tilde{\varepsilon}_{eff} = \tilde{\varepsilon}_{eff\perp} = (\tilde{\varepsilon}_{\perp} - j\tilde{\sigma}_{\perp}/\omega)$, from above] and $k_0$ is the free-space wave number. The cutoff wave number $k_c$ corresponds to the boundary condition $J_1'(k_c a) = 0$ ($E_\theta = 0$), where $k_c a = 1.841$ is the first root of $J_1'(k_c r)$ for TE$_{11}$ mode [15]. Here, $J_1(k_c r)$ is the Bessel function of the first kind with number of circumferential ($\theta$) variation $m = 1$, and $J_1'(k_c r)$ is its first derivative. Some typical values of $\tilde{\beta} = \gamma - j\alpha$ are given in Table 12.1 for three radial positions

**TABLE 12.1**

Typical Values of $\tilde{\beta} = \gamma - j\alpha$ at Three Radial Locations for $z = 15$ cm, at $p = 0.15$ mTorr

| $\tilde{\beta}$ | $P_{in}$(W) | Center ($r = 0.0$) | ECR ($r = 2.75$ cm) | Edge ($r = 4.0$ cm) |
|---|---|---|---|---|
| $\gamma$ (cm$^{-1}$) | 180 | 0.15240 | 0.00829 | 0.12439 |
| | 450 | 0.15238 | 0.00944 | 0.09828 |
| $\alpha$ (np $\cdot$ cm$^{-1}$) | 180 | $1.00 \times 10^{-10}$ | 1.43772 | $1.675 \times 10^{-5}$ |
| | 450 | $1.01 \times 10^{-10}$ | 1.74086 | $3.477 \times 10^{-5}$ |

at the center ($z = 15$ cm) of the MC, which correspond to different conditions of $N_i$, $T_e$, and $B$. It can be seen that the attenuation number ($\alpha$) is maximum and the wave number ($\gamma$) is minimum in the resonance region (ECR), indicating resonant power absorption. It may also be noted that the X-mode-like wave can propagate through the central plasma, which is overdense for an O-mode-like wave. The wave can also propagate freely along the peripheral plasma.

The formation of the SW pattern can be explained as follows. The forward wave propagates from the entrance toward the exit of the MC. At the exit, due to sudden change of the cross section, a fraction of the wave is reflected back inside the MC depending upon the reflection coefficient due to such a transition, thereby superposing with the incident wave and forming an SW. A part of the reflected wave inside the MC will again be reflected at the entrance side, and hence there will be multiple reflections between the two ends of the MC.

A comparison between the SW profile for the case where the MC exit is blanked off with a metal plate to provide a shorted boundary condition, and the normal condition where the exit (end) is kept open in which the MC normally operates is made. On an average, the SW amplitude in the blanked case is higher than the open case, indicating better reflection. Taking the blanked-off condition as a reference, the reflection coefficient is estimated to be $\bar{R} = 0.87 \exp(-j2.05)$.

Figure 12.8 shows the contour plot of the effective refractive index, $n = \mathrm{Re}\left(\sqrt{\varepsilon_{eff}/\varepsilon_0}\right)$, with (a) $\alpha_p^2$ and $\beta_c^2$ and (b) $r$ and $z$ in the MC. Figure 12.8a shows the conditions in which the X-mode-like wave used in the calculations will either propagate ($n \rightarrow 1$), be cutoff ($n \rightarrow 0$) or undergo resonant absorption ($n \rightarrow \infty$) as functions of the plasma density and magnetic field. An average collisionality ($\delta_{avg} = 2.5 \times 10^{-3}$) has been assumed for our calculation. It may be noted that the CMA diagram here is different from the conventional one, since the effect of the boundary (through $k_c$) has been incorporated in the refractive index calculation, in contrast to the unbounded calculation. Figure 12.8b shows the variation of $n$ with respect to the geometry of the MC. It can be clearly seen that the X-mode-like wave can propagate through the center ($r \sim 0$–0.75 cm) and the periphery ($r > 2.25$ cm) with little attenuation ($n \sim 1.0$). Near the entrance ($z \sim 0$–3 cm; $r \sim 0.5$–1.75 cm), a cusped

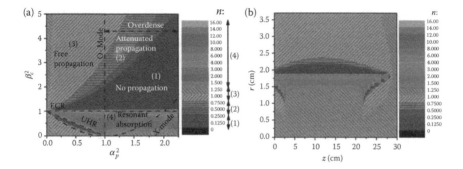

**FIGURE 12.8**
Contour plot of the refractive index as a function of (a) $\alpha_p^2$ and $\beta_c^2$ and (b) $r$ and $z$. (Reprinted with permission from I. Dey and S. Bhattacharjee. 2008. Experimental investigation of standing wave interaction with a magnetized plasma in a minimum B field, *Phys. Plasmas* **15**: 123502. Copyright 2008, American Institute of Physics.)

region is observed where the cutoff and the resonance conditions are satisfied very close to each other, and hence there will be damping of the SW near the entrance (Figure 12.5a). Another such cusp is also observed near the exit ($z \sim 22.5$–26 cm; $r \sim 0.5$–1.75 cm), but since the SW minima falls in that region, its effect is not observed. For $z > 26$ cm, there is again free propagation, where the peripheral waves and the central waves join up to enforce the SW pattern ($z \sim 27.5$ cm). The O-mode-like wave has no dependence on magnetic field and will propagate only when the plasma density is below cutoff, which is satisfied in the peripheral region ($r > 3$ cm) of the MC.

The variation of numerically evaluated $|E_n|$ with $\gamma$ is shown in Figure 12.6a for 180 W, 0.45 mTorr plasma condition. The wave number ($\gamma$) spectrum shows reasonable agreement with the spectrum obtained from the Fourier transform of the experimental data.

## Summary and Discussion

The interaction of microwave SWs with a plasma column confined in a minimum-B field is studied experimentally. The SW pattern is investigated as a function of input power and neutral density. It is found that the waves are damped toward the entrance of the MC, and undergo wave number downshift owing to the dispersive property of the plasma. The high average electron temperature ($T_{e\ avg} \cong 12$ eV) can be attributed to the interaction of the SW with the weakly collisional plasma, since the waves provide a heating of the electrons throughout the bulk of the plasma, apart from the ECR and UHR resonance zones occurring at the peripheral regions [9], thereby contributing to the axial plasma uniformity. The SW affects the plasma electron dynamics along the axial plasma and is demonstrated by the periodic undulation of $V_f$ since $V_s = V_f + 4.7T_e$ [10].

The analysis using the effective dielectric permittivity of the plasma is limited by the fact that Maxwell's equations were not solved to obtain the actual electric field amplitude and phase in the inhomogeneous and anisotropic magnetoplasma. The explanation provided here gives a preliminary understanding of the phenomenon, and a rigorous formulation is required to fully understand the physics involved. The observation of SWs in the overdense, minimally magnetized axial plasma cannot be explained by the conventional plane wave analysis, usually undertaken for preliminary analysis [9]. If the SW pattern was due to an evanescent plane wave penetrating axially into the plasma, then the SW intensity would have decayed toward the exit due to attenuation $\propto \exp(-\alpha z)$ by the plasma. However, it is observed that SW intensity profile is higher at the exit, implying that the pattern is getting reinforced along the axis as $N_i$ and $T_e$ decrease on an average toward the exit. From conventional theory of wave propagation in waveguides bounded by a conductor, it is known that the waves rise to various propagation modes.

However, there are several issues that need to be carefully addressed in regard to the above model, namely, why the screening by the axial plasma column is incomplete, what is the role of the radially varying magnetostatic field and plasma density, and how the waveguide mode actually gets modified in the anisotropic medium having regions where resonance criterion is satisfied. A detailed radial profiling of the wave field intensity in the MC cross section along with a theoretical model for wave dispersion in this unique geometry would be able to satisfactorily address the issues. Investigation in this regard is reported and discussed in the next chapter.

## References

1. S. Bhattacharjee and H. Amemiya. 1998. Microwave plasma in a multicusp circular waveguide with a dimension below cutoff, *Jpn. J. Appl. Phys.* **37**: 5742–5745.
2. J.V. Mathew, I. Dey, and S. Bhattacharjee. 2007. Microwave guiding and intense plasma generation at subcutoff dimensions for focused ion beams, *Appl. Phys. Lett.* **91**: 041503-3.
3. S. Bhattacharjee and H. Amemiya. 1999. Production of microwave plasma in narrow cross sectional tubes: Effect of the shape of cross section, *Rev. Sci. Instrum.* **70**: 3332–3337.
4. S. Bhattacharjee and H. Amemiya. 2000. Production of pulsed microwave plasma in a tube with a radius below the cut-off value, *J. Phys. D: Appl. Phys.* **33**: 1104–1116.
5. H. Amemiya, S. Ishii, and Y. Shigueoka. 1991. Multicusp type electron cyclotron resonance ion source for plasma processing, *Jpn. J. Appl. Phys.* **30**: 376–384.

6. J.L. Warren, G.P. Boicourt, M.T. Menzel, G.W. Rodenz, and M.C. Vasquez. 1985. Revision of and documentation for the standard version of the Poisson group codes, *IEEE Trans. Nucl. Sci.* **32**: 2870–2872.
7. K.N. Leung, N. Hershkowitz, and K.R. MacKenzie. 1976. Plasma confinement by localized cusps, *Phys. Fluids* **19**: 1045–1053.
8. J.H. Malmberg and C.B. Wharton. 1964. Collisionless damping of electrostatic plasma waves, *Phys. Rev. Lett.* **13**: 184–186.
9. I. Dey and S. Bhattacharjee. 2008. Experimental investigation of standing wave interactions with a magnetized plasma in a minimum-B field, *Phys. Plasmas* **15**: 123502.
10. M.A. Lieberman and A.J. Lichtenberg. 1994. *Principles of Plasma Discharges and Material Processing* (Wiley-Interscience, New York) p. 73.
11. J.A. Bittencourt. 2004. *Fundamentals of Plasma Physics* (3rd edition, Springer International Edition, New York) p. 400.
12. W.P. Allis, S.J. Buchsbaum, and A. Bers. 1963. *Waves in Anisotropic Plasmas* (M.I.T. Press, Cambridge, MA) pp. 133, 179.
13. D.K. Kalluri. 1998. *Electromagnetics of Complex Media* (CRC Press, Boca Raton) p. 105.
14. M.A. Lieberman and A.J. Lichtenberg. 1994. *Principles of Plasma Discharges and Material Processing* (Wiley-Interscience, New York) p. 73.
15. D.M. Pozar. 2006. *Microwave Engineering* (3rd edition, John Wiley and Sons Inc., Singapore) p. 118.
16. I. Dey and S. Bhattacharjee. 2011. Penetration and screening of perpendicularly launched electromagnetic waves through bounded supercritical plasma confined in multicusp magnetic field, *Phys. Plasmas* **18**: 22101.

# 13

## Penetration and Screening of Perpendicularly Launched Electromagnetic Waves

### Introduction

The physics of the interaction of electromagnetic (EM) waves with plasmas has been an active area of research over the years [1–12]. In the presence of a static magnetic field in the plasma, the interaction becomes all the more interesting due to the generation of a variety of wave modes, cutoffs, and resonances [1]. Most of the earlier experiments have focused on launching waves parallel to a nearly uniform axial magnetostatic ($B_o$) field produced by current-carrying coils ($k \parallel B_o$ case) [1–5]. The experiments on the Whistler [2,3] and the electron-cyclotron waves are well known [4,5]. A few investigations, where waves are launched perpendicular to the static magnetic field ($k \perp B_o$ case), have also been carried out by Sugai [6], Rypdal et al. [7], and Yadav and Bora [8], with emphasis on wave interactions in the UHR region.

However, not much work has been carried out to understand the wave interaction with plasmas operating in the $k \perp B_o$ configuration. An example of such a configuration includes the magnetic multicusp devices [9–11], where the magnetic field is predominantly transverse to the axially launched wave. In a multicusp device, supercritical plasmas ($\omega_p > \omega$) are generated in compact geometries, which interact self-consistently with the wave that generates the plasma [9–12].

Such wave-assisted plasma sources are a major requirement in emerging applications, such as high-current focused ion beams [13], miniplasma thrusters [14], broadband wave sources [15], and plasma wake field accelerators [16].

It is known that when an EM wave impinges on an infinitely long, unbounded, and uniform plasma, where $\alpha_p^2 > 1$ (supercritical) and $\beta_c^2 \ll 1$ (minimally magnetized), most of the EM wave is reflected from the plasma. The wave penetration is usually limited to the skin depth ($\delta_s$), which is of the order of a few centimeters, and depends on the wave number and the plasma density $[\delta_s = \{k \, \mathrm{Im}(1 - \alpha_p^2)^{1/4}\}^{-1}]$ (e.g., for a density of $10^{11}$ cm$^{-3}$ at 2.45 GHz, $\delta_s \sim 3$ cm).

The aim of this chapter is to investigate the issue of wave penetration and screening through bounded and radially inhomogeneous supercritical plasma for waves launched in the $k \perp B_o$ mode, which is realized experimentally in a compact multicusp device [9–12]. The reflection of waves needs to be minimized, while at the same time the plasma density is desired to be maintained above the critical density—two contradictory requirements. The conventional assumption of the plasma and the $B_o$ field being homogeneous and uniform [1–8] are not valid since $|B_o/(\partial B_o/\partial r)|_{r=33 \text{ mm}} \sim 7$ mm and $|n_e/(\partial n_e/\partial r)|_{r=33 \text{ mm}} \sim 11$ mm, $\ll \lambda_o \sim 122$ mm (free space wavelength at 2.45 GHz), thus making the plane wave approach inapplicable. A recent observation of large-amplitude EM-standing waves being supported in the central, minimally magnetized ($\beta_c^2 \ll 1$) plasma region [12] adds further to the motive behind this work.

## Experimental Setup

Figure 13.1 shows a schematic of the experimental setup [12], which consists of a stainless-steel vacuum chamber (VC) with a length of 50 cm and a diameter of 20 cm, with four cylindrical arms having numerous ports for pumping, gas inlet, vacuum gauges, and plasma diagnostics. The chamber is evacuated to a base pressure of $\sim 5 \times 10^{-7}$ Torr using a turbomolecular pump (TMP) backed by a rotary pump. Argon (zero grade) is used as the experimental gas. A 30-cm-long, eight-pole magnetic multicusp (MC) [9–12] having a pole-to-pole diameter of 84 mm is placed coaxially at the center of VC for plasma generation and confinement. A thin metallic cylinder of inner radius $a = 41$ mm is inserted in the MC to provide a uniform conducting boundary. Symmetric holes of 8 mm diameter are available for insertion of probes from the radial directions. The microwave generator (MWG) produces continuous-mode microwaves of 2.45 GHz, which are guided into the experimental chamber via standard WR340 waveguides. The water-cooled isolator (ISO) protects the magnetron from any reflected power, a dual directional coupler (DC) is used for measuring the forward and reflected powers, and a triple stub tuner (TST) for impedance matching between the source and the load; the TST is tuned to keep the reflected power below 5% of the incident power in all experiments. The waves are allowed to pass through a straight waveguide section (SS) and then through a high-grade quartz window (QW) into the vacuum chamber; QW maintains the vacuum integrity. A guiding cylinder (GC) of 104 mm inner diameter is placed coaxially at the entrance to guide the microwaves into MC.

Figure 13.2a shows an enlarged three-dimensional schematic of the plasma chamber MC in the larger VC. The wave polarization, vessel wall, and the antenna "pickup" probes are indicated for clarity. A nonuniform $B_o$ field is generated by the alternating arrangement of permanent magnets of the

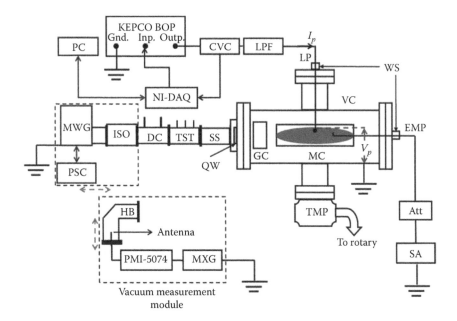

**FIGURE 13.1**
Schematic of the experimental setup. VC: vacuum chamber, MC: multicusp, GC: guiding cylinder, TMP: turbomolecular pump, QW: quartz window, SS: straight section, TST: triple stub tuner, DC: directional coupler, ISO: isolator, MWG: microwave generator, PSC: power supply and control for the MWG, WS: Wilson's seal, LP: Langmuir's probe, $V_p$: probe voltage, $I_p$: probe current, LPF: low-pass filter, CVC: current-to-voltage converter, KEPCO BOP: programmable bipolar power supply, Gnd.: common ground, Inp.: input (from −9 to +6 V slow ramp), Outp.: output −90 to +60 V slow ramp), NI-DAQ: National Instruments data acquisition device, PC: personal computer with LABVIEW, EMP: electromagnetic probe, Att.: 30 dB attenuator, SA: Agilent N1996A spectrum analyzer, HB: H-bend, PMI-5074: low-power isolator, and MXG: Agilent N5183A microwave signal generator. (Reprinted with permission from I. Dey and S. Bhattacharjee. 2011. Penetration and screening of perpendicularly launched electromagnetic waves through bounded supercritical plasma confined in multicusp magnetic field, *Phys. Plasmas* **18**: 22101. Copyright 2011, American Institute of Physics.)

eight-pole MC [9–12] (cf. Figure 13.2a and b). A Poisson Superfish (Ref. 17) simulation demonstrating the magnetic flux lines that lie primarily in the $(r, \theta)$ plane is shown in Figure 13.2b, where the reference polar angles as viewed from the axial probe entry side (opposite to microwave entrance side) of the MC are indicated. The measured radial variation of $B_0(r,\theta) = \sqrt{B_0(r)^2 + B_0(\theta)^2}$ from the center to the magnet gap along 270°, using a Lakeshore 421 gaussmeter, is shown in Figure 13.3a. The magnetostatic well with minimum-$B$ at the center is clearly realized, as demonstrated in Figure 13.2b and in the radial measurements of Figure 13.3. The waves are launched primarily in the $k \perp B_0$ mode [9–12].

A planar Langmuir probe having a 4 mm diameter stainless-steel tip is swept between −90 and +60 V to obtain the $I$–$V$ characteristics with the help of

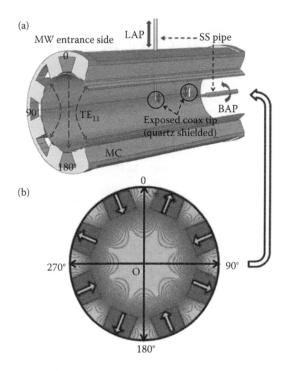

**FIGURE 13.2**

(a) Expanded view of the multicusp demonstrating the orientation of the electric field probes (linear antenna LAP and bent antenna BAP) with respect to the incident wave $TE_{11}$ mode polarization. (b) Poisson's simulation of the magnetic flux lines in the multicusp showing the angular scheme followed. Magnetization direction is indicated by arrows. (Reprinted with permission from I. Dey and S. Bhattacharjee. 2011. Penetration and screening of perpendicularly launched electromagnetic waves through bounded supercritical plasma confined in multicusp magnetic field, *Phys. Plasmas* **18**: 22101. Copyright 2011, American Institute of Physics.)

a National Instruments data acquisition system in association with a PC running LABVIEW 8.0. The total sweep time is set at 75 s in steps of 0.2 V, with 100 data points acquired and averaged at each point. The low-pass filter (LPF) prevents the interference of any microwave pickup in the measurements. The probe is inserted from the 270° gap for the radial measurement, and its plane is held parallel to the MC axis and to the wave vector of the microwaves so that the disturbance by the microwave is minimal. Also, the surface of the probe is maintained perpendicular to the $B_o$ field lines (Figure 13.2b). Hence, the standard method of evaluating the plasma parameters from the *I–V* characteristics may be employed without much error, as has been done by Amemiya, Maeda, Bhattacharjee, and Mathew previously [9–11].

The probes for wave electric and magnetic field measurements are constructed out of a Micro-Coax microwave grade (0–18 GHz) cable, having an inner conductor diameter of ~1 mm. A linear antenna probe (LAP) with a 2-mm-diameter exposed spherical tip, enclosed in a quartz glass tube, is

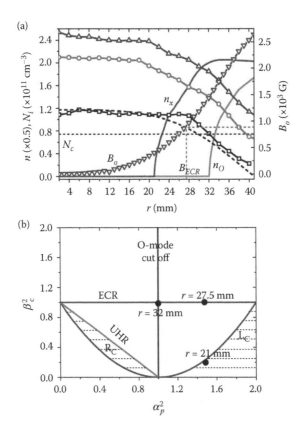

**FIGURE 13.3**

(a) Experimental variation of $N_i$ with $r$ at 180 W and 0.50 mTorr (–□–), 360 W and 0.35 mTorr (–○–), and 540 W and 0.35 mTorr (–△–). A Bessel fit (– – –) is superimposed on the 180 W and 0.50 mTorr data. Experimental variation of $B_o$ with $r$ is shown by (–▽–). Radial variation of $n_O$ and $n_x$ calculated at 180 W and 0.50 mTorr is shown by solid lines. (b) CMA diagram showing the propagation, left cutoff ($L_C$), right cutoff ($R_C$), and ECR regions. (Reprinted with permission from I. Dey and S. Bhattacharjee. 2011. Penetration and screening of perpendicularly launched electromagnetic waves through bounded supercritical plasma confined in multicusp magnetic field, *Phys. Plasmas* **18**: 22101. Copyright 2011, American Institute of Physics.)

used for the measurement of the total wave electric field along the radial direction. An L-shaped bent antenna probe (BAP) is constructed with an exposed tip length of 12 mm, oriented perpendicular to the axis of the probe for measurement of polar variation of the integrated average central wave electric field intensity along the axis of MC [18]. The orientation of the electric field probes with respect to the plasma chamber MC and the incident wave $TE_{11}$ mode polarization is shown in Figure 13.2a. A B-dot probe of 1 mm diameter, 2 mm length, and having three turns is employed to measure the wave $B$ field of the incident microwave. The probe signal is attenuated via a Weinschel 30 dB attenuator and measured with an Agilent N1996A spectrum analyzer by recording the peak power corresponding to 2.45 GHz, which is averaged over 100 shots.

All probes are vacuum-sealed inside a 5.5-mm stainless-steel (SS-304 L) pipe and inserted via Wilson's seals. The dimensions of the probes are chosen such that they cause minimal disturbance to the plasma during measurement. The radial measurements are carried out from the top (0°) and left (270°; see Figure 13.2b) at the center of the MC ($z = 15$ cm from the MC entrance). However, it is observed that the measurements from the top direction cause disturbance in the reflected power and the data. This may be attributed to the fact that the length of the probe is predominantly parallel to the electric field lines (Figure 13.2a) and hence disturbs the field distribution significantly, compared to the case when the probe is inserted perpendicularly from the left direction. Hence, the measurements by the probes inserted from the left (270°) are reported in this study.

The measurement of vacuum EM fields is necessary both for comparison of the spatial wave field distribution in the presence and absence of plasma and for measuring the vacuum polarization mode in MC. The vacuum EM field measurements are carried out using an Agilent N5183A microwave signal generator (MXG) (Figure 13.1), which provides a maximum output power of 20 dBm (100 mW) at 2.45 GHz. The signal is launched through the waveguide system into MC by a quarter-wave monopole antenna of 3 cm exposed length, mounted on an aluminum plate and attached to the WR 340 waveguide flange, as shown in Figure 13.1. The block containing MWG, ISO, and PSC is replaced with the "vacuum measurement module," as depicted in Figure 13.1. A PMI-5074 Western microwave ISO is used between the antenna and the MXG to isolate any reflected power picked up by the antenna.

The experiments are performed at a wave input power range $P_{in} = 180$–$540$ W and in the pressure range $p = 0.20$–$0.60$ mTorr corresponding to $(6$–$14) \times 10^{12}$ cm$^{-3}$ of Ar-neutral density at room temperature (25°C). Table 13.1 lists the typical plasma parameters that are realized in the experimental device and are reproduced here for completeness [12].

**TABLE 13.1**

Range of Values for the Calculated Plasma Parameters Obtained from Experiments

| Symbol | Parameter | Range of Value |
|--------|-----------|----------------|
| $\nu_{en}$ | Electron–neutral collision frequency | $(3$–$10) \times 10^6$ Hz |
| $\nu_{in}$ | Ion–neutral collision frequency | $(1$–$4) \times 10^4$ Hz |
| $f_{pe}$ | Electron–plasma frequency | $(0$–$4) \times 10^9$ Hz |
| $f_{ce}$ | Electron–cyclotron frequency | $(0$–$1) \times 10^{10}$ Hz |
| $\nu_e$ | Electron velocity | $(3$–$9) \times 10^8$ cm/s |
| $\nu_i$ | Ion velocity | $(3$–$4) \times 10^5$ cm/s |

*Source:* Adapted from I. Dey and S. Bhattacharjee. 2008. *Phys. Plasmas* **15**: 123502–7.

## Experimental Observation

Figure 13.3 shows the radial variation of plasma (ion) density $N_i$ at three combinations of power and pressure, namely, 180 W at 0.50 mTorr, and 360 and 540 W at 0.35 mTorr, at $z = 15$ cm (center of MC). The plasma is approximately uniform ($N_i \sim (1.11 \pm 0.03) \times 10^{11}$ cm$^{-3}$ for 180 W, 0.50 mTorr) up to about $r \sim 25$ mm from the center, which corresponds to a length parameter $\rho(= r/a) \sim 0.6$, where $a$ is the radius. At 540 W and 0.35 mTorr, a central density of about $2.5 \times 10^{11}$ cm$^{-3}$ is obtained. It is observed that the central density is supercritical, that is, $N_i >$ critical density, $N_c (= 7.44 \times 10^{10}$ cm$^{-3})$ with respect to the wave frequency of 2.45 GHz for a major part of MC. In previous works [7,8], the plane wave dispersion relation in the $k \perp B_o$ mode for infinite plasma was used for explaining wave propagation in the MC. The real part of the refractive index $n$ (indicating propagation) for the ordinary (O) $\left( n_O = \sqrt{1 - \alpha_p^2} \right)$ and extraordinary $(X) \left[ n_X = \sqrt{(1 + \beta_c - \alpha_p^2)(1 - \beta_c - \alpha_p^2)(1 - \alpha_p^2 - \beta_c^2)} \right]$ modes is calculated from the experimental $N_i$ data at 180 W at 0.50 mTorr and $|B_o|$,

with the plane wave dispersion relations [7,8] and plotted in the same graph (Figure 13.3a). The values for $n_O$ and $n_X$ indicate that the wave propagation through the central plasma region $r \sim (0–21)$ mm is forbidden, and most of the wave propagation would take place through the peripheral regions, as indicated in Figure 13.3a [9–11]. To relate the propagation conditions with that predicted by the CMA diagram [7], the O-mode cutoff condition ($\alpha_p^2 = 1$ at $r = 32$ mm), the X-mode left cutoff ($1 + \beta_c - \alpha_p^2 = 0$ at $r = 21$ mm), and the ECR condition ($\beta_c^2 = 1$ at $r = 27.5$ mm) obtained in the MC are mapped onto a CMA plot (Figure 13.3b).

Figure 13.4a shows the radial variation of the experimentally measured total electric field intensity proportional to $|E_{tot}|^2 (= E_r^2 + E_\theta^2 + E_z^2)$ at 180 W, 0.50 mTorr and 360 W, 0.35 mTorr, which corresponds to average central densities of $\sim 1.1 \times 10^{11}$ and $2.1 \times 10^{11}$ cm$^{-3}$, respectively. The data at 360 W have been downscaled by 13 dB for comparison with the 180 W data. Contrary to the predictions of the plane wave dispersion theory (cf. Figure 13.3a and b) for the bounded and inhomogeneous plasma in the minimum-B field, the $E_{tot}$ field data indicate the presence of the wave field in the central region ($r \sim 0–28$ mm), which decreases to a minimum at $r \sim 28$ mm and then rises again toward the boundary. For the input power of 180 W, the central $|E_{tot}|^2$ detected by the probe system in the absence of plasma was found to be $\sim -10$ dBm, whereas in the presence of plasma (0.50 mTorr), it was $\sim -39$ dBm, with the noise floor being $\sim -75$ dBm, indicating that there is a substantial amount of remnant intensity present in the central plasma after the generation of the supercritical plasma by the microwave. It may be noted that the central wave field intensity is of the same order as the peripheral values.

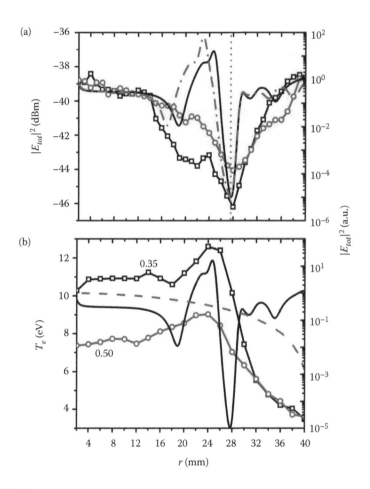

**FIGURE 13.4**

(a) Experimental variation of $|E_{tot}|^2$ with $r$ at 180 W and 0.50 mTorr (–☐–) and 360 W and 0.35 mTorr (–O–). Variation of $|E_{tot}|^2$ (model) with $r$ at $n_o = 1.1 \times 10^{11}$ cm$^{-3}$ (—) and $2.1 \times 10^{11}$ cm$^{-3}$ (– · –) for comparison with experimental data at 180 and 360 W, respectively. (b) Experimental variation of $T_e$ with $r$ with 180 W at 0.35 mTorr (–☐–) and 0.50 mTorr (–O–). Radial variation of $|E_{tot}|^2$ (model) at $n_o = 1.1 \times 10^{11}$ cm$^{-3}$ (–). The radial variation of $|E_{tot}|^2$ (model) in vacuum case is shown by (– –). (Reprinted with permission from I. Dey and S. Bhattacharjee. 2011. Penetration and screening of perpendicularly launched electromagnetic waves through bounded supercritical plasma confined in multicusp magnetic field, *Phys. Plasmas* **18**: 22101. Copyright 2011, American Institute of Physics.)

Small discernable peaks are observed at $r \sim 24$ and 22 mm for 180 and 360 W, respectively, indicating that the peaks shift inward with an increase in the plasma density. The minimum value of the electric field for the two plasma conditions approximately coincides at 28 mm and lies between the X-mode left cutoff (L$_C$) $\omega_L = 0.5\left[-\omega_c + \sqrt{\left(\omega_c^2 + 4\omega_p^2\right)}\right]$ at 21 mm and the O-mode cutoff $\omega_O = \omega_p$ at 32 mm (Figure 13.3a and b). The minimum is observed to occur close to the ECR point ($r_{ECR} = 27.5$ mm) (Figure 13.3a and b) and thus may be related to wave absorption at ECR.

The radial variation of plasma (electron) temperature $T_e$ at 0.35 and 0.50 mTorr pressures and 180 W microwave power is shown in Figure 13.4b. A peak in the temperature profile is observed at $r \sim 24$ mm from the center, thereby indicating the occurrence of a possible wave magnetoplasma resonant interaction. The plane wave dispersion analysis employed in earlier works [7,8] does not predict such a resonance, since $n_X \sim 0.5$ at $r \sim 24$ mm (cf. Figure 13.3a), whereas, at resonance, $n_X$ must be $\gg 1$ in the weakly collisional regime of our experiments [12], thereby demonstrating the shortcoming of the plane wave analysis. At a higher pressure (0.50 mTorr), the peak value of the temperature decreases 1.5 times and the spatial spread increases, $\Delta r \sim 2$ mm, due to the enhanced electron–neutral collisions.

In Figure 13.5a, the radial variation of the axial component of the wave magnetic field intensity proportional to $|B_z|^2$ is shown for 360 W, at pressures of 0.20 and 0.35 mTorr corresponding to central densities of $\sim 1.2 \times 10^{11}$ and $2.1 \times 10^{11}$ cm$^{-3}$, respectively. The measurements are carried out by the B-dot probe described in the section "Experimental Setup," with its plane perpendicular to the axis of MC. The intensity is minimum at the center and maximum at the edges, with a discernable dip being observed at $r \sim 27$ mm for the 0.20 mTorr data. The existence of $B_z$ component indicates that the wave mode sustained in the plasma is predominantly transverse electric (TE) modified by the plasma. This is further substantiated by the fact that the measured intensity of the axial component of the wave electric field ($|E_z|^2$) is an order of magnitude less than $|B_z|^2$ and 72% smaller than the intensity corresponding to the radial ($|E_r|^2$) and polar ($|E_\theta|^2$) electric field intensities [19].

To further confirm the waveguide mode, a polar plot of the normalized average central $|E_{tot}|^2$ is obtained at the center of MC by the L-shaped bent antenna probe, both in vacuum and in the presence of plasma (360 W, 0.35 mTorr), and is shown in Figure 13.5b. The exposed probe tip is 12 mm long and records the average integrated wave electric field intensity of the central region (0–12 mm) as a function of $\theta$. The nature of the plot further confirms the phenomenon of wave field penetration in the central supercritical plasma. On fitting the experimental data sets with a $\cos^2\{m(\theta + \theta_c)\}$ curve gives $m \sim 1$, and $\theta_c$ is the angle of rotation of the polarization axis from the vacuum value ($\theta_{c\ vacuum} = 0$). The bilobed ($\cos^2 \theta$-like) nature of the profile clearly indicates that the sustained wave fields inside the plasma are in the dominant transverse electric TE$_{11}$ mode for cylindrical waveguides. The measurement is performed in the central region to know the polarization state of the penetrating field. Polarization measurements are also performed at other radii (not reported here), and it is observed that the primary pattern remains dipole in nature, but the distortion of the lobes increases as the resonance region is approached. The typical cross-sectional electric wave vector profile in a cylindrical waveguide for the TE$_{11}$ mode is schematically shown in Figure 13.5c, where the dipole nature is clearly demonstrated. The rotation in the polarization angle ($\theta_c \sim 20°$) with respect to vacuum angle at $z = 15$ cm,

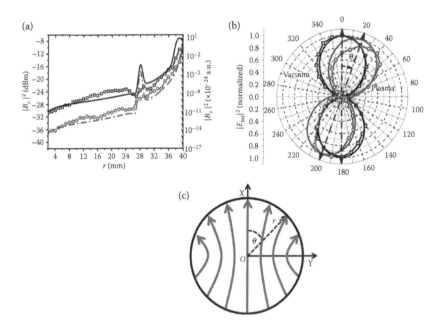

**FIGURE 13.5**

(a) Experimental variation of $|B_z|^2$ with $r$ at 360 W for 0.20 mTorr (–□–) and 0.35 mTorr (–O–). Variation of $|B_z|^2$ (model) at $n_o = 1.2 \times 10^{11}$ and $2.1 \times 10^{11}$ cm$^{-3}$ is shown by solid lines superimposed on the 0.20 mTorr (—) and 0.35 mTorr (– · –) data. (b) Experimental variation of $|E_{tot}|^2$ with polar angle $\theta$ in vacuum (–□–) and at 360 W and 0.35 mTorr (–O–). Cosine-squared fits of the data (—) are superimposed. (c) Schematic view of the cross-sectional wave electric field lines for the TE$_{11}$ mode in a circular waveguide. (Reprinted with permission from I. Dey and S. Bhattacharjee. 2011. Penetration and screening of perpendicularly launched electromagnetic waves through bounded supercritical plasma confined in multicusp magnetic field, *Phys. Plasmas* **18**: 22101. Copyright 2011, American Institute of Physics.)

as shown in the plot (Figure 13.5b), may be attributed to the phenomenon of birefringence in the anisotropic plasma medium. The polarization axis rotates by ~12° over a distance of 20 cm ($z = 5$–25 cm) along the axial plasma. These are interesting observations and will be discussed in a future article.

The experimentally obtained $|E_{tot}|^2$ and $T_e$ profiles are interesting because wave propagation through the central supercritical plasma is unexpected from the plane wave theory [7] and as reported in a recent experiment [11] in a similar geometry, but without the inner conducting boundary. The observed results therefore motivate us to consider the effect of the metallic boundary with a finite conductivity, with all the plasma and magnetic field nonuniformities taken into account. Allis et al. [20] looked at the wave field solutions in bounded plasmas where the plasma and the confining magnetic field are homogeneous. Similarly, Takahashi et al. [21] solved the wave equations in the $k||B_o$ mode where the plasma density and the static magnetic field are uniform inside the boundary. However, the solution of the wave equations in a compact geometry, having strong magnetoplasma inhomogeneities with the wave launch belonging to the $k \perp B_o$ mode, has not been reported earlier in such geometries. In the subsequent sections, we attempt to obtain numerical solutions in such a configuration based on the experimental observations and investigate the physics of the interaction.

## Modeling

In general, there is no perfect TE and TM mode in plasma-filled waveguides [20,21]. However, from the experimental results, the following points may be noted: (i) The dipole structure in Figure 13.5b indicates that the electric field has a $\theta$ dependence. For a TM mode, $E_\theta = 0$. (ii) A finite $B_z$ is obtained, as shown in Figure 13.5a. Ideally, for TM mode, $B_z = 0$. (iii) The measured value of $|E_z|^2$ is smaller by an order of magnitude compared to $|B_z|^2$ and also 72% smaller than $E_r$ and $E_\theta$ components.

Therefore, it is seen that $TE_{11}$ mode is predominant in the presence of plasma, and thus Maxwell's equations are solved numerically in the $TE_{11}$ mode. The highlights of the computation are that the equations are solved with the magnetized plasma in the multicusp as a dispersive medium characterized by a radially inhomogeneous and anisotropic permittivity, enclosed by a conducting boundary. The static cusped magnetic field can be resolved into $r$ and $\theta$ components, and for an octupole they are given by [22] $B_{or} = -B_p (r/a)^3 \sin(4\theta)$, $B_{o\theta} = -B_P(r/a)^3 \cos(4\theta)$ defined with respect to the polar angle scheme demonstrated in Figure 13.2b. Here, $B_p$ is the magnetic field strength at the pole of the magnet and $(B_p/a^3) = 0.049$ G/mm$^3$.

Maxwell's equations for a general medium are given by

$$\vec{\nabla} \cdot \varepsilon_0 \bar{\bar{K}} \tilde{\vec{E}} = 0, \tag{13.1a}$$

$$\vec{\nabla} \cdot \tilde{\vec{B}} = 0, \tag{13.1b}$$

$$\vec{\nabla} \times \tilde{\vec{E}} = i\omega \tilde{\vec{B}}, \tag{13.1c}$$

and

$$\vec{\nabla} \times \tilde{\vec{B}} = -(i\omega/c^2)\bar{\bar{K}}\tilde{\vec{E}}, \tag{13.1d}$$

where

$$\bar{K} = \begin{bmatrix} K_{11} & K_{12} & K_{13} \\ K_{21} & K_{22} & K_{23} \\ K_{31} & K_{32} & K_{33} \end{bmatrix} \tag{13.2}$$

is the general dielectric tensor.

In the $k \perp B_o$ mode, the tensor components $K_{ij}$ are given by [19] $K_{11} = 1 - \{(\alpha_p^2/U_o)(U^2 - \beta_{cr}^2)\}$, $K_{12} = K_{21} = \{(\alpha_p^2/U_o) \times (\beta_{cr}\beta_{c\theta})\}$, $K_{13} = -K_{31} = -i\{(\alpha_p^2/U_o)(U\beta_{c\theta})\}$, $K_{22} = 1 - \{(\alpha_p^2/U_o)(U^2 - \beta_{c\theta}^2)\}$, $K_{23} = -K_{32} = i\{(\alpha_p^2/U_o)(U\beta_{cr})\}$ and $K_{33} = 1 - \{(\alpha_p^2/U_o)(U^2)\}$ in terms of the normalized plasma parameters, which have been derived under the cold plasma approximation. The experimentally obtained electron velocities (cf. Table 13.1) [12] $v_e$ (~$10^6$ m/s) being much smaller than $v_{ph}$ (~$10^8$ m/s), the wave phase velocity, in the bulk plasma allow the use of cold plasma approximation [4,7,8,20]. It is expected that any deviation from the model would be reflected in the comparison of simulation with the experimental results. Here, $U = 1 + i\delta$, $\alpha_p = \omega_p/\omega$, $\beta_{cr} = \omega_{cer}/\omega$, $\beta_{c\theta} = \omega_{ce\theta}/\omega$, $U_o = U(U^2 - \beta_{cr}^2 - \beta_{c\theta}^2)$, and $\delta = v_{coll}/\omega$, where $\omega_p$, $\omega_{cer}$ and $\omega_{ce\theta}$, and $v_{coll}$ are the electron–plasma frequency, the electron–cyclotron frequency corresponding to the $r$ and $\theta$ components of the $B_o$ field, and the electron–neutral collision frequency, respectively. The expressions relating the frequencies to the spatially varying plasma parameters and magnetostatic field are [12] $\omega_p(r,\theta) = \sqrt{n_e(r,\theta)e^2/m_e\varepsilon_0}$ (rad/s), $\omega_{ce}(r,\theta) = eB_o(r,\theta)/m_e$ (rad/s), and $v_{coll}(r,\theta) = 4\pi \times 10^{28} p \times \sqrt{T_e(r,\theta)}\, \sigma_{coll}$ (rad/s), where $\sigma_{coll}$ is the electron–neutral collision cross section. The finite conductivity of the metal boundary (stainless steel, $\sigma_{coll} = 9.8 \times 10^5$ S/m) is taken into account, while imposing the grounded boundary condition on Maxwell's equations.

A cylindrical coordinate system is assumed in the MC geometry. The problem is formulated as follows. The plasma density is assumed to be symmetric in the $\theta$ direction, varies only along $r$, and is represented by a Bessel fit $n_e(r) = n_o J_o\{2.405(r/a)^2\}$, where $n_o$ is the central density at $r = 0$ [9–12], the fitting being reasonably good, as shown in Figure 13.3a for the 180 W case. The Bessel function ensures the boundary condition $n_e(a) = 0$. The collisionality parameter is calculated assuming an average $T_e = 10$ eV [12] over the MC cross section. It is of the order of $10^{-4}$ in the pressure range of our experiments.

The solution procedure is on similar lines as outlined in Ref. [23]. Let the solution of the wave electric and magnetic fields in the cylindrical coordinate system be

$$\tilde{E}(r,\theta,z,t) = \tilde{E}_r(r,\theta,z,t)\hat{r} + \tilde{E}_\theta(r,\theta,z,t)\hat{\theta} + \tilde{E}_z(r,\theta,z,t)\hat{z}$$

$$= \vec{E}_c(r,\theta)\exp\{i(k_z z + m\theta - \omega t)\}, \tag{13.3a}$$

where

$$\vec{E}_c(r,\theta) = E_{cr}\hat{r} + E_{c\theta}\hat{\theta} + E_{cz}\hat{z}$$

and similarly

$$\tilde{B}(r,\theta,z,t) = \tilde{B}_r(r,\theta,z,t)\hat{r} + \tilde{B}_\theta(r,\theta,z,t)\hat{\theta} + \tilde{B}_z(r,\theta,z,t)\hat{z}$$

$$= \vec{B}_c(r,\theta)\exp\{i(k_z z + m\theta - \omega t)\}, \tag{13.3b}$$

where

$$\vec{B}_c(r,\theta) = B_{cr}\,\hat{r} + B_{c\theta}\,\hat{\theta} + B_{cz}\,\hat{z}.$$

$\vec{E}_c$ and $\vec{B}_c$ are the complex wave electric and magnetic field amplitudes, respectively, which are functions of $r$ and $\theta$ only, incorporating all nonharmonic variations of the fields. It may be noted that the magnetoplasma is taken as uniform in the axial direction, and hence variation along $z$ occurs only in the harmonic phase term. Here, $m$ is the polar mode number and $k_z$ is the magnitude of the wave vector for $z$ axis propagation in the waveguide (MC). It is related to the free space wave vector ($k_o = \omega/c$) and the cutoff wave number for the cylindrical waveguide ($k_c$) by $k_z(r,\theta)^2 = k_o^2 K_{22}(r,\theta) - k_c^2$ [24], where it is assumed that $k_c$ is not much affected by the plasma, and from the experimental observations, the TE$_{11}$ waveguide mode dominates and is given by $k_c = 1.841/a$. The $k_z$ obtained from the above relation is used in the calculations.

From Equations 13.3a and 13.3b, six wave field components are obtained, namely

$$\tilde{E}_r(r,\theta,z,t) = E_{cr}(r,\theta)\exp\{i(k_z z + m\theta - \omega t)\},$$

$$\tilde{E}_\theta(r,\theta,z,t) = E_{c\theta}(r,\theta)\exp\{i(k_z z + m\theta - \omega t)\},$$

$$\tilde{E}_z(r,\theta,z,t) = E_{cz}(r,\theta)\exp\{i(k_z z + m\theta - \omega t)\},$$

$$\tilde{B}_r(r,\theta,z,t) = B_{cr}(r,\theta)\exp\{i(k_z z + m\theta - \omega t)\},$$

$$\tilde{B}_\theta(r,\theta,z,t) = B_{c\theta}(r,\theta)\exp\{i(k_z z + m\theta - \omega t)\},$$

and

$$\tilde{B}_z(r,\theta,z,t) = B_{cz}(r,\theta)\exp\{i(k_z z + m\theta - \omega t)\}.$$

Employing the above six wave field components obtained from equation set 13.3 along with the two curl Equations 13.1c and 13.1d and Equation 13.2, and noting that $m = 1$ and $E_z = 0$ for TE$_{11}$ mode [24], six equations involving the cylindrical ($r$, $\theta$, $z$) components for the $\tilde{\vec{E}}$ and $\tilde{\vec{B}}$ oscillating fields are obtained, namely

$$B_{cr} = -\frac{k_z}{\omega}E_{c\theta}, \qquad (13.4a)$$

$$B_{c\theta} = \frac{k_z}{\omega} E_{cr}, \tag{13.4b}$$

$$\frac{1}{r}\left[\frac{\partial}{\partial r}(rE_{c\theta}) - \left(i + \frac{\partial}{\partial\theta}\right)E_{cr}\right] - i\omega B_{cz} = 0, \tag{13.4c}$$

$$\frac{1}{r}\left(1 - i\frac{\partial}{\partial\theta}\right)B_{cz} - k_z B_{c\theta} = -\frac{k_o^2}{\omega}(K_{11}E_{cr} + K_{12}E_{c\theta}), \tag{13.4d}$$

$$k_z B_{cr} + i\frac{\partial B_{cz}}{\partial r} = -\frac{k_o^2}{\omega}(K_{21}E_{cr} + K_{22}E_{c\theta}), \tag{13.4e}$$

and

$$\frac{1}{r}\left[\frac{\partial}{\partial r}(rB_{c\theta}) - \left(i + \frac{\partial}{\partial\theta}\right)B_{cr}\right] + \frac{ik_o^2}{\omega}(K_{31}E_{cr} + K_{32}E_{c\theta}) = 0. \tag{13.4f}$$

From Equations 13.4a, 13.4b, 13.4d, and 13.4e, the $E_{cr}$, $E_{c\theta}$, $B_{cr}$, and $B_{c\theta}$ components can be expressed in terms of the $B_{cz}$ component as

$$E_{cr} = \frac{i\omega}{D_{00}}\left\{\frac{iD_{22}}{r} + D_{12}\frac{\partial}{\partial r} + \frac{D_{22}}{r}\frac{\partial}{\partial\theta}\right\}B_{cz}, \tag{13.5a}$$

$$E_{c\theta} = \frac{i\omega}{D_{00}}\left\{\frac{iD_{21}}{r} + D_{11}\frac{\partial}{\partial r} + \frac{D_{21}}{r}\frac{\partial}{\partial\theta}\right\}B_{cz}, \tag{13.5b}$$

$$B_{cr} = \frac{ik_z}{D_{00}}\left\{\frac{iD_{21}}{r} + D_{11}\frac{\partial}{\partial r} + \frac{D_{21}}{r}\frac{\partial}{\partial\theta}\right\}B_{cz}, \tag{13.5c}$$

and

$$B_{c\theta} = \frac{ik_z}{D_{00}}\left\{\frac{iD_{22}}{r} + D_{12}\frac{\partial}{\partial r} + \frac{D_{22}}{r}\frac{\partial}{\partial\theta}\right\}B_{cz}. \tag{13.5d}$$

The coefficients are functions of $r$ and $\theta$ and are related to the dielectric tensor components through the following transformations: $D_{ii} = k_o^2 K_{ii} - k_z^2$ $(i \rightarrow 1, 2, 3)$, $D_{ij} = k_o^2 K_{ij}(i \neq j)$ $(i, j \rightarrow 1, 2, 3)$, and $D_{00} = D_{11}D_{22} - D_{12}^2$.

Replacing $E_{cr}$, $E_{c\theta}$, $B_{cr}$, and $B_{c\theta}$ from equation set 13.5 in Equations 13.4c and 13.4f, we obtain a pair of coupled radial differential equations for $B_{cz}$ at each $\theta$, namely

$$P_1 B''_{cz} + Q_1 B'_{cz} + R_1 B_{cz} = 0 \tag{13.6a}$$

and

$$P_2 B''_{cz} + Q_2 B'_{cz} + R_2 B_{cz} = 0, \tag{13.6b}$$

where the primes indicate differentiation with respect to $r$, and $P_1 = i\omega r^2$, $P_2 = ir^2 F_{12}$,

$$Q_1 = -r\omega\{C_{12} - 2iC_{11} + C_{21} - ir(\partial C_{11}/\partial r) - i(\partial C_{12}/\partial \theta)\},$$

$$Q_2 = r\{2iF_{12} + F_{11} - F_{22} - i(\partial F_{11}/\partial \theta) + ir(\partial F_{12}/\partial r) + r(D_{32}C_{11} - D_{31}C_{12})\},$$

$$R_1 = -\omega\{C_{12} + iC_{22} - ir + r(\partial C_{12}/\partial r) - i(\partial C_{22}/\partial \theta)\},$$

and

$$R_2 = \{iF_{12} - F_{22} + (\partial F_{12}/\partial \theta) - r(\partial F_{22}/\partial r) + ir(D_{32}C_{12} - D_{31}C_{22})\}.$$

The $C$ and $F$ coefficients are given by $C_{ij} = D_{ij}/(rD_{00})$ and $F_{ij} = k_z D_{ij}/(rD_{00})$ ($i, j \rightarrow 1, 2, 3$).

The boundary conditions applicable are $(\varepsilon_{medium}E_{\perp medium} - \varepsilon_{metal}E_{\perp metal}) = \rho_s$, $(B_{\perp medium} - B_{\perp metal}) = 0$, $(E_{\parallel medium} - E_{\parallel metal}) = 0$, and $(B_{\parallel medium} - B_{\parallel metal}) = \mu_0 J_s$, where $E_\perp = E_{cr}$, $B_\perp = B_{cr}$, $E_\parallel = E_{c\theta}$, $B_\parallel = \sqrt{B_{c\theta}^2 + B_{cz}^2}$, $\rho_s$ is the surface charge density, and $J_s$ is the surface current density.

The radial differential Equations 13.6a and 13.6b are solved simultaneously using the finite difference method [25] in MATLAB® R2009A over the whole polar range (1°–360°) in steps of 1°. The resulting $B_{cz}$ obtained from Equations 13.6a and 13.6b is substituted into Equations 13.5a and 13.5b to obtain $E_{cr}$ and $E_{c\theta}$ from which $\vec{E}_c(r,\theta) = E_{cr}\hat{r} + E_{c\theta}\hat{\theta}$ can be calculated ($E_{cz} = 0$). Using the value of $\vec{E}_c$ thus obtained, the wave electric field $\tilde{E}(r,\theta,z,t)$ can be calculated from Equation 13.3a. The axial magnetic field component can be obtained from the relation $\tilde{B}_z = B_{cz}\exp\{i(k_z z + m\theta - \omega t)\}$.

## Modeling Results

The wave field solution varies both with radius and the polar angle. The wave mode formation due to the radial boundary reflections is taken care by the formulation in the section "Modeling." The standing wave formation due to axial reflections occurring at the terminals of the MC [12] is taken into account by considering the superposition of a forward and a reflected wave. The forward wave field is given by Equation 13.3a, namely,

$$\tilde{\vec{E}}_{fwd}(r,\theta,z,t) = \tilde{E}_{rfwd}\,\hat{r} + \tilde{E}_{\theta fwd}\,\hat{\theta} + \tilde{E}_{zfwd}\,\hat{z} = \tilde{E}_{zref}\,\hat{z} = \tilde{E}_{cref}(r,\theta)\,\exp\left\{(k_z z + m\theta - \omega t)\right\},$$

and the counter propagating reflected wave is given by $\tilde{\vec{E}}_{ref}(r,\theta,z,t) = \tilde{E}_{rref}\,\hat{r} +$

$\tilde{E}_{\theta ref}\,\hat{\theta} + \tilde{E}_{zref}\,\hat{z} = \tilde{E}_{cref}(r,\theta)\exp\left\{i(-k_z z + m\theta - \omega t)\right\}$, which can be obtained by the same procedure in Sec. IV. Here, $\tilde{E}_{cref} = \bar{R}\tilde{E}_{cfwd}$ is the complex amplitude of the reflected wave, in which the reflection coefficient was estimated to be $\bar{R} = 0.87\exp(-j2.05)$ from experiments in Ref. [12]. The total wave electric field in the MC is obtained from $\tilde{\vec{E}}_{tot} = \tilde{\vec{E}}_{fwd} + \tilde{\vec{E}}_{ref}$, where the radial, polar, and axial variations, along with the radial and axial reflections, have been taken into account and describe the behavior of the wave electric field in the MC in the presence of the inhomogeneous magnetoplasma. The intensity proportional to $|E_{tot}|^2$ (model) is evaluated from $\tilde{\vec{E}}_{tot}$ and compared with the experimental results. A similar procedure is applied for the axial component of the wave magnetic field to obtain $|B_z|^2$.

A typical polar contour plot showing the normalized total wave electric field intensity $|E_{tot}|^2$ in arbitrary units, in the $(r - \theta)$ plane at $z = 15$ cm for a central density $n_o = 1.1 \times 10^{11}$ cm$^{-3}$, is shown in Figure 13.6. Major features such as the wave field penetration at the center (yellow-orange, $|E_{tot}|^2 \sim 0.02$), damping at the ECR ($|E_{tot}|^2 \sim 10^{-8}$), wave plasma resonance regions ($|E_{tot}|^2 \sim 1$), and peripheral propagation ($|E_{tot}|^2 \sim 0.1$) are clearly realized. The effect of

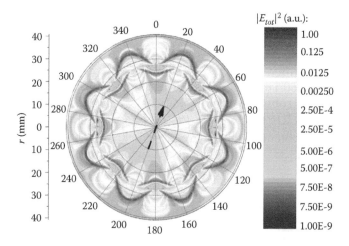

**FIGURE 13.6**

Polar contour plot of the numerical solution for $|E_{tot}|^2$ in the $r - \theta$ plane of the MC at $z = 15$ cm with $n_o = 1.1 \times 10^{11}$ cm$^{-3}$. (Reprinted with permission from I. Dey and S. Bhattacharjee. 2011. Penetration and screening of perpendicularly launched electromagnetic waves through bounded supercritical plasma confined in multicusp magnetic field, *Phys. Plasmas* **18**: 22101. Copyright 2011, American Institute of Physics.)

the cusped magnetic field profile is clearly manifested in the resonance regions ($r \sim$ 22–24 mm) and closely resembles bright cusp-like regions that are observed visually in the MC cross section (cf. Figure 2b of Ref. [12]) due to enhanced ionization in these regions. At the center (0–12 mm), a distorted dipole-like feature is visible, which resembles the experimental data in Figure 13.5b. The average rotation ($\sim 20°$) compared to the vacuum case is shown by an arrow in the central region.

Since $\vec{E}_{tot}$ incorporates the polar term exp $i(m\theta)$, the polar variation will arise as $\cos^2(m\theta)$ in $|E_{tot}|^2$, which gives the primary dipole pattern. The rotation of the polarization axis is due to the orientation-dependent interaction of the $E_r$ and $E_\theta$, with $B_0(r, \theta)$ in the presence of $N_i(r)$, leading to the phenomenon of birefringence, as can be seen from Equations 13.5a and 13.5b. The distortion of the dipole is due to the derivatives of the tensor components $K_{ij}$ with respect to $r$ and $\theta$, which are incorporated in $E_r$ and $E_\theta$ by solution of equation set 13.6 and subsequent evaluation from Equations 13.5a and 13.5b. The distortion becomes pronounced for $r > 20$ mm since the tensor components are highly nonuniform due to the radially varying magnetoplasma and the presence of resonance and damping regions close to each other. Detailed analysis will be presented in a future article dedicated to the phenomenon of polarization axis rotation.

Since the experimental results have been taken from the 270° direction, the wave solutions are compared at $\theta = 270°$. The electric field intensity $|E_{tot}|^2$ in arbitrary units is obtained for comparison with experimental results for different plasma density profiles. The comparison of the radial variation of the profiles rather than the actual magnitudes, which are in arbitrary units, is of interest. From Figure 13.4a, it is noted that the numerically calculated radial $|E_{tot}|^2$ profile in the TE$_{11}$ mode shown for two different conditions of central plasma density ($1.1 \times 10^{11}$ and $2.1 \times 10^{11}$ cm$^{-3}$), as obtained from the model, has a qualitative agreement with the experimental plots of $|E_{tot}|^2$. The distinctive features of the existence of the wave fields in the central plasma region, the minimum at $r \sim 28$ mm, and the propagation at the periphery are reproduced reasonably well. The resonance-like peaks that occur at $r \sim 22$–24 mm are much larger than the small peak-like features obtained experimentally. The discrepancy in the relative magnitudes of resonance peaks between the experiment and the model, as observed in Figure 13.4a, may be due to different scales of $y$ axis (experiment [dBm] and model [AU]) used in the plot. Also, it is expected that the absorption of wave energy by the plasma particles at resonance [7] can contribute to the difference. The electron temperature ($T_e$) is seen to have a peak in that region, which matches quite well with the peak of $|E_{tot}|^2$ (Figure 13.4b). The wave energy absorption by the plasma has not been taken into account, since we use the cold plasma model, which does not consider the effect of the electron kinetics. The assumption of cold plasma is valid for most of the plasma region, where electron velocity $v_e \ll$ wave velocity $v_{ph}$. However, at resonance, the electrons gain high kinetic energies and $v_e$ becomes comparable to the $v_{ph}$ and may undergo damping [7,26]. The discrepancy between the

wider minima obtained experimentally (Figure 13.4a) compared to the numerical solution may also be attributed to the effect of electron kinetics.

The variation of $|E_{tot}|^2$ in vacuum for the $TE_{11}$ mode, obtained by solv-
ing Maxwell's equation in the MC with $K = 1$, is shown in Figure 13.4b for
comparison. In Figure 13.5a, the radial variation of $|B_z|^2$ from the model is

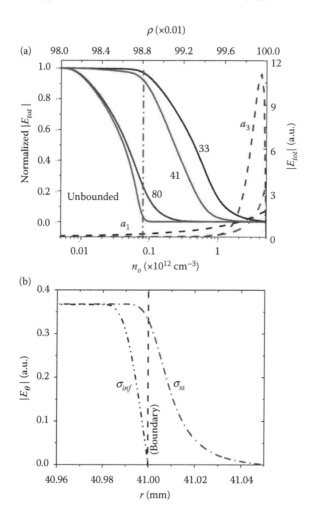

**FIGURE 13.7**
(a) Variation of the normalized $|E_{tot}|$ (model) with $n_o$ for three different radii, namely, 33, 41,
and 80 mm, and the unbounded case (solid lines). Theoretical variation of $|E_{tot}|$ with the nor-
malized radius $\rho$ near the boundary for bounding radius of $a_1 = 205$ mm, $a_2 = 410$ mm, and
$a_3 = 4100$ mm (dashed lines). (b) Behavior of $|E_\theta|$ (model) with infinite ($\sigma_{inf}$) and finite ($\sigma_{SS}$)
conductivities at the boundary. (Reprinted with permission from I. Dey and S. Bhattacharjee.
2011. Penetration and screening of perpendicularly launched electromagnetic waves through
bounded supercritical plasma confined in multicusp magnetic field, *Phys. Plasmas* **18**: 22101.
Copyright 2011, American Institute of Physics.)

superimposed on the experimental data. The theoretical peaks at $r \sim 28$ mm are not very well resolved in the experimental data, but discernable humps are observed. The reason for better matching of the $|B_z|^2$ data with the numerical results compared to the $|E_{tot}|^2$ data may be attributed to the fact that the mutual interaction of the wave electric field with the plasma electrons is much more than the wave magnetic field, thereby affecting the features of $|E_{tot}|^2$.

In Figure 13.7a, the variation of the normalized $|E_{tot}|^2$ (model) at the center of MC is plotted against the central density $n_o$ to understand the effect of plasma on wave field screening. It is observed that with an increase in $n_o$, the wave field penetration at the center decreases exponentially. The decay rate of the field amplitude becomes faster for larger radii. For a density of $8 \times 10^{10}$ cm$^{-3}$ (vertical dashed-dotted line in Figure 13.7a) just above $N_c$ ($7.44 \times 10^{10}$ cm$^{-3}$), $|E_{tot}|$ decreases by $\sim 30$ times as the radius goes from 33 mm to the *unbounded* case. We found that for the bounded case only at densities $n_o \sim 10^{13}$ cm$^{-3}$, the wave field is 99% screened and then most of the wave propagation happens through the peripheral plasma. Figure 13.7a also demonstrates the effect of the bounding radius on peripheral wave propagation. The theoretical $|E_{tot}|$ fields with $n_o = 1.1 \times 10^{11}$cm$^{-3}$ are plotted with respect to a normalized radius $\rho = r/a$ for three bounding radii, $a_1 = 205$ mm, $a_2 = 410$ mm, and $a_3 = 4100$ mm, that is, 5, 10, and 100 times larger than the radius $a = 41$ mm used in the experiments. It is clearly seen that for very large radii, the wave propagation primarily takes place through the peripheral region and the field in the central region is nearly zero, resembling closely the infinite boundary case.

The boundary behavior of the wave fields is further investigated in Figure 13.7b, where the effect of the boundary for finite conductivity ($\sigma_{SS}$) is compared to the case of infinite conductivity ($\sigma_{inf}$) for $n_o = 1.1 \times 10^{11}$ cm$^{-3}$. It is noted that the $|E_\theta|$ falls off at the conducting boundary within a very small distance of $\sim 50$ μm, whereas for the infinite boundary case, the drop is sharp and at the boundary as expected.

## Discussion and Summary

The wave field penetration in bounded supercritical plasma in a minimum-B field is attributed to the inability of the central plasma column in effectively screening the wave field modes, which are sustained by reflection of the waves from the conducting boundary. The wave modes sustained at the center are not evanescent since the length of the plasma traversed by the reflected waves is smaller than the corresponding e-folding length for infinite plasma (cf. Figure 13.7a) and the waveguide mode modified by the plasma dominates. The maxima and minima along the axial direction are due to the superposition and phase mixing of the wave fields on reflection

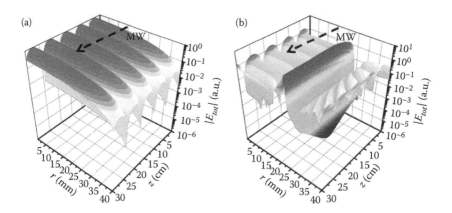

**FIGURE 13.8**
Numerically generated variation of $|E_{tot}|$ with $r$ and $z$ in the multicusp for (a) vacuum and (b) plasma with $n_o = 1.1 \times 10^{11}$ cm$^{-3}$. (Reprinted with permission from I. Dey and S. Bhattacharjee. 2011. Penetration and screening of perpendicularly launched electromagnetic waves through bounded supercritical plasma confined in multicusp magnetic field, *Phys. Plasmas* **18**: 22101. Copyright 2011, American Institute of Physics.)

from the conducting boundary and the MC terminals due to sudden geometrical transition [12,27]. This is demonstrated by the three-dimensional plots in Figure 13.8a and b, where a variation of $|E_{tot}|$ along the radial and axial directions has been shown both in vacuum and in the presence of plasma ($n_o = 1.1 \times 10^{11}$ cm$^{-3}$) in MC. In the vacuum case, it is observed that the general feature of the TE$_{11}$ mode is obtained [24].

In most of the work related to wave plasma interactions [1–12], the measurement of wave number [28] is considered to be important to describe wave propagations. However, in the current experiment, the radial inhomogeneity of the magnetoplasma medium causes the wave number to be a function of $r$ and $\theta$, and hence must be measured along multiple directions in the plasma, which is difficult in a compact source. As the axial plasma is approximately uniform leaving the edges, an effective wave number along the central axis may be estimated by performing a fast Fourier transform (FFT) of a $|E_{tot}|^2$ versus $z$ standing wave data similar to the one obtained in Ref. [12] and can be compared with the model. From the FFT for the 180 W, 0.50 mTorr data, $kz$ (expt) ~ 0.1501 cm$^{-1}$ and $kz$ (model) ~ 0.1484 cm$^{-1}$, which is similar to the results obtained in Ref. 12. Similarly, an FFT of the $|E_{tot}|^2$ versus $r$ data gives the effective radial wave number ($k_r$). The effective $k_r$ from experiment and model are 0.2451 and 0.2507 cm$^{-1}$, respectively, at 180 W, 0.50 mTorr (Figure 13.3a). The above wave number comparisons show reasonably good agreement between the experimental results and model.

It may be noted that the radial profile of the $|E_{tot}|^2$ and $|B_z|^2$ (Figures 13.4a and 13.5a) is maintained approximately similar over the axial distance, but their magnitudes vary due to the axial standing wave profile and rotation of polarization axis. Therefore, in the presence of plasma confined in the minimum $B_o$ field, the vacuum wave mode is modified by the inhomogeneous and anisotropic dielectric permittivity $K$. The wave polarization is predominantly $TE_{11}$ and undergoes rotation.

The peaks observed in the wave field intensity and electron temperature profile may be attributed to an upper hybrid-like resonance mechanism. Since the plasma is bounded, the plane wave expression for the UHR condition, namely, $\alpha_p^2 + \beta_c^2 = 1$, is modified by equating the denominator of Equation 13.5a or Equation 13.5b to zero

$$\alpha_p^2 \left( \beta_{cr}^2 - \beta_{c\theta}^2 \right) + \left( k_c / k_o \right)^2 U \left( U^2 - \beta_c^2 \right)$$

$$- \left( k_o / k_c \right)^2 \left[ \left( \alpha_p^2 \beta_{cr} \beta_{c\theta} \right)^2 / \left\{ U \left( U^2 - \beta_c^2 \right) \right\} \right] = 0, \tag{13.7}$$

where $\beta_c^2 = \beta_{cr}^2 + \beta_{c\theta}^2$. For $\theta = 270°$ and Equation 13.7 reduces to $(\alpha_p \beta_c k_o / k_c)^2 = U(U^2 - \beta_c^2)$ since $\beta_{cr} \rightarrow 0$, which is satisfied at a radial point $r = r_{reso}$, where $0 < r_{reso} < r_{ECR}$. For $n_o = 10^{11}$ cm$^{-3}$, $r = r_{reso}$ mm. From the model, the group velocity $(v_g)$ of the wave in the resonance region is found to be $\sim 1 \times 10^6$ ms$^{-1}$, which is two orders of magnitude smaller than $v_g$ in the central region ($\sim 1 \times 10^8$ ms$^{-1}$) and signifies resonant wave plasma coupling, which is manifested by the $T_e$ peaks obtained experimentally in Figure 13.4b. The results are consistent with the ones obtained by us in Ref. [11].

In summary, the penetration of EM waves through the central supercritical $(\alpha_p^2 \geq 1)$ plasma in a minimum-B field with a conducting boundary is studied experimentally and by numerical modeling using a cold plasma theory. It is observed that (a) there is finite wave field penetration through the central plasma region unexplained by the plane wave theory and (b) wave plasma resonant interaction occurs at a radial location satisfied by a modified UHR relation; (c) damping due to wave energy absorption occurs at the ECR location satisfied by $\omega = \omega_c$; (d) phase mixing and superposition of the wave fields reflected from the metallic boundary lead to the formation of EM-standing waves modes along the central axis of MC; and (e) the wave polarization undergoes a rotation of $\sim 20°$ due to birefringence as it traverses the plasma.

# References

1. T.H. Stix. 1990. Waves in plasmas: Highlights from the past and present, *Phys. Fluids B* **2**: 1729–1743.
2. R.L. Stenzel. 1976. Whistler wave propagation in a large magnetoplasma, *Phys. Fluids* **19**: 857–864.

3. R.W. Boswell. 1984. Very efficient plasma generation by whistler waves near the lower hybrid frequency, *Plasma Phys. Controlled Fusion* **26**: 1147–1162.

4. M. Sugimoto, M. Tanaka, and Y. Kawai. 1996. Electron cyclotron wave plasma production using a concave lens, *Jpn. J. Appl. Phys. Part 1* **35**: 2803–2807.

5. O. Sauter, M.A. Henderson, F. Hofmann, T. Goodman, S. Alberti, C. Angioni, K. Appert et al. 2000. Steady-state fully noninductive current driven by electron cyclotron waves in a magnetically confined plasma, *Phys. Rev. Lett.* **84**: 3322–3325.

6. H. Sugai. 1981. Mode conversion and local heating below the second electron cyclotron harmonic, *Phys. Rev. Lett.* **47**: 1899–1902.

7. K. Rypdal, A. Fredriksen, O.M. Olsen, and K.G. Hellblom. 1997. Microwave-plasma in a simple magnetized torus, *Phys. Plasmas* **4**: 1468–1480.

8. V.K. Yadav and D. Bora. 2004. Electron Bernstein wave generation in a linear plasma system, *Phys. Plasmas* **11**: 4582–4588.

9. M. Maeda and H. Amemiya. 1994. Electron cyclotron resonance plasma in multicusp magnets with a checkered pattern, *Jpn. J. Appl. Phys. Part 1* **33**: 5032–5037.

10. S. Bhattacharjee and H. Amemiya. 1999. Production of microwave plasma in narrow cross-sectional tubes; effect of the shape of cross section, *Rev. Sci. Instrum.* **70**: 3332–3337.

11. J.V. Mathew, I. Dey, and S. Bhattacharjee. 2007. Quasi steady state interpulse plasmas, *Appl. Phys. Lett.* **91**: 041503–3.

12. I. Dey and S. Bhattacharjee. 2008. Experimental investigation of standing wave interaction with a magnetized plasma in a minimum B field, *Phys. Plasmas* **15**: 123502–7.

13. J.V. Mathew, A. Chowdhury, and S. Bhattacharjee. 2008. Subcutoff microwave driven plasma ion sources for multi elemental focused ion beam systems, *Rev. Sci. Instrum.* **79**: 063504–5.

14. J. Yang, Y. Xu, Z. Meng, and T. Yang. 2008. Effect of applied magnetic field on a microwave plasma thruster, *Phys. Plasmas* **15**: 023503–7.

15. M.I. Bakunov, A.M. Bystrov, and V.B. Gildenburg. 2002. Frequency self-upshifting of intense microwave radiation producing ionization in a thin gaseous layer, *Phys. Plasmas* **9**: 2803–2811.

16. H. Ito, C. Rajyaguru, N. Yugami, and Y. Nishida. 2004. Propagation characteristics and guiding of a high-power microwave in plasma waveguide, *Phys. Rev. E* **69**: 066406–5.

17. J.L. Warren, G.P. Boicourt, M.T. Menzel, G.W. Rodenz, and M.C. Vasquez. 1985. Revision of and documentation for the standard version of the Poisson Group Codes, *IEEE Trans. Nucl. Sci.* **32**: 2870–2872.

18. S.R. Douglas, C. Eddy Jr., and B.V. Weber. 1996. Faraday rotation of microwave fields in an electron cyclotron resonance plasma, *IEEE Trans. Plasma Sci.* **24**: 16–17.

19. F.M. Aghamir and M. Abbas-nejad. 2007. Electron temperature effects on the eigenmodes of a plasma waveguide, *Phys. Plasmas* **14**: 062110–7.

20. W.P. Allis, S.J. Buchsbaum, and A. Bers. 1963. *Waves in Anisotropic Plasmas* (MIT Press, Cambridge, MA) pp. 133, 179.

21. K. Takahashi, T. Kaneko, and R. Hatakeyama. 2005. Effects of polarization reversal on localized-absorption characteristics of electron cyclotron wave in bounded plasmas, *Phys. Plasmas* **12**: 102107–7.

22. F. K. Azadboni, M. Sedaghatizade, and K. Sepanloo. 2010. Design Studies of a Multicusp Ion Source with FEMLAB Simulation, *J. Fusion Energy* **29**: 5–12.
23. D.G. Swanson. 2003. *Plasma Waves* (2nd edition, IOP, Bristol) p. 230.
24. D.M. Pozar. 2006. *Microwave Engineering* (3rd edition, Wiley, Singapore), p. 208.
25. J.H. Mathews and K.D. Finks. 2005. *Numerical Methods Using MATLAB* (4th edition, Pearson Education, Singapore) p. 546.
26. T. Maekawa, S. Tanaka, Y. Terumichi, and Y. Hamada. 1978. Wave trajectory and electron-cyclotron heating in toroidal plasmas, *Phys. Rev. Lett.* **40**: 1379–1383.
27. S. Bhattacharjee and H. Amemiya. 2000. Production of pulsed microwave plasma in a tube with a radius below the cutoff value, *J. Phys. D: Appl. Phys.* **33**: 1104–1116.
28. G.S. Eom, G.C. Kwon, I.D. Bae, G. Cho, and W. Choe. 2001. Heterodyne wave number measurement using a double B-dot probe, *Rev. Sci. Instrum.* **72**: 410–412.
29. I. Dey and S. Bhattacharjee. 2011. Penetration and screening of perpendicularly launched electromagnetic waves through bounded supercritical plasma confined in multicusp magnetic field, *Phys. Plasmas* **18**: 22101.

# 14

## Wave Birefringence in Perpendicular Propagation

### Introduction

In the previous chapter, the experimental observation of a dipole-like pattern in the polar variation of the wave electric field intensity in the central plasma region was reported. The observation provided a major support for the development of the wave–plasma interaction model in the bounded, inhomogeneous magnetoplasma in the multicusp (MC). Rotation of the polarization axis with respect to vacuum case and distortion of the dipole pattern was observed in the presence of plasma. In this chapter, the polar wave electric field measurements are carried out in detail and compared to the wave–plasma interaction model developed in Chapter 13.

When a polarized electromagnetic (EM) wave interacts with a magneto-optic medium, a rotation in the axis of polarization of the incident wave is observed due to birefringence [1,2]. For waves launched with $k \parallel B_o$, where $k$ is the wave vector and $B_o$ is the applied static magnetic field, the Faraday effect [3,4] is observed, which arises due to the difference in the refractive index of the right and the left circularly polarized waves. For $k \perp B_o$ propagation, the Cotton–Mouton effect [3,5] is observed, which is due to the difference in the refractive index of the ordinary and extraordinary waves. An observable consequence of both these effects is a rotation in the axis of polarization. A plasma permeated by a magnetostatic field behaves similarly to a magneto-optic medium, and causes such a rotation [3–10]. In the case of plasma, the rotation is due to the magnetostatically induced anisotropic response of the plasma electrons to the wave electric field, and depends primarily on the electron density, the $B_o$, and the distance traversed by the waves in the plasma. Knowledge of the polarization state of an EM wave in a plasma is of prime importance and has been utilized as a powerful tool for studying various plasmas ranging from laboratory to the space [6–10]. However, the question of how the wave polarization state is affected during its passage through a plasma confined in a nonuniform magnetic field such as the one generated in an MC with a conducting boundary has not been addressed earlier.

The details of the experimental setup have been given earlier in Chapters 4 and 5. A brief schematic is shown in Figure 14.1a, where the polar wave electric field measurement scheme is demonstrated. The octupole field lines [11–15] in the MC cross section, along with the wave electric field lines in the vacuum $TE_{11}$ mode, are shown in Figure 14.1b. The experiments are carried out with an input power range $P_{in} = 180$–$540$ W and neutral pressure $p = 0.20$–$0.50$ mTorr. The measurement of the polar variation of wave electric field intensity in the MC cross section is performed by two "L-shaped" bent antenna probes (BAP), constructed from microwave-grade coaxial cables, one with an exposed tip of 12 mm for average integrated measurement over the central region (region C in Figure 14.1b) and another of radius 23 mm with an exposed tip of 2 mm for average integrated measurement in the annular ring (region C in Figure 14.1b). The probes are inserted axially into the MC (Figure 14.1a), opposite the microwave input side for measurement of polar wave electric field variation along the axis of MC. Figure 14.1b also shows the polar angle convention followed, as viewed from the probe entry side of the MC.

Plasma density measurement in the MC cross section is carried out by three bent Langmuir probes inserted from the axial direction and rotated about the axis. The three probes scan an annular region of 4 mm (planar tip diameter) at radii of 5, 10, and 15 mm from the MC axis, covering the central minimally magnetized region C. Measurements at $r = 0$ are obtained from a straight Langmuir probe. Polar Langmuir probe measurement is not carried out at higher radii since the magnetic field becomes stronger at larger radii, which affects the measurements because the probe plane is not always maintained perpendicular to the cusped magnetostatic field lines while in rotation.

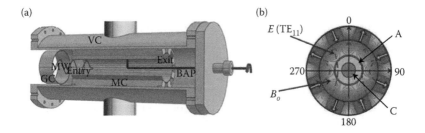

**FIGURE 14.1**

(a) Three-dimensional schematic of the setup. The orientation of the bent antenna probe (BAP) is indicated. (b) Schematic of the exit side of the MC showing the magnetostatic flux lines ($B_o$), $TE_{11}$ electric field lines, and the angular scheme followed. The central region (C) and the annular region (A) are shaded. Magnetization direction of the permanent magnets is indicated by broad arrows. (Reprinted with permission from I. Dey and S. Bhattacharjee. 2011. Anisotropy induced wave birefringence in bounded supercritical plasma confined in a multicusp magnetic field, *Appl. Phys. Lett.* **98**: 151501. Copyright 2011, American Institute of Physics.)

## Polar Plasma Characteristics

At 180 W, 0.50 mTorr: $N_{i\ avg} = (1.4 \pm 0.10) \times 10^{11}$ cm$^{-3}$ and at 360 W, 0.35 mTorr: $N_{i\ avg} = (1.9 \pm 0.14) \times 10^{11}$ cm$^{-3}$ are the average $N_i$ over the 15 mm radius of plasma cross section. Thus, there are small variations of $N_i$ over the central region, but they are uniform on an average, with a deviation of about 7% of the average value. The variations may be attributed to the small asymmetry in the minimum-$B$ magnetic field since the confining magnets are not exactly identical. Asymmetric interaction of the penetrating wave electric field with the central plasma may also give rise to the observed local "hotspots." From the radial plots of plasma density (cf. Figure 14.2a) [15], it may be noted that the plasma is supercritical ($N_i > N_c \sim 0.74 \times 10^{17}$ m$^{-3}$), up to $r \sim 24$ mm of the MC, and is uniform over the central region C. From radial variation of $|B_o|$ (Figure. 14.2a) from the center to the magnet gap ($\theta = 270°$), it is seen that near the central region (0–12 mm), $B_o|_{avg}(\sim 0.002$ T) is quite small and $\beta_{ce}^2 = (\omega_{ce}/\omega)^2 \sim 10^{-4}$ (where $\omega_{ce}$ is the electron–cyclotron frequency).

## Polar Wave Field Intensity Measurements

The measurement of the remnant wave electric field intensity $|E_{tot}|^2$ (Figure 14.3a), carried out in the previous chapter, reveals that there is finite wave penetration in the central plasma region, which has been attributed to inadequate screening of the reflected wave fields from the conducting boundary.

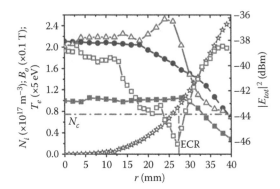

**FIGURE 14.2**
$N_i$ versus $r$ at 180 W, 0.50 mTorr (solid square); 360 W, 0.35 mTorr (solid circle). $|E_{tot}|^2$ versus $r$ at 180 W, 0.50 mTorr (open square). $T_e$ versus $r$ at 180 W, 0.50 mTorr (open triangle). $B_o(r)$ is shown by stars. (Reprinted with permission from I. Dey and S. Bhattacharjee. 2011. Anisotropy induced wave birefringence in bounded supercritical plasma confined in a multicusp magnetic field, *Appl. Phys. Lett.* **98**: 151501. Copyright 2011, American Institute of Physics.)

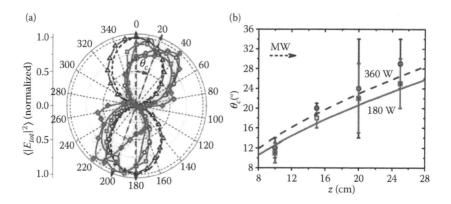

**FIGURE 14.3**

(a) Experimental variation of normalized $\langle|E_{tot}|^2\rangle$ versus $\theta$ (i) at the center, under vacuum (open triangle), and in plasma at 180 W, 0.50 mTorr (open square); 360 W, 0.35 mTorr (open circle) and (ii) at $r \sim 23$ mm with 180 W, 0.50 mTorr (diamond). Theoretical $\cos^2(\theta + \theta_c)$ fitting for the vacuum (dashed line) and numerical plot for plasma with $n_o = 1.0 \times 10^{11}$ cm$^{-3}$ (solid line) is shown. Dashed arrows indicate rotation. (b) Experimental variation of polarization angle $\theta_c$ versus $z$ at 180 W, 0.50 mTorr (open square); 360 W, 0.35 mTorr (open circle). Theoretical axial variation of average $\theta_c$ for $n_o = 1.0 \times 10^{11}$ cm$^{-3}$ (solid line) and $2.0 \times 10^{11}$ cm$^{-3}$ (dashed line) from the model. (Reprinted with permission from I. Dey and S. Bhattacharjee. 2011. Anisotropy induced wave birefringence in bounded supercritical plasma confined in a multicusp magnetic field, *Appl. Phys. Lett.* **98**: 151501. Copyright 2011, American Institute of Physics.)

Additionally, a small discernable peak indicating wave–plasma resonance at $r \sim 24$ mm (180 W), and minima at $r \sim 28$ mm (ECR region), is observed. The radial variation of electron temperature ($T_e$) (Figure 14.3b) shows a distinct peak corresponding to the local plasma resonance at $r \sim 24$ mm [15]. Since the plasma medium is anisotropic, it is expected to affect the polarization state of the wave due to birefringence.

To study the effect of the plasma on the polarization state of the wave, a measurement of the variation of average integrated electric field intensity $\langle|E_{tot}|^2\rangle$ in the central region (C in Figure 14.1b) and in the annular ring (A in Figure 14.1b) with the polar angle $\theta$ in the MC cross section is performed. It may be noted that the plasma density is uniform on an average in the central region (Figure 14.2a and b). The L-shaped antennas are inserted axially into the MC and rotated in steps of 10° with an accuracy of ±1°, and the $\langle|E_{tot}|^2\rangle$ as a function of $\theta$ is recorded both in the presence and absence of plasma at various axial positions. The probes are scanned both clockwise and counterclockwise and the average of the two measurements is taken.

The measured $\langle|E_{tot}|^2\rangle$ normalized with its maximum value versus $\theta$ at the central region (C) of the MC for the vacuum, and two different plasma conditions (180 W, 0.50 mTorr and 360 W, 0.35 mTorr) are shown in Figure 14.3a. The polar variation of normalized $\langle|E_{tot}|^2\rangle$ in the 2 mm annular region (A) at $r = 23$ mm for microwave power of 180 W and pressure of 0.50 mTorr is shown in Figure 14.3a.

In vacuum, the dipole-like profile expected from a $TE_{11}$-mode electric field distribution (Figure 14.1b) is obtained at the central region. A theoretical fitting of the form $\cos^2\{m(\theta + \theta_c)\}$, where $\theta_c$ is the polarization axis rotation angle with respect to the central maximum of the dipole pattern in vacuum, $m = 1$ for $TE_{11}$ mode [15] is superimposed as dashed lines on the vacuum polar plot and shows a very good agreement, with $\theta_c \pm \Delta\theta_c = (0.0 \pm 0.6)°$. Here, $\Delta\theta_c$ is the fitting error and is a measure of the mismatch in perfect fitting with an ideal dipole profile. In the presence of plasma, two major effects are observed in the nature of the polar plots: (i) there is a rotation in $\theta_c$ for both the plasma conditions, $\theta_c \sim 18°(180 \text{ W})$ and $20°(360 \text{ W})$ in region C, with respect to the vacuum condition, and (ii) the shapes of the dipoles are distorted ($\Delta\theta_c \sim \pm 1.5°$) with $\sim 4\%$ asymmetry between the peaks of the top and the bottom lobes, and differ from the $\cos^2(\theta - \theta_c)$ profile. In the annular region, the polarization angle $\theta_c \sim 40°$ for 180 W, 0.50 mTorr, and the distortion is much more pronounced, evidently due to the proximity to the wave–plasma resonance region at $r \sim 24 \text{ mm}$ (Figure 14.3a and b). The asymmetry is attributed to the small fluctuations in the plasma density during the time taken to acquire the profile ($\sim 10 \text{ min}$). The distortion is attributed to the effect of wave birefringence due to the inhomogeneous anisotropic plasma having resonance regions in its cross section, and is discussed in the section "Modeling of the Wave Birefringence."

Figure 14.3b shows the axial variation of the polarization angle ($\theta_c$) in the central region C, for the two plasma conditions of Figure 14.2. It is observed that $\theta_c$ increases with the distance $\Delta z$ traversed in the plasma, which is expected from a medium exhibiting birefringence. The error bars indicate $\Delta\theta_c$. The error at $z \sim 20 \text{ cm}$ may be attributed to a large distortion of the profile due to the fact that the EM–SW pattern has a minimum in that region [12].

## Modeling of the Wave Birefringence

The experimental observations discussed in Chapter 5 and this chapter suggest that the predominant wave mode present in the plasma-filled MC is the $TE_{11}$ mode. The highlights of the computation are that Maxwell's equations are solved numerically with the magnetized plasma in the MC as a dispersive medium characterized by a radially inhomogeneous and anisotropic permittivity, enclosed by a conducting boundary. The spatial derivates of the density and the magnetic field are taken into account. The static cusped magnetic field can be resolved into $r$ and $\theta$ components, and for an octupole they are given by $B_{o\theta} = -B_p(/a)^3 \cos(4\theta)$ [15], defined with respect to the polar angle scheme demonstrated in Figure 14.1b. Here, $B_p$ is the magnetic field strength at the pole of the magnet, and $(B_p/a^3) = 0.048 \text{ G/mm}^3$. Different plasma conditions are incorporated by providing the azimuthally symmetric average central plasma density ($n_0$) (Figure 14.2a and b) with the experimental radial variation as input. The solution procedure and evaluation of the total electric field intensity ($|E_{tot}|^2$) are detailed in Ref. 15.

From Equations 13.5a and 13.5b in Chapter 13, the $r$ and $\theta$ components for the wave electric fields can be written as

$$E_{rc} = \left(\frac{i\omega}{D_{00}}\right)\left[\frac{D_{22}}{r}\left(i + \frac{\partial}{\partial\theta}\right) + D_{12}\frac{\partial}{\partial r}\right]B_{zc},$$

(14.1a)

and

$$E_{\theta c} = \left(\frac{-i\omega}{D_{00}}\right)\left[\frac{D_{21}}{r}\left(i + \frac{\partial}{\partial\theta}\right) + D_{11}\frac{\partial}{\partial r}\right]B_{zc},$$

(14.1b)

where $D_{ii} = k_0^2 K_{ii} - k_z^2$ (with $i \rightarrow 1, 2, 3$), $D_{ij} = k_0^2 K_{ij}$ (with $i \neq j$, $i, j \rightarrow 1, 2, 3$), and $D_{00} = D_{11}D_{22} - D_{12}^2$. Here, $k_z^2 = k_0^2 K_{22} - k_c^2$ is the propagation vector component along $z$-direction, free-space wave vector $k_0 = 2\pi/\lambda_0$, and waveguide cutoff wave vector $k_c = 1.841/a$ in $TE_{11}$ mode, with $a$ being the MC radius (= 41 mm). The dielectric tensor components $K_{ij}$ (cf. Equation 13.2) are given in the "Modeling" section of Chapter 13 [15].

The differential Equations 13.6a and 13.6b are solved to obtain the axial wave magnetic field amplitude $B_{zc}$, subject to the boundary and magnetoplasma conditions in the MC using a finite difference method MATLAB® code (cf. "Modeling" section of Chapter 13). The total electric field $\vec{E}_{tot}(r,\theta,z,t) = \tilde{E}_{rtot}(r,\theta,z,t)\hat{r} + \tilde{E}_{\theta tot}(r,\theta,z,t)\hat{\theta}$ is obtained as discussed in Chapter 13, from which the field intensity $|E_{tot}|^2$ can be obtained. Here, $\tilde{E}_{rtot} = \tilde{E}_{rfwd} + \tilde{E}_{rref}$ and $\tilde{E}_{\theta tot} = \tilde{E}_{\theta fwd} + \tilde{E}_{\theta ref}$ are the total radial and polar standing wave components obtained by superposition of forward and reflected wave electric fields in the MC (cf. Equations 13.7a and 13.7b) [15].

The polarization rotation angle $\theta_c$ can be expressed as [16]

$$\theta_c = \tan^{-1}\left[\left\{Im(E_{rtot})/Im(E_{\theta tot})\right\}\left(2\pi \cdot \Delta z/\lambda_g\right)\right] (\text{radians})$$

(14.2)

where $\tilde{E}_{rtot} = E_{rtot}\exp(-\omega t)$ and $\tilde{E}_{\theta tot}\exp(-\omega t)$ as given above, $\Delta z$ is the axial distance traversed in the MC, and $\lambda_g = 2\pi/\sqrt{(k_0^2 - k_c^2)}$ is the guided wavelength in the MC. From Equations 14.1a and 14.1b, it can be seen that the $E_{rtot}$ and $E_{\theta tot}$ will have different dispersion at various $r$ and $\theta$. In vacuum, $E_{rtot}$ and $E_{\theta tot}$ are at quadrature [16] and $Im(E_{rtot}) = 0$, hence $\theta_c = 0$. In the presence of plasma permeated by $B_o$, $Im(E_{rtot})$ is finite, and $\theta_c$ varies with distance and the radial cross section. Toward the center ($r < 22$ mm), where the polar component of the magnetostatic field $B_{o\theta}$ component is stronger, the radial wave electric field component, which is perpendicular to it, interacts more with the magnetoplasma and has a higher degree of penetration. To compare with the experimental results obtained in the central region, the calculated value of $\theta_c$ with $n_o = 1.0 \times 10^{17}$ m$^{-3}$ (180 W) and $2.0 \times 10^{17}$ m$^{-3}$ (360 W) is superimposed on the experimental data points for the two plasma conditions in Figure 14.3b, and a good agreement is obtained. In the calculation, the plasma has been assumed to be axially uniform and to be present between $z = 0$ cm and $z = 30$ cm. The deviation between the theory and the experiments is within error limits and may be attributed to the small nonuniformity in the axial plasma variation and effect of end plugging [11–14].

In Figure 14.4, the numerically calculated, normalized $|E_{tot}|^2$ (AU) is plotted as a function of both $r$ and $\theta$ in a contour-polar plot for the vacuum (Figure 14.4a) and the plasma case corresponding to the central density $n_o \sim 1.0 \times 10^{11}$ cm$^{-3}$ (Figure 14.4b), as seen from the exit side of the MC. The color map is the same for the two plots to enable comparison of relative magnitudes. In Figure 14.4a, the vacuum TE$_{11}$ field variation is clearly realized, with an arrow in the central region showing the polarization angle. In the presence of the plasma (Figure 14.4b), the electric field intensity profile is modified. The general features of the numerical plot have been discussed in Chapter 13. Major features like the wave field penetration (Figure 14.4b) at the center (P), damping at the ECR (D), and wave–plasma resonance regions (R) are clearly realized. At the center, the dipole-like feature is visible, which is quite distorted compared to the vacuum. The average polarization axis rotation from the vacuum case is shown by arrows in the central region (12 mm radius). The numerically obtained $\langle|E_{tot}|^2\rangle$ versus $\theta$ for the central region C

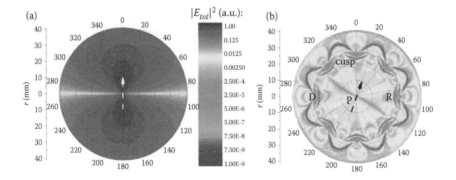

**FIGURE 14.4**
Numerically generated contour-polar plot of $|E_{tot}|^2$ versus $r - \theta$ in (a) vacuum and (b) plasma with $n_o = 1.0 \times 10^{11}$ cm$^{-3}$. The average polarization angle rotation at the center is shown by dashed arrows. Penetration (P), damping (D), and resonance (R) regions are marked. (Reprinted with permission from I. Dey and S. Bhattacharjee. 2011. Anisotropy induced wave birefringence in bounded supercritical plasma confined in a multicusp magnetic field, *Appl. Phys. Lett.* **98**: 151501. Copyright 2011, American Institute of Physics.)

is superimposed on the experimental data at 180 W ($n_o = 1.0 \times 10^{17}$ m$^{-3}$) and shows a very good agreement, clearly demonstrating the distortion of the dipole feature (Figure 14.3a).

## Discussion and Summary

In order to understand how the bounded plasma and magnetostatic field affect the polarization state, an average value of $\theta_c$ at the central region ($\theta_{c\,avg}$) is evaluated under various conditions as a function of $n_o$. It was seen that in total absence of any magnetostatic field, there is no rotation in the presence of plasma ($\theta_{c\,avg} = 0$). For the octupole configuration used, the experiment, at low $n_o (<10^9$ cm$^{-3}$) gives $\theta_{c\,avg} \sim 10^{-3\circ}$ since $E_r$ and $E_\theta$ both penetrate into the plasma, making their ratio in Equation 14.2 small. As $n_o$ increases, $E_\theta$ decreases more rapidly and hence $\theta_{c\,avg}$ increases. As $n_o > N_c$, the modified UHR-like condition $[(k_o/k_c)^2 \{(\alpha_p^2 \beta_c^2)/(U^2 - \beta_c^2)\} = 1]$ from Equation 13.2 shifts toward the center, thereby enhancing the wave interaction with the magnetoplasma ($E_r$ increases), which results in a rapid increase in the value of $\theta_{c\,avg}$. At still higher densities ($>10^{12}$ cm$^{-3}$), the angle decreases as the effect of the plasma screening becomes dominant over the field penetration.

In summary, rotation of the polarization axis of $|E_{tot}|^2$ penetrating the axial supercritical plasma has been observed experimentally. $\theta_c$ is affected by the plasma density and $B_o$ field, and increases with distance inside the plasma. Solution of Maxwell's equations in the experimental configuration has identified birefringence of the wave electric field components ($E_r$ and $E_\theta$) of the modified TE$_{11}$ profile as the reason for the rotation of polarization axis. Even though the central plasma is minimally magnetized, the presence of $B_o$ field and resonance regions in the cross section causes rotation at the center, since the entire bounded electric field profile is affected. The results generated by the model have reasonably good agreement with the experimental observations.

## References

1. E. Hecht. 2006. *Optics* (4th edition, Pearson Education, New Delhi) pp. 372–376.
2. M. Born and E. Wolf. 1959. *Principles of Optics* (Pergamon, London, New York, Paris, Los Angeles) pp. 700–705.
3. S.E. Segre. 1995. Plasma polarimetry for large Cotton–Mouton and Faraday effects, *Phys. Plasmas* **2**: 2908.
4. J. Howard. 1995. Quadrature polarimetry for plasma Faraday rotation measurements, *Rev. Sci. Instrum.* **66**: 383.

5. V.I.V. Kocharovsky, V.V. Kocharovsky, and V.V. Zheleznyakov. 2002. Microwave polarization diagnostics of solar current sheets with transverse component of magnetic field, *Adv. Space Res.* **29**: 1107.

6. B.W. Reinisch, D.M. Haines, K. Bibl, G. Cheney, I.A. Galkin, X. Huang, S.H. Myers, G.S. Sales, R.F. Benson, and S.F. Fung. 2000. The Radio Plasma Imager investigation on the IMAGE spacecraft, *Space Sci. Rev.* **91**: 319.

7. J.M. Kovac, E.M. Leitch, C. Pryke, J.E. Carlstrom, N.W. Halverson, and W.L. Holzapfel. 2002. Detection of polarization in the cosmic microwave background using DASI, *Nature (London)* **420**: 772.

8. M.A. Van Zeeland, R.L. Boivin, T.N. Carlstrom, and T.M. Deterly. 2008. $CO_2$ laser polarimeter for Faraday rotation measurements in the DIII-D tokamak, *Rev. Sci. Instrum.* **79**: 10E719.

9. K. Nagasaki, A. Isayama, and A. Ejiri. 1995. Application of a grating polarizer to the 106.4 GHz ECH system on Heliotron-E, *Rev. Sci. Instrum.* **66**: 3432.

10. S.R. Douglass, C. Eddy, Jr., and B.V. Weber. 1996. Faraday rotation of microwave fields in an electron cyclotron resonance plasma, *IEEE Trans. Plasma Sci.* **24**: 16.

11. S. Bhattacharjee and H. Amemiya. 1998. Microwave plasma in a multicusp circular waveguide with a dimension below Cutoff, *Jpn. J. Appl. Phys. Part 1* **37**: 5742.

12. I. Dey and S. Bhattacharjee. 2008. Experimental investigation of standing wave interactions with a magnetized plasma in a minimum-B field, *Phys. Plasmas* **15**: 123502.

13. J.V. Mathew, I. Dey, and S. Bhattacharjee. 2007. Microwave guiding and intense plasma generation at subcutoff dimensions for focused ion beams, *Appl. Phys. Lett.* **91**: 041503.

14. J.V. Mathew, A. Chowdhury, and S. Bhattacharjee. 2008. Subcutoff microwave driven plasma ion sources for multielemental focused ion beam systems, *Rev. Sci. Instrum.* **79**: 063504.

15. I. Dey and S. Bhattacharjee. 2011. Penetration and screening of perpendicularly launched electromagnetic waves through bounded supercritical plasma confined in multicusp magnetic field, *Phys. Plasmas* **18**: 022101.

16. W.P. Allis, S.J. Buchsbaum, and A. Bers. 1963. *Waves in Anisotropic Plasmas* (MIT Press, Cambridge, MA) p. 55.

17. I. Dey and S. Bhattacharjee. 2011. Anisotropy induced wave birefringence in bounded supercritical plasma confined in a multicusp magnetic field, *Appl. Phys. Lett.* **98**: 151501.

# 15

## Electron Localization and Trapping Physics Revisited

## Introduction

In recent investigations of wave interaction with intense plasmas sustained in magnetic multicusps (MCs) [1–4], it was observed that electromagnetic standing waves (SW) are realized along the axis of the minimally magnetized ($|B_o| < 1$ gauss) plasma column [2]. The origin of the SW was attributed to the penetration and inadequate screening of the wave fields reflected from the conducting radial boundary [3,4] and from the axial plasma–vacuum interface at the ends [2,3]. During the investigation of the SW phenomenon, measurement of the axial plasma floating potential was performed [2], which indicated that the variation of the potential closely follows the SW pattern, leading to the possibility of localization of plasma electrons in the SW intensity minima (troughs) [2]. However, detailed experimental confirmation of the localization, including investigation of the origin and nature of the interaction leading to axial localization of the electrons in the presence of a predominantly transversely polarized ($TE_{11}$) [3,4] wave fields, has not been carried out earlier. Such a study is useful not only from the basic physics point of view but also from the viewpoint of emerging applications like wave plasma-based focused ion beams [5,6], plasma thrusters for space propulsion [7,8], and so on, many of which employ compact MC [9–11] for generation, sustenance, and confinement of wave-induced plasmas.

The phenomenon of plasma electron localization or trapping has been earlier investigated in large-amplitude electrostatic waves (~100 V/cm) sustained in uniform subcritical plasma columns [12–14]. Wakeren et al. studied the trapping of beam electrons in a potential well of electrostatic waves resulting from beam–plasma interaction [12]. Franklin et al. investigated parametric coupling of noise-induced sideband frequencies with oscillating plasma electrons trapped in large-amplitude electrostatic waves [13]. In a recent work, Danielson et al. observed linear Landau damping and nonlinear wave-particle trapping oscillations with standing plasma waves in a trapped pure electron plasma [14]. However, the possibility of plasma electron localization in the intensity minimum of *electromagnetic standing waves*, which is different from trapping in electrostatic waves of earlier works [12–14], has not been explored. The current experiment investigates the phenomena through measurements of the plasma parameters, wave electric field intensity, and optical intensity of lines emitted from the plasma along the axial (z) direction where electron localization is observed. The spectral broadening of the injected wave is investigated using electromagnetic probes and a high resolution spectrum analyzer (SA) using frequency domain analysis. The phenomenon is verified using Monte Carlo simulations, where the temporal evolution of the plasma to steady state in the MC geometry is followed in space and time and many of the results observed experimentally are confirmed.

---

## Experimental Setup

A schematic of the experimental setup is shown in Figure 15.1a, where argon plasma is generated by continuous-mode microwaves (MWs) of 2.45 GHz in an octupole MC of radius $a = 41$ mm and length $L = 30$ cm [2–4]. The octupole is composed of alternating linear arrays of NdFeB permanent magnet blocks of surface magnetic field $B_p \sim 0.4$ T. The resultant field lines in the MC cross section are shown in Figure 15.1c using the Poisson simulation [15], where the minimum-$B$ structure is clearly demonstrated. The minimum-$B$ magnetic field structure helps in radial confinement of the plasma and the MC is end-plugged for axial confinement [11–16]. The superposed-dashed lines in Figure 15.1c indicate the wave electric field. MWs are launched directly into the MC [9–11], and the wave–plasma interaction occurs predominantly in the $k \perp B_o$ mode, where $k$ is the wave vector and $B_o$ is the radially varying magnetostatic field. The details of the experimental setup are given in Refs. [2] and [3].

Axial measurements of plasma parameters and the wave electric field are carried out by a planar Langmuir probe (LP) and an electric field probe (EP), respectively [2,3]. Plasma ion density ($N_i$), electron temperature ($T_e$), space potential ($V_s$), and floating potential ($V_f$) are obtained from the Langmuir characteristics where magnetostatic field correction is incorporated. $V_f$ is also measured separately by a Keithley-2001 multimeter with the LP in the floating mode [2]. The wave electric field intensity spectra detected by the antenna probe are recorded by an Agilent E4408B SA having a bandwidth of ~26 GHz [3,4].

An optical probe (OP), consisting of an optical fiber, is inserted through a radial port at the boundary of MC. The probe can be made to collect the light of a particular wavelength from different axial positions by moving it axially. Light collected by the optical fiber is allowed to fall on a spectrometer (Ocean Optics USB4000) capable of detecting emission spectra in the wavelength range of 200–1100 nm with an optical resolution of 0.22 nm. The

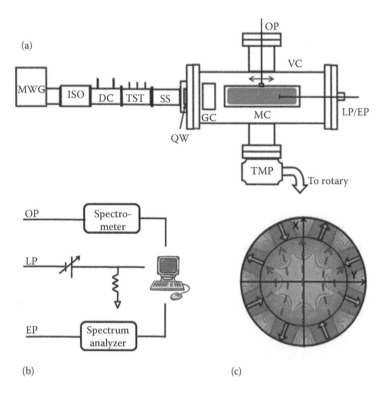

**FIGURE 15.1**
(a) Schematic of the experimental setup; VC: vacuum chamber, MC: multicusp, GC: guiding cylinder, TMP: turbomolecular pump, QW: quartz window, SSC: straight section, TST: triple stub tuner, DC: directional coupler, ISO: isolator, MWG: microwave generator, PSC: power supply and control for the MWG, OP: optical probe, LP: Langmuir's probe, EP: electric field probe. (b) Measurement circuit for OP, LP, and EP. (c) Poisson's simulation of an octupole (solid curves) with $TE_{11}$ wave electric fields (dashed curves) superposed.

spectrometer integration time was fixed at 100 ms, and the spectrum data were saved by averaging over five scans.

All measuring equipments are computer-interfaced using GPIB protocol in tandem with appropriate LABVIEW programs, as shown in Figure 15.1b. This ensures faster and steady measurements along with lower random error and white noise reduction by the averaging of data points over a large number of acquisitions.

## Experimental Results

Plasma is generated and sustained for MW input power in the range $P_{in} = 180$–270 W and neutral pressure range of $p = 0.10$–0.55 mTorr in MC via ECR, and

UHR occurring near the peripheral plasma [1–4,9–11]. It is observed that stable plasma is not obtained below 0.10 mTorr, and above 0.55 mTorr, the plasma density decreases. The axial variation of $N_i$, $T_e$, and the SW electric field intensity $|E_{tot}|^2$ has been reported in details in Ref. [2], and the data obtained in the current experiment has a similar trend. The electromagnetic character of the SW was confirmed by performing an axial B-dot probe measurement, where the SW variation of the wave magnetic field was observed [2].

The plot of $\ln(I_e)$ versus $V$ of the LP characteristics has two different slopes, corresponding to two electron temperatures of $T_e \sim$ 6–10 eV and a lower one of $T_e \sim$ 12–14 eV. The weighted average $T_{e\text{-}avg}$ is obtained by employing a linear fitting over both the slopes. It is of interest to investigate axial evolution of the "hot" electrons (lower slope), which are expected to be affected maximally by the SW pattern. Figure 15.2a shows the axial variation of $T_{e\text{-}hot}$ at 0.15 and 0.25 mTorr with 180 W, and at 0.25 mTorr with 234 W. The axial variation of the SW intensity (in dBm) under the same experimental conditions is plotted in Figure 15.2a. Small undulations (~±1 eV) are observed with

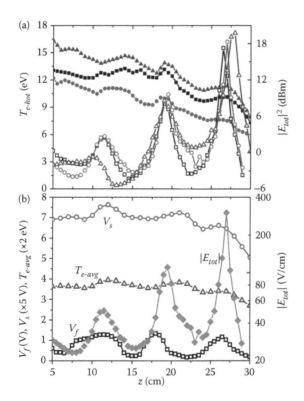

**FIGURE 15.2**
(a) Axial variation of SW intensity $|E_{tot}|^2$ (hollow symbol) and $T_{e\text{-}hot}$ (solid symbol) at 0.15 mTorr (square), 0.25 mTorr (circle) for 180 W, and with 234 W at 0.25 mTorr (triangle). (b) Axial variation of calibrated $|E_{tot}|$ (diamond), $V_f$ (square), $V_s$ (circle), and $T_{e\text{-}avg}$ (triangle) with 180 W, at 0.25 mTorr.

the temperature maxima corresponding to the SW maxima. $T_{e\text{-}hot}$ maximizes near the SW maxima, signifying that the SW intensity, which is stationary in space and oscillates in time, accelerates the electrons near the maxima to a greater extent than the ones near the minima.

The variation of $V_f$, $V_s$, and $T_{e\text{-}avg}$ is plotted, along with SW amplitude at 0.25 mTorr, 180 W in Figure 15.2b. The SW amplitude (in V/cm) is obtained from the intensity (Figure 15.2a) by using the EP calibration factor, 0 dBm (probe) ~37 V/cm (actual). The average electron temperature $T_{e\text{-}avg}$ (down-scaled by 2) shows variation similar to $T_{e\text{-}hot}$ (cf. Figure 15.2a), with the value maximizing at the SW maxima due to higher acceleration of the electrons. $V_f$ shows a strong correlation with the SW following the same trend, especially around $z \sim 15$–30 cm, where the pattern is well pronounced. The variation of $V_s$ (downscaled by 5) is loosely correlated to the SW amplitude (Figure 15.2b), where it is noted that $V_s$ is higher at SW maxima. For argon plasma, the difference between $V_s$ and $V_f$ can be written as $V_s - V_f = 4.7T_{e\text{-}avg}$. Hence, it is expected that the difference between $V_s$ and $V_f$ would follow the same trend as the electron temperature, although not explicitly shown in Figure 15.2b.

Figure 15.3 shows the axial variation of normalized intensities of two Ar (II) lines: (i) 426.48 nm, and (ii) 488.0 nm, taken at 0.25 mTorr of argon pressure, with the MW power set at 180 W. From Figure 15.3, it is observed that corresponding to the SW minima, there is a maximum in the intensity profile of the two optical wavelengths.

Figure 15.4 shows typical wave spectrum screenshots averaged over 50 acquisitions, taken at $z \sim 18$ cm (maxima) and 180 W for vacuum and

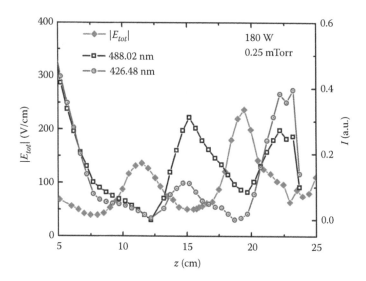

**FIGURE 15.3**
Axial profile of intensities of 426.6 and 488.0 nm Ar II lines at 180 W microwave power and 0.25 mTorr gas pressure, along with the corresponding SW pattern of E-field, $|E_{tot}|$.

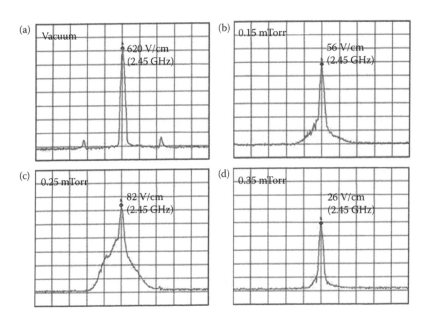

**FIGURE 15.4**
Wave spectrum screenshot at 180 W, (a) vacuum, (b) 0.15, (c) 0.25, and (d) 0.35 mTorr, at z = 18 cm. Central frequency = 2.45 GHz, and span = 300 MHz.

plasma conditions at 0.15, 0.25, and 0.35 mTorr. Each division along the $x$ axis is 30 MHz. In vacuum, the spectrum is sharp with typical broadening of ~0.015 GHz, defined at 5% of the peak intensity ($P_{peak}$ ~ 620 V/cm). Two small peaks with intensity 100 times lower than the primary peak (2.45 GHz) are observed at 2.38 and 2.52 GHz, and may be attributed to recirculating electrons in the magnetron, which is often a consequence of aging [17]. In the presence of plasma, the primary peak is reduced by an order of magnitude due to utilization of the field energy for plasma generation and sustenance. The small peaks are suppressed and an asymmetric broadening toward the lower frequency is observed. Maximum broadening is observed at 0.25 mTorr (~0.09 GHz) (Figure 15.4c), and the broadening decreases below ~0.05 GHz at 0.15 mTorr (Figure 15.4b) and to ~0.04 GHz at 0.35 mTorr (Figure 15.4d). At 0.25 mTorr, the broadening leads to the formation of a sideband-like structure with frequency (~2.43 GHz) very close to the injected wave. At 0.25 mTorr, there appear to be small peaks at 2.442 GHz, 2.423 GHz, and 2.414 GHz, the average of these frequencies being 2.43 GHz.

Figure 15.5a–c shows the contour plot of the wave spectrum (amplitude vs. frequency) as a function of the axial distance $z$ in the MC at (a) 0.15 mTorr, (b) 0.25 mTorr, and (c) 0.35 mTorr with 180 W of MW power. The contour plot provides a much clearer picture of the asymmetry and spectral spread. The SW pattern is clearly observed and it may be noted that the spectral broadening is more near the maxima, where electrons are accelerated more.

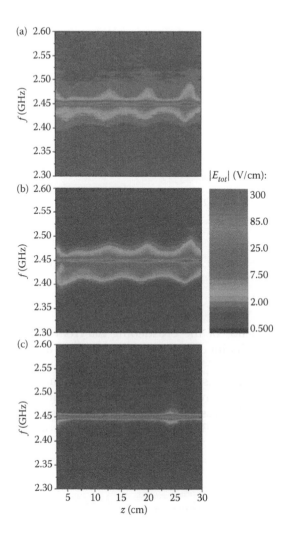

**FIGURE 15.5**
Asymmetric spectral broadening of central frequency (2.45 GHz) at (a) 0.15, (b) 0.25, and (c) 0.35 mTorr, at 180 W.

Maximum spectral broadening is clearly observed for 0.25 mTorr (Figure 15.5b) along z. The frequency broadening is clearly more toward the lower frequency side of 2.45 GHz than the higher frequency side, and hence the term asymmetric broadening is used to indicate that the integrated spectral power on the lower frequency is greater than the integrated power on the higher frequency side of 2.45 GHz. The integrated power on the lower frequency side is calculated to be ~88% of the total power at 0.25 mTorr pressure as calculated from Figure 15.4c.

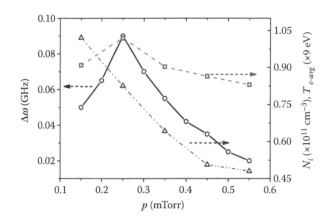

**FIGURE 15.6**
Pressure variation of the central frequency spread ($\Delta\omega$) from experimental electromagnetic spectrum (open circle), $N_i$ (open square), and $T_{e\text{-avg}}$ (open triangle) at $z = 18$ cm for 180 W.

Pressure dependence of the broadening $\Delta\omega$ from the center frequency (2.45 GHz) to the midpoint of asymmetric sideband at a fixed input power of 180 W for the maxima at $z \sim 18$ cm is shown in Figure 15.6. The trend follows the variation of $N_i$ with pressure [12], plotted in the same graph (Figure 15.6). The variation of $T_{e\text{-avg}}$ (downscaled by 9), also plotted in Figure 15.6, shows a decreasing trend with pressure as expected [12]. Thus, the degree of broadening is more sensitive to the number of charged particles interacting with the wave.

## Modeling

In order to understand the phenomenon of electron localization, a Monte Carlo simulation is carried out to investigate the electron dynamics in the presence of bounded $TE_{11}$-mode MWs by considering the plasma and magnetic field inhomogeneities. The details of the code and formulation of the problem are described in Refs. [18] and [19]. We briefly present here the essential features and the experimental results.

The equation of motion for an electron subject to the electromagnetic SW and magnetostatic force fields modified by the presence of the temporally growing plasma is given by

$$\ddot{\vec{r}} = -(e/m_e)[\bar{E}_P(\vec{r})\{\exp i(\vec{k} \cdot \vec{r} - \omega t) + \bar{R}_o \exp i(-\vec{k} \cdot \vec{r} - \omega t)\} + \dot{\vec{r}} \times \bar{B}_o(\vec{r}) + \vec{E}_{int}(\vec{r})]$$

or

$$\ddot{\vec{r}} = -[\vec{C}_o(\vec{r})\{\exp i(\vec{k}\cdot\vec{r} - \omega t) + \vec{R}_o \exp i(-\vec{k}\cdot\vec{r} - \omega t)\} + \dot{\vec{r}} \times \vec{\omega}_{ce}(\vec{r}) + \vec{C}_{int}(\vec{r})], \quad (15.1)$$

where $\vec{r}$ is the position vector of the electron, $\dot{\vec{r}}$ is the velocity, $\ddot{\vec{r}}$ is the accelera-tion, $m_e$ is the electron mass, $e$ is the electronic charge, $\vec{C}_o(\vec{r}) = (e\vec{E}_P(\vec{r})/m_e)$ is the acceleration due to the wave electric field $\vec{E}_P(\vec{r})$, $\vec{k}$ is the wave vector, $\vec{R}_o$ is the reflection coefficient of the backward moving wave, $\vec{\omega}_{ce}(\vec{r}) = e\vec{B}_o(\vec{r})/m_e$ is the electron–cyclotron frequency corresponding to the MC magnetostatic field $\vec{B}_o(\vec{r})$, and $\vec{C}_{int}(\vec{r}) = (e\vec{E}_{int}(\vec{r})/m_e)$ is the acceleration due to the internal electric field $\vec{E}_{int}(\vec{r})$ generated due to the presence of electrons and ions in the volume. The MC magnetic field in the cylindrical coordinates is implemented using the equations [20] $B_{or} = -B_p(r/a)^3 \sin(4\theta)$, $B_{o\theta} = -B_p(r/a)^3\cos(4\theta)$, and $B_{oz} = 0$, where $B_p$ (~0.4 T) is the field at the pole (surface) of the magnet.

Separating out the time variation, the spatial electric field amplitude of the $TE_{11}$ mode ($E_z = 0$) modified by an inhomogeneous anisotropic plasma is given by [3]

$$\vec{E}_P(\vec{r}) = E_{rP}(r,\theta)\hat{r} + E_{\theta P}(r,\theta)\hat{\theta}, \quad (15.2)$$

where

$$\vec{E}_{rP}(r,\theta) = \left(\frac{i\omega}{D_o}\right)\left[\frac{i}{r}(k_o^2 K_{22} - k_z^2) + \frac{1}{r}(k_o^2 K_{22} - k_z^2)\frac{\partial}{\partial\theta} + k_o^2 K_{12}\frac{\partial}{\partial r}\right]B_z(r,\theta) \quad (15.3a)$$

and

$$\vec{E}_{\theta P}(r,\theta) = \left(-\frac{i\omega}{D_o}\right)\left[\frac{i}{r}k_o^2 K_{21} + \frac{1}{r}k_o^2 K_{21}\frac{\partial}{\partial\theta} + (k_o^2 K_{11} - k_z^2)\frac{\partial}{\partial r}\right]B_z(r,\theta) \quad (15.3b)$$

are the radial and polar components. Here, $K_{ij}$ are the dielectric tensor ele-ments derived for the $k{\perp}B_o$ mode of propagation [3].

Assuming that the vacuum $TE_{11}$ mode is modified due to the plasma disper-sion, the axial wave magnetic field amplitude can be expressed in terms of vac-uum electric field amplitude $E_o$ for a cylindrical waveguide of radius $a$ as [21]

$$B_z(r,\theta) = \frac{E_o}{c} J_1(k_c r)\sin\theta, \quad (15.4)$$

where the dipole polar variation is explicitly implemented as $\sin\theta$, and $k_c$ (= $1.841/a$) is the cutoff wave number for the $TE_{11}$ mode [21], which is related to $k_o$ and $k_z$ as

$$k_z(r,\theta)^2 = k_o^2 K_{22}(r,\theta) - k_c^2. \quad (15.5)$$

In vacuum, $K_{11} = K_{22} = 1$ and $K_{12} = K_{21} = 0$, and Equations 15.5, 15.3a, and 15.3b reduce to the vacuum cases given in Ref. [21]. In the presence of a radially varying plasma $n_e(r) = n_o J_o[2.405(r/a)^2]$ ($n_o$, maximum density at $r = 0$) [2–4], Equations 15.3a and 15.3b give the modified $TE_{11}$-mode SW pattern in the MC geometry [3,4].

The electric field amplitude $E_o$ is related to the peak power $P_o$ in vacuum by $P_o = \pi k_o E_o^2 k_z (p_{11}' - 1) J_1^2(k_c a)/(4\mu_o c k_c^4)$, where $p_{11}' = 1.841$ for $TE_{11}$-mode propagation [21]. An input of $P_o = 180$ W would correspond to an SW field of ~620 V/cm.

The above are implemented in a FORTRAN-90 code, where a fourth-order Runge–Kutta method is used for solving the electron equation of motion, and a random number generator is used wherever probabilistic (Monte Carlo) determination is required. The collisional scattering of an electron with the neutral has been implemented using Monte Carlo collisions, and occurs in three categories depending on the energy and cross section of the process, namely, elastic scattering ($\sigma_{el}$), excitation ($\sigma_{exc}$), and ionization ($\sigma_{inz}$) [18,19,22]. Accurate cross sections are employed to ensure realistic simulations [18,22].

To account for the enhancement of the electron numbers by several orders of magnitude (density change of $10^3$ cm$^{-3}$ to $10^{11}$ cm$^{-3}$) due to ionization, dynamic charge scaling is implemented [23]. The scaling factor continuously increases with the generation of electrons, keeping the number of superparticles constant. Correspondingly, the charge scaling is accounted for in the equations of motion, and plasma density relations. Separate scaling is used for electrons and ions since electrons have higher escape probability due to their larger mobility. The simulation is started with $N_e = 10^4$ seed superelectrons and zero superions, in the cylindrical volume defined by $a = 4.1$ cm and $L = 30$ cm. The electrons evolve together in time steps of $dt$ (~0.02 ns) $\ll \tau_p$ (~0.36 ns), with a characteristic plasma electron oscillation time at $n_e \sim 10^{11}$ cm$^{-3}$, and undergo collisions, gain energy, excite, ionize, and escape. The primary and secondary electrons and generated ions are tracked, and the corresponding densities updated after each time step, which in turn influences the wave electric field. The evolution is studied up to $T \sim 50$ μs, which is more than twice the steady-state time, and the position and velocity vectors of the electrons are stored after certain time intervals ($\Delta T \sim 5$ μs). The generated data are used to study the distribution of the particles, evolution of phase space, and other parameters with time in the MC. The results are discussed in the next section.

---

## Modeling Results

The growth of plasma is studied, and it is observed that the plasma parameter variations become approximately steady in about 20 μs. Figure 15.7 shows the one-dimensional $v_z$–$z$ phase space plots evolving in time up to 20 μs for

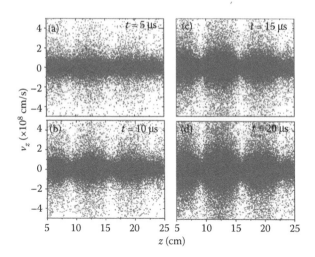

**FIGURE 15.7**
Time shots of $v_z$ versus $z$ phase plot in the MC for 180 W, 0.25 mTorr at (a) 5, (b) 10, (c) 15, and (d) 20 μs.

180 W, 0.25 mTorr. At initial times (~5 μs), the electrons are nearly uniformly distributed. With time, the electrons gradually become phase-locked with the SW pattern, and at ~20 μs, well-defined striations are obtained. At SW maxima ($z \sim 5$, 11.5, 18 cm), the electrons are accelerated more and hence gain higher velocity, which is clearly observed in the plots (Figure 15.7a through d), and also demonstrate higher velocity spread compared to the minima. The higher velocity spread at SW maxima is consistent with the experimental variation of electron energy spread (Figure 15.3a).

Figure 15.8a–c shows the spatial distribution of electrons in the MC geometry (transverse and longitudinal section) after the steady state has been achieved. The color map represents the energy of the electrons. The mean energy comes out to be $U_{mean} \sim 12$ eV, which gives $T_{e-avg} \sim 8$ eV, consistent with the experimentally obtained results. Figure 15.8a and b shows the electron distribution in the transverse plane of the MC at a maxima ($z \sim 18$ cm) and minima ($z \sim 22$ cm), respectively. In the transverse section (Figure 15.8a), the heating of the electrons near the cusped ECR and UHR regions is clearly observed, and complements earlier wave–plasma interaction results [3,4]. The electrons are comparatively cooler in the minima (Figure 15.8b) as expected. The asymmetry in heating along the $x$–$y$ plane is due to the $TE_{11}$-mode electric field polarization directed primarily along the $x$ axis as shown superposed in Figure 15.1b [4]. In the longitudinal section (Figure 15.8b), the striations in the electron density are clearly visible, with low-energy electrons localized in the SW minima regions ($z \sim 2.5$, 9, 15.5, 22 cm) and the regions of intense ionization are observed between $|x| \sim 2$–3 cm, where the resonance conditions are satisfied [2–4].

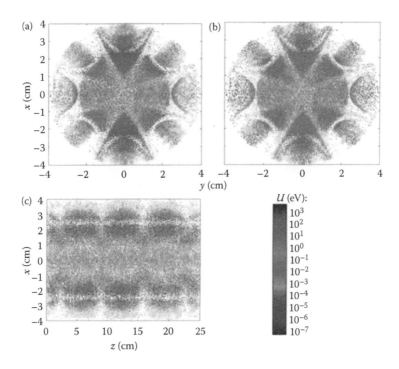

**FIGURE 15.8**

Spatial distribution of electrons in the MC at $t = 20\,\mu s$ for 180 W, 0.25 mTorr. Cross-sectional view (a) at $z \sim 18$ cm, (b) at $z \sim 22$ cm, and (c) longitudinal view. The energy of the electrons is shown in gray scale.

## Summary and Discussion

The phenomenon of plasma electron localization and asymmetric frequency sideband generation in the interaction of MW SWs sustained in a plasma column confined in a minimum-$B$ magnetic field is studied experimentally and through Monte Carlo simulations.

It is observed that the axial variation of electron temperature, floating, and plasma potential closely follow the SW pattern. The observations indicate localization of plasma electrons in the SW minima, which is further verified from optical measurements.

The corona balance equation suggests that $n_e\, N_1 k_{1j} = N_j\, \Sigma A_{jk}$, where $n_e$, $N_1$, and $N_j$ are, respectively, the electron, ground level, and excited state of argon population densities; $k_{1j}$ is the electron-impact excitation rate coefficient from ground state 1 to level $j$; and $A_{jk}$ is the transition probability for spontaneous radiative emission from level $j$ to the lower level $k$. The emitted light intensity $I_{jk} = (h\nu_{jk}\, A_{jk}\, N_j\, L)/4\pi$, where $h\nu_{jk}$ and $L$ are the energy separation between levels, and $j$ and $k$ and the plasma length that the emitted light has to pass through, respectively. Substituting for $N_j$ from the corona balance equation, $I_{jk}$ can be written as $I_{jk} = (hv_{jk} \, A_{jk} \, n_e \, N_1 \, k_{1j} L \,)/(4\pi \sum A_{jk})$. Since $hv_{jk}, A_{jk}$, and $L$ are constants, and as the electron temperature (Figure 15.2b) does not vary much (~1 eV) over the axial length ($z$), $N_1$ and $k_{1j}$ may also be taken as constants. This suggests that the increase in optical intensity $I_{jk}$ near the minima of the SWs (E-field minima location) is due to an increase in the local electron concentration and has been observed in the optical measurements, confirming electron localization in the SW troughs.

The local plasma frequency $f_p$ is ~2.83 GHz and the bounce frequency of the electrons $f_b$ may be estimated from the relation $\omega_b = k_o v_b$, where $v_b = \sqrt{2eV/m_e}$, $V$ is the potential difference between the crest and the trough, and $m_e$ is the mass of the electron in the potential well of the SWs: $f_b$ is estimated to be ~0.13 GHz. It is possible that $f_b$ beats with $f_o$ the drive frequency as one of the beat frequencies $f_o - f_b \sim 2.33$ GHz shows up as the asymmetric frequency observed in the lower-frequency side of $f_o$ and the higher beat $f_o + f_b \sim 2.58$ GHz is not observed or has a very small amplitude. The broadening of the frequencies is considered to occur due to initial phase variation of the electrons, finite temperature, and energy distribution of the oscillating electrons. Electrons with different initial phase and energy would oscillate at slightly different frequencies than $f_o$ and hence broaden it. Since the number of electrons with energy ($U$) less than the mean electron energy ($U_{mean}$) is greater than electrons with energy higher than the mean energy (typical Maxwell–Boltzmann-type distribution, $f(U) \propto \sqrt{U} \exp(-U/U_{mean})$, the broadening is expected to be asymmetric (Figure 15.4a–d).

At lower neutral pressure, there is less number of plasma electrons responding to the oscillation, which are more or less coherent with $f_o$ (Figures 15.4b, 15.5a, and 15.6). As the neutral density increases, collisions increase and so does the plasma density, and an optimum condition is achieved at ~0.25 mTorr, where $N_i$ is maximum, after which the density decreases again due to frequent elastic collisions (cf. Figure 15.6). The existence of an optimum pressure in MC has been explained earlier on the basis of particle and energy balance [9]. The large ensemble of plasma electrons results in greater variation in initial phase and thermal velocities, thereby broadening the wave spectrum to a higher degree (Figures 15.4c, 15.5b, and 15.6).

Finally, a Monte Carlo simulation is developed to understand the phenomenon, which successfully reproduces many of the experimental observations. It is intriguing that electron localization occurs along the axial direction in the predominantly transverse electromagnetic fields [2–4]. This may be understood in terms of the $\vec{E} \times \vec{B}_o$ force acting on the plasma electrons. The transverse electric and magnetic field components couple with each other ($E_r$ with $B_{o\theta}$, and $E_\theta$ with $B_{or}$) and generate force that pushes plasma electrons in the axial direction and helps them get trapped in the SW minimas as observed in Figure 15.3b.

Electrons, while oscillating near the maxima, collide with the neutrals, whereby their directions are randomized. Some of these randomly directed electrons would be pushed into the SW minima. The electron may also move out of the minima by collisions, but with a lesser probability since it has lower energy ($kT_e \ll$ barrier height $V$).

## References

1. J.V. Mathew, I. Dey, and S. Bhattacharjee. 2007. Microwave guiding and intense plasma generation at subcutoff dimensions for focused ion beams, *Appl. Phys. Lett.* **91**: 041503.
2. I. Dey and S. Bhattacharjee. 2008. Experimental investigation of standing wave interactions with a magnetized plasma in a minimum-B field, *Phys. Plasmas* **15**: 123502.
3. I. Dey and S. Bhattacharjee. 2011. Penetration and screening of perpendicularly launched electromagnetic waves through bounded supercritical plasma confined in multicusp magnetic field, *Phys. Plasmas* **18**: 022101.
4. I. Dey and S. Bhattacharjee. 2011. Anisotropy induced wave birefringence in bounded supercritical plasma confined in a multicusp magnetic field, *Appl. Phys. Lett.* **98**: 151501.
5. N.S. Smith, W.P. Skoczylas, S.M. Kellogg, D.E. Kinion, P.P. Tesch, O. Sutherland, A. Aanesland, and R.W. Boswell. 2006. High brightness inductively coupled plasma source for high current focused ion beam applications, *J. Vac. Sci. Technol. B* **24**: 2902.
6. J.V. Mathew, A. Chowdhury, and S. Bhattacharjee. 2008. Subcutoff microwave driven plasma ion sources for multielemental focused ion beam systems, *Rev. Sci. Instrum.* **79**: 063504.
7. J. Yang, Y. Xu, Z. Meng, and T. Yang. 2008. Effect of applied magnetic field on a microwave plasma thruster, *Phys. Plasmas* **15**: 023503.
8. N. Yamamoto, S. Kondo, T. Chikaoka, H. Nakashima, and H. Masui. 2007. Effects of magnetic field configuration on thrust performance in a miniature microwave discharge ion thruster, *J. Appl. Phys.* **102**: 123304.
9. S. Bhattacharjee and H. Amemiya. 1998. Microwave plasma in a multicusp circular waveguide with a dimension below cutoff, *Jpn. J. Appl. Phys.* **37**: 5742.
10. S. Bhattacharjee and H. Amemiya. 1997. Transversely magnetized microwave plasma in a rectangular waveguide under cutoff conditions, *Rev. Sci. Instrum.* **68**: 3061.
11. S. Bhattacharjee and H. Amemiya. 1999. Production of microwave plasma in narrow cross-sectional tubes; effect of the shape of cross section, *Rev. Sci. Instrum.* **70**: 3332.
12. J.H.A. van Wakeren and H.J. Hopman. 1972. Trapping of electrons in large-amplitude electrostatic fields resulting from beam-plasma interaction, *Phys. Rev Lett.* **28**: 295–298.
13. R.N. Franklin, S.M. Hamberger, H. Ikezi, G. Lampis, and G.J. Smith. 1972. Nature of the instability caused by electrons trapped by an electron plasma wave, *Phys. Rev. Lett.* **28**: 1114–1117.

14. J.R. Danielson, F. Anderegg, and C.F. Driscoll. 2004. Measurement of Landau damping and the evolution to a BGK equilibrium, *Phys. Rev. Lett.* **92**: 245003.
15. J.L. Warren, G.P. Boicourt, M.T. Menzel, G.W. Rodenz, and M.C. Vasquez. 1985. Revision of and documentation for the standard version of the Poisson group codes, *IEEE Trans. Nucl. Sci.* **32**: 2870.
16. M.A. Lieberman and A.J. Lichtenberg. 1994. *Principles of Plasma Discharges and Material Processing* (Wiley-Interscience, New York) p. 73.
17. I. Dey, J.V. Mathew, S. Bhattacharjee, and S. Jain. 2008. Subnanosecond electron transport in a gas in the presence of polarized electromagnetic waves, *J. Appl. Phys.* **103**: 083305.
18. S. Bhattacharjee and S. Paul. 2009. Random walk of electrons in a gas in the presence of polarized electromagnetic waves: Genesis of a wave induced discharge, *Phys. Plasmas* **16**: 104502.
19. D.M. Pozar. 2006. *Microwave Engineering* (3rd edition, John Wiley and Sons Inc., Singapore) p. 118.
20. D. Vender, H.B. Smith, and R.W. Boswell. 1996. Simulations of multipactor-assisted breakdown in radio frequency plasmas, *J. Appl. Phys.* **80**: 4292.
21. F.K. Azadboni, M. Sedaghatizade, and K. Sepanloo. 2010. Design studies of a multicusp ion source with FEMLAB simulation, *J. Fusion Energ.* **29**: 5.
22. V.B. Neculaes, R.M. Gilgenbach, and Y.Y. Lau. 2003. Low-noise microwave magnetrons by azimuthally varying axial magnetic field, *Appl. Phys. Lett.* **83**: 1938.
23. J.J. Seough and P.H. Yoon. 2009. Analytic models of warm plasma dispersion relations, *Phys. Plasmas* **16**: 092103.

# 16

## New Experiments in Pulsed Microwaves: Quasisteady-State Interpulse Plasmas

### Introduction

Pulsed plasmas have found wide applications both in industry and for basic studies [1–5]. The prime advantages include additional control over the plasma properties by being able to vary the pulse duration, duty cycle, and the repetition frequency. The pulsing parameters are known to provide control over the electron energy distribution function [6] and gas-heating effects can be avoided [7]. Pulsed plasmas have therefore found attractive applications that include generation of metastables [8], radicals [9], and negative ions [10].

Previously, pulsed discharges that have been widely investigated belong to low-to-moderate powers (few 100 W–2 kW) and relatively wide duration (10 µs–100 ms) pulses [11–13]. In most of these studies, it has been reported that the plasma attains a steady state within the pulse, and after the end of the pulse, the plasma decays temporally with the electron temperature attaining close to thermal values. Such a behavior is typical of afterglow plasmas.

However, if high power (60–100 kW) is supplied to the plasma in small-duration pulses ($\tau_p$ = 0.5–1.2 µs), the parent plasma is quite active and is followed by a comparatively long-lived plasma in the power-off phase, with electrons still having a high average energy ($3/2kT_e$). Such an "interpulse plasma" [14] in the power-off phase has interesting properties [15–17] that are distinctly different from afterglows. Some of these experimentally determined differences from our earlier work on microwaves of 3 GHz are as follows: (a) the plasma buildup is extended beyond the pulse duration, with the peak density attained a few to tens of microseconds later in the power-off phase; (b) a high value of electron temperature (~6–10 eV) continues for a time comparable to the buildup time; and (c) enhanced interpulse plasma densities could be obtained at a pressure ($\omega/v_c \ll 1$) where resonant wave absorption such as ECR may be considered negligible (here, $\omega$ is the wave frequency and $v_c$ is the electron–neutral collision frequency).

In this chapter, we report experimental observations on the generation of a quasisteady state in the power-off phase of a plasma produced by high-frequency (9.45 GHz) pulsed-mode microwaves in the $X$ band. For the time between the pulses (interpulse time) where the steady state is obtained, the plasma is basically without an external energy source. The dependence of the steady state on the plasma collisionality is studied over a pressure range from 1 mTorr to 1 Torr. The influences of microwave pulse parameters such as the pulse amplitude and repetition frequency and the characteristic diffusion length on the steady state are investigated. The experimental results are complemented by theoretical model calculations that are based on the coupled rate equations of the plasma density and the electron temperature. A reasonably good agreement is found between the experimental observations and the model calculations.

## Experiment

The compact plasma source employs microwaves in the $X$ band (9.45 GHz), which are launched into a circular plasma chamber about 30 cm long. It may be noted that in this experiment the microwave frequency is much higher than the 2.45 GHz frequency used in the other experiments. The plasma chamber is kept inside a larger vacuum chamber of about 60 cm diameter and 250 cm length. The plasma is confined by a minimum-$B$ field generated by permanent magnets [Nd–Fe–B, surface magnetic field $(B_{max}) \sim 1$ T] surrounding the chamber. The resulting field structure with a null at the center of the chamber provides plasma confinement. A schematic diagram of the experimental setup is shown in Figure 16.1 and the radial magnetic field profile for the hexapole is shown in Figure 16.2a, including a Poisson simulation field plot in Figure 16.2b.

The wave launch mode belongs to $k \perp B$, where $k$ is the wave vector and $B$ is the static magnetic field. Argon is used as the test gas. The experiments were carried out on a few different plasma chambers that were designed with the radius at integer multiples of the quarter wavelength of the wave $n\lambda/4$ ($n \leq 4$ has been used). The small chamber cross section enhances the microwave power density ($\sim$10 kW/cm$^2$), which is considerably larger than conventional moderate-power, large-diameter plasma sources (e.g., for power $\sim$500 W and plasma chamber radius $\sim$3 cm, the power density is $\sim$0.01 kW/cm$^2$).

Plasma current measurements have been made at the center of the plasma chamber, where the magnetic field is almost zero. A circular planar Langmuir probe with a diameter of 4 mm, made of stainless steel with a thickness of 0.1 mm, is used in the fixed voltage (20 V for electron current and –70 V for ion current) mode to study the temporal profiles of the particle (electron and ion) saturation currents and in a sweep voltage mode (–100 to 100 V) to obtain the probe ($I$–$V$) characteristics. Both sides of the planar probe were used for charge

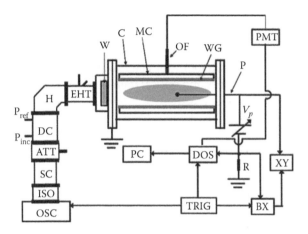

**FIGURE 16.1**
Schematic of the experimental apparatus; OSC: magnetron oscillator, ISO: isolator, SC: straight section, ATT: attenuator, DC: directional coupler, H: rectangular bend, EHT: E–H tuner, W: quartz window, C: chamber, MC: multicusp, OF: optical fiber, WG: waveguide, P: Langmuir probe, PMT: photomultiplier tube, $V_p$: probe voltage, PC: personal computer, DOS: digital oscilloscope, R: resistor, XY: X–Y chart recorder, TRIG: trigger signal, BX: Boxcar's integrator. (Reprinted with permission from S. Bhattacharjee et al. 2007. Quasisteady state interpulse plasmas, *J. Appl. Phys.* **101**: 113311. Copyright 2007, American Institute of Physics.)

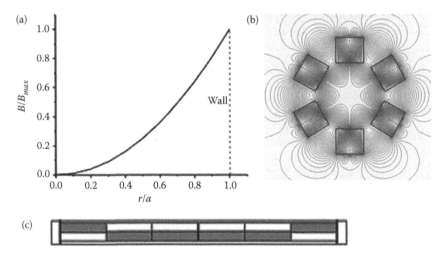

**FIGURE 16.2**
(a) Radial variation of the static B field of the magnetic multicusp in the normalized scale. (b) Two-dimensional Poisson field plot. (c) Longitudinal magnetic arrangement demonstrating axial-end plugging. (Reprinted with permission from S. Bhattacharjee et al. 2007. Quasisteady state interpulse plasmas, *J. Appl. Phys.* **101**: 113311. Copyright 2007, American Institute of Physics.)

particle collection, and the probe surface was kept parallel to the wave vector. The time-varying particle currents (electron and ion) were measured using a fast digital oscilloscope (Tektronix model TDS2014). A Boxcar integrator (Stanford Research Systems, model SR200) was used to make time-resolved measurements of probe characteristics and optical data. For the optical studies, the light emitted by the pulsed argon plasma was obtained by an optical probe, which consists of an optical fiber and a lens attached at the tip. The probe is inserted through a radial port onto a tiny gap in the multicusp and the plasma chamber. The obtained light is passed onto a monochromator (CVI model CM110), and the signal is amplified by a photomultiplier tube (CVI model AD110). We have studied the transition $3s^23p^4\,(^1D)\,4s \rightarrow 3s^23p^4(^1D)\,4p$, which is the most intense in the visible region of the spectrum of the pulsed argon plasma.

## Experimental Results

Figure 16.3 shows the buildup of the plasma electron current density. It is seen that the buildup is extended beyond the end of the pulse (which ends at ~1.2 µs), with the peak density attained about 20–30 µs later in the power-off phase. In general, the plasma buildup time $\tau_b$, defined as the time to obtain the peak density from the end of the pulse, increases with increase in pressure. The trend agrees reasonably well with what was seen earlier in the case of a 3 GHz discharge [15]. It may be noted in Figure 16.3 that at 40 mTorr the buildup and the decay profiles tend to become flatter.

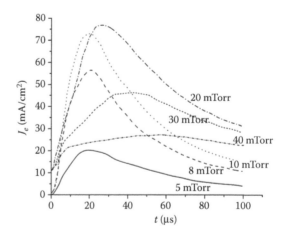

**FIGURE 16.3**

Temporal variation of the electron current density $J_e$ with pressure $p$ as a parameter. (Reprinted with permission from S. Bhattacharjee et al. 2007. Quasisteady state interpulse plasmas, *J. Appl. Phys.* **101**: 113311. Copyright 2007, American Institute of Physics.)

Figure 16.4 shows a typical plasma quasisteady state in the electron current density that is maintained for more than 30 µs in the interpulse phase. The result is remarkable in that the steady state is maintained even in the absence of any microwave input power. The end of the microwave pulse is clearly indicated in Figure 16.4. The temporal variation of the electron current density $J_e$ is shown at a discharge pressure $p$ of 36 mTorr and a pulse repetition frequency $f_r$ of 100 Hz.

Figure 16.5 shows the variation of the peak electron current density $J_{me}$ with pressure $p$ and pulse repetition frequency $f_r$ as a parameter. $J_{me}$ corresponds to the maximum value of the current profiles shown in Figure 16.3. The pulse repetition frequency has only a weak influence on $J_{me}$. The plasma density decreases below 4 mTorr and above 100 mTorr, and the pressure where the steady state is observed (36–40 mTorr) lies approximately between these limits. The plasma can be sustained over a wide pressure range of $10^{-3}$–1 Torr, and for this condition a maximum current density of ~100 mA/cm² is obtained.

Figure 16.6 shows the variation of the electron confinement time $\tau_c$ with pressure $p$, and the pulse repetition frequency $f_r$ as a parameter. The electron confinement time in the discharge may be obtained as the sum of the buildup time $\tau_b$, the quasisteady state time $\tau_{flat}$, and the decay time $\tau_d$, where $\tau_d$ is defined as the e-folding decay time required for the electron current to decay from its peak value to ~ $1/10e$ when measured with a highest sensitivity of ~0.1 µA. The electron confinement time may be considered a qualitative measure of the amount of time the plasma is sustained in the multicusp magnetic field. At 100 Hz, $\tau_c$ is found to increase steadily from ~0.1 ms at 3 mTorr to ~10 ms at ~40 mTorr, which is the maximum value and

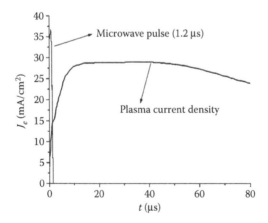

**FIGURE 16.4**
Temporal variation of the electron current density $J_e$ showing the development of a plasma steady-state density flattop in the interpulse regime. (Reprinted with permission from S. Bhattacharjee et al. 2007. Quasisteady state interpulse plasmas, *J. Appl. Phys.* **101**: 113311. Copyright 2007, American Institute of Physics.)

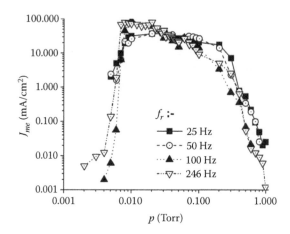

**FIGURE 16.5**
Variation of the peak electron current density $J_{me}$ with pressure $p$ and pulse repetition frequency $f_r$ as a parameter. (Reprinted with permission from S. Bhattacharjee et al. 2007. Quasisteady state interpulse plasmas, *J. Appl. Phys.* **101**: 113311. Copyright 2007, American Institute of Physics.)

corresponds to the pressure at which the plasma steady state is observed in Figure 16.4. As we will soon see, from these data we may infer that the flattop duration follows the simulation trend.

Figure 16.7 shows the variation of the optical intensity at 404 nm at 60 mTorr in comparison with the microwave pulse. It may be noted that the intensity of the 404 nm optical line has a delayed buildup and remains

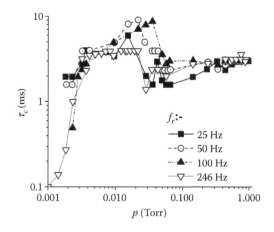

**FIGURE 16.6**
Variation of the electron confinement time $\tau_c$ with pressure $p$ and pulse repetition frequency $f_r$ as a parameter. (Reprinted with permission from S. Bhattacharjee et al. 2007. Quasisteady state interpulse plasmas, *J. Appl. Phys.* **101**: 113311. Copyright 2007, American Institute of Physics.)

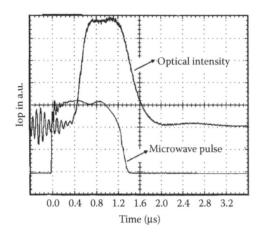

**FIGURE 16.7**
Variation of the 404 nm optical intensity with time at a fixed microwave power for $a = 2.2$ cm. (Reprinted with permission from S. Bhattacharjee et al. 2007. Quasisteady state interpulse plasmas, *J. Appl. Phys.* **101**: 113311. Copyright 2007, American Institute of Physics.)

constant for ~400 ns before undergoing decay. The 404 nm line corresponds to the transition between the states of Ar II (ion) with energy levels, $3s^23p^4$ $(^1D)4p$ (21.4980485 eV) and $3s^23p^4(^1D)4s$ (18.4265479 eV). Thus, we can see that although the energy difference between the levels is around 3.07 eV, the excitation energy needed to cause this transition is about 21.5 eV. In order to have this level populated sufficiently for an easily detectable output, a high value of average electron energy is required ($\geq 21.5$ eV). This is provided primarily during the pulse on time, as shown in Figure 16.8 for the 0.98 cm radius.

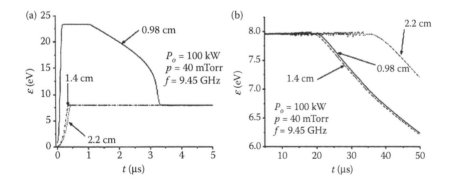

**FIGURE 16.8**
Temporal variation of electron energy for different tube radii at a pressure of 40 mTorr, power of 100 kW, and input frequency of 9.45 GHz. (a) For radii 0.98 cm, 1.4 cm, and 2.2 cm until 5 μs and (b) for radii 0.98 cm, 1.4 cm, and 2.2 cm until 50 μs. (Reprinted with permission from S. Bhattacharjee et al. 2007. Quasisteady state interpulse plasmas, *J. Appl. Phys.* **101**: 113311. Copyright 2007, American Institute of Physics.)

## Model Calculation

### Temporal Evolution of the Electron Energy

The plasma steady state is considered to be a result of a balance between the production and loss mechanisms occurring in the power-off phase of the discharge. During the pulse, strong electric fields of short duration accelerate electrons to high energies.

The temporal evolution of the electron thermal energy ($\varepsilon$) is determined by a balance between the heating and the loss rates given by

$$\frac{\partial \varepsilon}{\partial t} = \frac{v_c \left[eE_P(t)\sin \omega t\right]^2}{2m_e \left(\omega^2 + v_c^2\right)} - L_c(\varepsilon), \tag{16.1}$$

where $v_c$ is the elastic collision frequency, $\omega$ is the angular wave frequency, $e$ and $m_e$ are the electron charge and mass, respectively, and $E_P(t)$ is the time-dependent amplitude of the electric field in the plasma chamber, which depends upon the nature of the microwave pulse.

Several improvements have been made to the model described in Ref. 15. First, a time-dependent electric field has been incorporated based upon the shape of the microwave pulse obtained experimentally. The pulsing scheme is implemented in the numerical calculations using the following conditions (for a pulse on time of ~1.2 μs):

$$E_P(t) = E_{P_0} \exp\left[(t - t_1)/t_1\right] \quad (\text{for } t < t_1), \tag{16.2}$$

$$= E_{P_0} \quad (\text{for } t_1 \le t \le t_2) \tag{16.3}$$

$$= E_{P_0} \exp\left[-(t - t_2)/\tau\right] \quad (\text{for } t > t_2), \tag{16.4}$$

where $E_{P_0}$ is the amplitude of the electric field, $t_1$ is the rise time of the pulse (~150 ns), and $\tau$ is the fall time of the electromagnetic field at the end of the pulse (~175 ns). The flat region of the pulse extends from $t = t_1$ to $t = t_2$ (~1.05 μs).

Second, when the microwave pulse switches off, the electromagnetic field energy does not become zero at once, but it decays slowly with a characteristic time $\tau_d$. This can be attributed to the decay of the wave in the plasma, which is proportional to the electron collision time and possible multiple reflections inside the chamber, which results in the elongation of the time taken by the electromagnetic energy to die out completely. We obtain the total decay time $\tau_d$ to be ~3 μs. Therefore, we replace the $\tau$ of Equation 16.4 with $\tau_f = \tau + \tau_d$ as the total decay time of the electromagnetic energy in the plasma.

For an input frequency of $f = 9.45$ GHz and a pressure range of $p = (10–100)$ mTorr with peak microwave power $P_0 = 25–100$ kW, we have performed numerical simulations to obtain the plots of $\varepsilon$ versus time $t$. For a chamber radius $a = 0.98$ cm, the electron energy was found to saturate within the pulse (1.2 μs) with the maximum energy $\varepsilon_{max} \sim 24–32$ eV, after which it dropped to about 8 eV just after the end of the pulse. It may be noted that this drop was not exponential. It then decayed with a weak exponential and around 100 μs was found to be ~5 eV. The weak exponential decay after about 2–3 μs agrees well with our earlier electron temperature measurements in a pulsed 3 GHz microwave discharge [17]. Thus, it can be seen that electron energies much above the thermal values can be maintained in the interpulse plasma. At larger radii (1.4 cm, 2.2 cm), the power density decreases and, as a result, the maximum electron energy decreases, as shown in Figure 16.8a, which is an expanded view. It is seen from Figure 16.8b that a steady value of the electron energy stays on for a longer time in the case of a larger radius, which indicates that the energy diffusion time is longer in the case of a larger radius.

## Temporal Evolution of the Plasma

The plasma evolution may be modeled by jointly considering the rate equations for the plasma (electron) density and the electron energy (Equation 16.1)

$$\frac{\partial n_e}{\partial t} = N_g \langle \sigma v \rangle n_e - D(1 - \delta)\nabla^2 n_e \qquad (16.5)$$

where $n_e$ is the electron density, $N_g\langle\sigma v\rangle$ is the ionization rate of the background gas given by $N_g \langle \sigma v \rangle = N_g \int_{E_i}^{\infty} \sqrt{2E/m_e}\,\sigma(E)F(E)\,dE$, $\sigma(E)$ being the collision cross section, $N_g$ is the neutral density, the electron energy distribution function $F(E)$ is assumed to be Maxwellian [15], $D$ is the diffusion rate of the generated plasma, and $\delta$ is the secondary electron coefficient for the electrons that are either reflected back or generated due to the primary plasma electrons impacting the chamber wall. The effect of the secondaries from the metal boundary is an addition to the previous model.

We can write Equation 16.5 as

$$\frac{\partial n_e}{\partial t} = N_g \langle \sigma v \rangle n_e - \frac{D}{\Lambda^2}(1 - \delta)n_e, \qquad (16.6)$$

where we have approximated $\nabla^2$ by $1/\Lambda^2$ and $\Lambda$ is the characteristic diffusion length, which we consider to be the scale length in our problem. We may write $D/\Lambda^2 = \rho$ as the decay rate due to radial diffusion.

In addition, when we include the effects of the static radial magnetic field, the diffusion coefficient $D$ is given by the modified diffusion coefficient $D_\perp$ given by the equation

$$D_\perp = \frac{D}{\left(1 + \omega_c^2/v_c^2\right)}, \tag{16.7}$$

where $D$ is the diffusion coefficient for the case when $B = 0$ and $\omega_c$, is the electron–cyclotron frequency. $D$ may be taken as the free electron diffusion given by $D = \kappa T_e/m_e v_e$. We take an average value of $B \sim 0.6$ T for our calculations. The characteristic diffusion length $\Lambda$ can be estimated by including the effect of the radial magnetic field as

$$\frac{1}{\Lambda^2} = \left(\frac{2.4}{R}\right)^2 + \left(\frac{\pi}{L}\right)^2 \chi, \tag{16.8}$$

where $\chi = v_c^2/(v_c^2 + \omega_c^2)$, $R$ is the effective plasma radius, and $L$ is the length of the plasma chamber ($\sim$30 cm).

The secondary electron coefficient $\delta$ can be estimated using the form [18]

$$\delta = \delta_s (\varepsilon/\varepsilon_m) \exp\left[2\left(1 - \sqrt{\varepsilon/\varepsilon_m}\right)\right]. \tag{16.9}$$

We use $\delta_s = 1.4$ and $\varepsilon_m = 400$ V for the stainless-steel surface of the multicusp that bounds the plasma.

Taking $\xi = \varepsilon/E_i$, the rate of change of the electron density can finally be written as

$$\frac{\partial n_e}{\partial t} = C\sqrt{\pi}\left(2 + \frac{1}{\xi}\right)\exp\left(-\frac{1}{\xi}\right)n_e - \frac{D}{\Lambda^2}(1 - \delta)n_e, \tag{16.10}$$

where the terms on the right-hand side are a function of the electron thermal energy ($\varepsilon$). Therefore, $\varepsilon(t)$ evaluated from Equation 16.1 by the numerical simulation can be fed to Equation 16.10 to obtain $n_e$ as a function of time $t$.

To compare our results with the experimental data, where the evolution of the plasma (electron) current density ($J_e$) with time is obtained, we evaluate $J_e$ using

$$J_e = \sqrt{\frac{e^3}{M}}\exp\left(-\frac{1}{2}\right)n_e\sqrt{\varepsilon}. \tag{16.11}$$

The simulation results have been plotted in Figures 16.9 through 16.12. In Figure 16.9, we see the variation of $J_e$ with pressure for a fixed power and radius. In conformity with the experimental result (Figure 16.3), the plasma tends to a quasisteady state at a lower pressure ($\sim$36 mTorr).

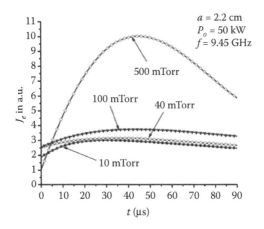

**FIGURE 16.9**
Temporal variation of the plasma (electron) current density for different gas pressures at a peak power of 50 kW, radius of 2.2 cm, and frequency of 9.45 GHz. (Reprinted with permission from S. Bhattacharjee et al. 2007. Quasisteady state interpulse plasmas, *J. Appl. Phys.* **101**: 113311. Copyright 2007, American Institute of Physics.)

Figure 16.10 shows the variation of $J_e$ with the radius of the multicusp at a fixed power and pressure. We see that $J_e$ is higher and there is a better flattop at a larger radius. This can be understood from the decay rate due to radial diffusion (Equations 16.6 through 16.8). Since $\rho = D/\Lambda^2$, for a larger radius (hence larger $\Lambda$), $\rho$ will be smaller and hence the plasma will diffuse out

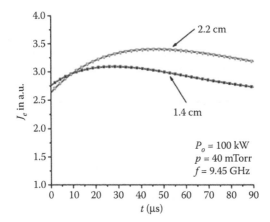

**FIGURE 16.10**
Temporal variation of the plasma (electron) current density for different tube radii at a pressure of 40 mTorr, peak power of 100 kW, and frequency of 9.45 GHz. (Reprinted with permission from S. Bhattacharjee et al. 2007. Quasisteady state interpulse plasmas, *J. Appl. Phys.* **101**: 113311. Copyright 2007, American Institute of Physics.)

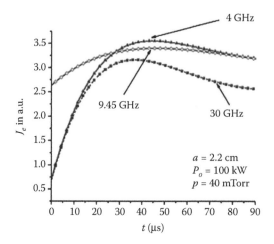

**FIGURE 16.11**

Temporal variation of the plasma (electron) current density for different input frequencies at a pressure of 40 mTorr, peak power of 100 kW, and radius of 2.2 cm. (Reprinted with permission from S. Bhattacharjee et al. 2007. Quasisteady state interpulse plasmas, *J. Appl. Phys.* **101**: 113311. Copyright 2007, American Institute of Physics.)

more slowly, thereby giving a higher current density and better flattop. Also, the field decay time $\tau$ will be longer for the larger radius, which will result in a slower decay of microwave energy from the plasma, thereby enhancing the plasma production.

Figure 16.11 shows the dependence of $J_e$ on the frequency of the incident microwave radiation at fixed radius, power, and pressure. This was not possible to do experimentally because of a fixed frequency source. As the frequency is increased from 4 to 30 GHz, we see that $J_e$ decreases, which is evident from Equation 16.1, as the energy gain dependence on the wave frequency goes as $1/\omega^2$. Although at 4 GHz the wave frequency nearly equals the cutoff frequency of the waveguide, from our earlier experiments [19], we have found that microwaves can be made to propagate and sustain a plasma even in a waveguide with a dimension slightly below cutoff with minimum reflection (~5%). We note that there is a better quasisteady state for $f = 9.45$ GHz, which can be attributed to the fact that the radius $a$ of the multicusp becomes comparable to $\lambda(n\lambda/4$, where $n = 4)$, which results in better power coupling between the field and the plasma.

Figure 16.12 shows the plot of $J_e$ versus $t$ for different peak powers of the incident microwave radiation at a fixed pressure and radius. We can see that the value of $J_e$ increases with the increase in microwave power. This is evident, from Equation 16.1, that the electron energy is directly dependent on the peak microwave power of the pulse.

Figure 16.13 shows the variation of the flattop duration with pressure. It is seen that for a fixed power, there is an optimum pressure at which the flattop

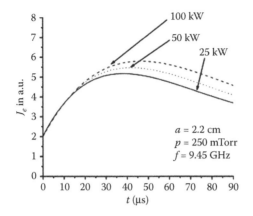

**FIGURE 16.12**
Temporal variation of plasma (electron) current density for different peak microwave powers at a pressure of 250 mTorr, radius of 2.2 cm, and input frequency of 9.45 GHz. (Reprinted with permission from S. Bhattacharjee et al. 2007. Quasisteady state interpulse plasmas, *J. Appl. Phys.* **101**: 113311. Copyright 2007, American Institute of Physics.)

duration is maximum. The simulation results are in good agreement with the experimental ones, particularly in the higher-pressure regime.

Figure 16.14 shows the variation of the flattop time $\tau_{flat}$ with microwave power at pressures of 10 and 60 mTorr. A good agreement is obtained with the experimental results. $\tau_{flat}$ shows a gradual increase with increase in microwave power and has a saturation tendency at a higher power ~100 kW.

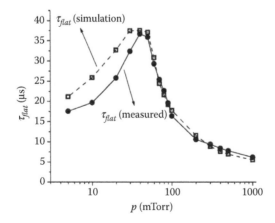

**FIGURE 16.13**
Variation of the flattop duration $\tau_{flat}$ with pressure. Dotted line and open squares, simulation; solid line and circles, experiment. The results are for a peak power of 100 kW. (Reprinted with permission from S. Bhattacharjee et al. 2007. Quasisteady state interpulse plasmas, *J. Appl. Phys.* **101**: 113311. Copyright 2007, American Institute of Physics.)

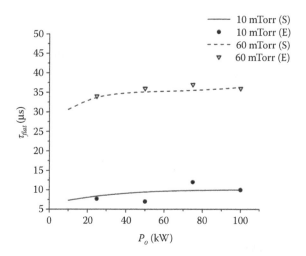

**FIGURE 16.14**

Variation of the flattop duration $\tau_{flat}$ with microwave powers at 10 and 60 mTorr. E: experimental values, S: simulation results. (Reprinted with permission from S. Bhattacharjee et al. 2007. Quasisteady state interpulse plasmas, *J. Appl. Phys.* **101**: 113311. Copyright 2007, American Institute of Physics.)

## Discussion and Conclusion

The simulation results are in good agreement with the experimental observations. The delayed plasma buildup tends to a steady state at lower pressures (~40 mTorr). The generation of the quasisteady state after the end of the microwave pulse depends upon the ionization processes in the interpulse phase, brought about by the energetic primary electrons generated during the pulse-on time by the waves of large amplitude. The steady-state duration depends primarily upon the discharge pressure and the decay rate $\rho(= D/\Lambda^2)$, which is related to the chamber size (Equation 16.6). The simulation results have predicted the existence of an optimum frequency, which couples to the plasma chamber more efficiently. In Figure 16.12, we see that the wave amplitude has an influence on the peak value of the current, although the peak current density appears after the microwave pulse is long gone. This is because the plasma current density depends jointly on the plasma electron density and the electron thermal energy. Higher wave amplitudes will therefore give rise to higher electron thermal energies (as seen from Equation 16.1). This effect will evidently be reflected in the peak value of the plasma current density. It was not possible to individually delineate the role played by the radiation decay time $\tau_d$, because a change in $\tau_d$ is intrinsically related to a change in the chamber radius, which in turn changes $\rho$ and the peak electric field $E_{R_0}$. However, we have found that its inclusion is significant as seen in

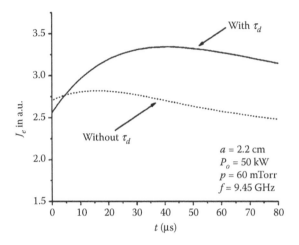

**FIGURE 16.15**

Comparison of the plasma (electron) current density evolution in the presence and in the absence of field diffusion time $\tau_d$ for $a = 2.2$ cm, peak power of 50 kW, at a pressure of 60 mTorr. (Reprinted with permission from S. Bhattacharjee et al. 2007. Quasisteady state interpulse plasmas, *J. Appl. Phys.* **101**: 113311. Copyright 2007, American Institute of Physics.)

Figure 16.15. Its effect is currently under further investigation. The secondary electron coefficient $\delta$ was found to extend the duration of the steady state to some extent in the case of chambers with smaller radius (~1 cm), but its effect at larger radii was found to be small.

The possibility of achieving a quasisteady state is interesting from the point of view of applications. From the experiments, we have seen that the duration of the steady state is controllable by a proper choice of background pressure, chamber size, and frequency. The pulse repetition frequency is another important factor that can provide control. Thus, one can produce a quasicontinuous plasma with high-frequency-pulsed sources of large amplitude by adjusting the pulse repetition parameters in such a fashion that before the plasma due to the first pulse decays to 90% of its maximum density, a second pulse arrives and reinforces the plasma. This may be quite economical, as one saves not only the expenses of acquiring high-frequency cw sources, but on the power budget as well. Furthermore, this scheme may offer a different method of auxiliary plasma heating in fusion plasmas or ion sources.

## References

1. A. Mizuno, R. Shimizu, A. Chakrabarti, L. Dascalescu, and S. Furuta. 1995. NOx removal process using pulsed discharge plasma, *IEEE Trans. Ind. Appl.* **31**(5): 957–963.

2. Y. Xu, P.R. Berger, J. Cho, and R.B. Timmons. 2006. Pulsed plasma polymerized dichlorotetramethyldisiloxane high-k gate dielectrics for polymer field effect transistors, *J. Appl. Phys.* **99**: 014104–09.

3. P. Subramonium and M.J. Kushner. 2004. Pulsed plasmas as a method to improve uniformity during materials processing, *J. Appl. Phys.* **96**: 82.

4. C.A. Sullivan, W.W. Destler, J. Rodgers, and Z. Segalov. 1988. Short-pulse high-power microwave propagation in the atmosphere, *J. Appl. Phys.* **63**: 5228.

5. S.P. Kuo, Y.S. Zhang, and P. Kossey. 1990. Propagation of high power microwave pulses in air breakdown environment, *J. Appl. Phys.* **67**(6): 2762–2766.

6. G.D. Conway, A.J. Perry, and R.W. Boswell. 1998. Evolution of ion and electron energy distributions in pulsed helicon plasma discharges, *Plasma Sources Sci. Technol.* **7**: 337.

7. B. Larisch, U. Brusky, and H.J. Spies. 1999. Plasma nitriding of stainless steel at low temperature, *Surf. Coat. Technol.* **116–119**: 205–211.

8. St. Behle, A. Brockhaus, and J. Engemann. 2000. Time-resolved investigations of pulsed microwave excited plasmas, *Plasma Sources Sci. Technol.* **9**: 57.

9. M.J. Kushner. 1993. Pulsed plasma-pulsed injection sources for remote plasma activated chemical vapor deposition, *J. Appl. Phys.* **73**: 4098.

10. T. Mieno and S. Samukawa. 1995. Time-variation of plasma properties in a pulse-time-modulated electron-cyclotron-resonance discharge of chlorine gas, *Jpn. J. Appl. Phys. Part 2* **34**: L1079–L1082.

11. S. Samukawa, H. Ohtake, and T. Mieno. 1996. Pulse-time-modulated electron cyclotron resonance plasma discharge for highly selective, highly anisotropic, and charge-free etching, *J. Vac. Sci. Technol. A, Vac. Surf. Films* **14**(6): 3049–3058.

12. V. Rousseau, S. Pasquiers, C. Boisse-Laporte, G. Callède, P. Leprince, J. Marec, and V. Peuch. 1992. Efficient pulsed microwave excitation of a high-pressure excimer discharge, *J. Appl. Phys.* **71**: 5712.

13. P. Sortias. 1992. Pulsed ECR ion source using the afterglow mode, *Rev. Sci. Instrum.* **63**: 2801.

14. S. Bhattacharjee and H. Amemiya. 1998. Interpulse plasma of a high-power narrow-bandwidth pulsed microwave discharge, *J. Appl. Phys.* **84**: 115.

15. S. Bhattacharjee, H. Amemiya, and Y. Yano. 2001. Plasma build-up by short-pulse high-power microwaves, *J. Appl. Phys.* **89**: 3573.

16. S. Bhattacharjee and H. Amemiya. 2000. Pulsed microwave plasma production in a conducting tube with a radius below cutoff, *Vacuum* **58**: 222.

17. S. Bhattacharjee and H. Amemiya. 2000. Production of pulsed microwave plasma in a tube with a radius below the cutoff value, *J. Phys. D* **33**: 1104.

18. D. Vender, H.B. Smith, and R.W. Boswell. 1996. Simulations of multipactor-assisted breakdown in radio frequency plasmas, *J. Appl. Phys.* **80**: 4292.

19. S. Bhattacharjee and H. Amemiya. 1999. Production of microwave plasma in narrow cross-sectional tubes; effect of the shape of cross section, *Rev. Sci. Instrum.* **70**: 3332.

20. S. Bhattacharjee, I. Dey, A. Sen, and H. Amemiya. 2007. Quasisteady state inter-pulse plasmas, *J. Appl. Phys.* **101**: 113311.

# 17

## Electron Plasma Waves Inside Large-Amplitude Electromagnetic Pulses

### Introduction

Wave plasma interaction has been widely investigated, especially with plasmas that are in steady state, inhomogeneous, and nonmagnetic in nature [1–7]. Some of the observed consequences are generation of high-energy electron bunches [1,2], excitation of temporal plasma-wave echoes in the ion wave regime [3], excitation of ion-wave wake field [4], large-amplitude electron plasma wave (EPW)-based plasma accelerators [5,6], and self-generation of magnetic fields [7]. The propagation of electromagnetic (EM) waves through temporally growing plasmas has also been analyzed, mainly in theory [8–10], including the study of dynamic behavior of microwaves through a gaseous medium that get ionized in the incident wave field [10]. It is shown that the waves can propagate through plasmas that are overdense in the steady-state approximation [9–11]. There are reports on the observation of charge oscillations in preformed steady-state plasmas subjected to an external electric field [12–14] where the plasma is produced by a different source. The plasma oscillations are analyzed in collision-free and fully ionized regimes in understanding the underlying phenomena [12–14].

This chapter is on the observation of EPW excited during interaction of high-power (~10 kW), short-pulse (~20 μs) EM waves in the microwave regime with a gaseous medium. The phenomena reported here occur within the pulses of the EM waves and in a temporally growing plasma where the plasma electron and ion densities $n_e(t)$ and $n_i(t)$ are both functions of time. In the ambient state before plasma initiation, the electron (ion) numbers are typically $10^3$–$10^4$ cm$^{-3}$. Moreover, different from earlier works of exciting EPW in a preformed plasma, here measurements were mainly carried out after the end of the pulse in the power-off phase, in the present experiment, the temporal evolution of the excited EPW is investigated inside the EM pulses by taking time-resolved measurements and looking at the spatial evolution of the waves.

As soon as the microwave pulse is launched into the confinement chamber containing argon at low pressure (~mTorr), EPWs are excited, which last for

a few microseconds before getting damped, and thereafter intense ionization and rapid growth of the plasma are observed. Three distinct phases
of plasma evolution can be identified: initial oscillatory phase of the EPW,
growth of the resulting ionization beyond the waves, and finally a decay
phase of the plasma. The later growth and decay phases have been investigated in earlier works [15–18]. The focus of this chapter is on the phenomena
occurring inside the pulses of the EM waves, in particular the origin and
characteristics of the EPW. We attempt to understand the physics behind the
generation of the plasma waves in the growing plasma, and address ques-
tions such as frequency and wavelength of the waves. The possible mechanisms of damping are also investigated through modeling and comparison
of experimental results with prediction from the model.

## Experimental Setup

A schematic of the experimental setup is shown in Figure 17.1a and consists of a stainless-steel vacuum chamber (VC), of length 50 cm and diameter 20 cm. Pulsed microwaves of 2.45 GHz with a peak output power of
~8–10 kW, pulse duration 20 μs, and repetition period variable in the range
500–2500 μs are produced by a magnetron generator (MWG) controlled by
a control unit (CU) and a power supply and controller (PSC; Richardson
Electronics PM740). In the present experiment, the pulse period is fixed at
2500 μs. The pulsed microwaves are guided into the experimental chamber
via standard WR340 waveguides. The water-cooled isolator (ISO) protects
the magnetron from the reflected power. The triple stub tuner (TST) is used
for impedance matching between the source and the load. In the experiment,
it is noted that the time period of the microwave $T(\sim 10^{-9}$ s$) \ll$ the duration of
the pulse ($\sim 10^{-5}$ s).

A 12-pole magnetic multicusp (MC) [19,20] with an inner diameter 12 cm
and length 25 cm is placed coaxially at a distance of 4.5 cm from the microwave entrance window (quartz window) of the VC for confining the plasma
with a minimum-$B$ field. Figure 17.1b shows the radial profile of the magnetic
field in the 12-pole MC. The magnetic field varies as $B_0(r) = B_p(r/a)^{(n/2-1)}$, where
$n$ (= 12) and $a$ are number of poles and the radius of the MC and $B_p$ is the
magnetic field at the pole (surface) of the magnet. The MC helps in the generation of the plasma by near-boundary resonances [19,21–23] and also provides radial confinement of the plasma. Moreover, both ends of the MC are
end-plugged for confining the plasma in the axial direction as well [21,22].

The base pressure in the VC is maintained below $10^{-6}$ Torr before the experiment. Argon is used as the experimental gas at an operating pressure of 0.2–
2.5 mTorr and is controlled with a mass flow controller (MKS Type 246). A
planar Langmuir probe (LP) of diameter 4 mm (made of stainless steel) can

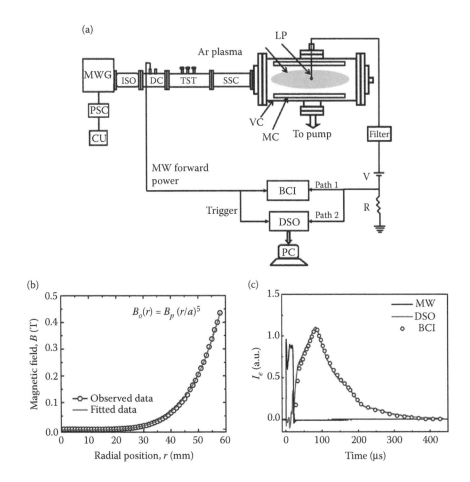

**FIGURE 17.1**

(a) Schematic of the experimental setup: CU: control unit, PSC: power supply and controller, MWG: microwave generator, ISO: isolator, DC: directional coupler, TST: triple stub tuner, SSC: straight section, VC: vacuum chamber, MC: multicusp, LP: Langmuir's probe, BCI: Boxcar's integrator, and DSO: digital oscilloscope. (b) Radial profile of a magnetic field in the multicusp shown in open circles along with the fitting given by $B_0(r) = B_p(r/a)^5$ with continuous line. (c) The electron current $I_e$ (t) as observed in the DSO (continuous line) and that taken from BCI (open circle) on adjusting its gate width to 100 ns shows close agreement with each other. The microwave pulse of 20 μs duration is also indicated in the plot. (Reprinted with permission from S. Pandey, D. Sahu, and S. Bhattacharjee. 2012. Transition from interpulse to afterglow plasmas driven by repetitive short-pulse microwaves in a multicusp magnetic field, *Phys. Plasmas Lett.* **19**: 80703. Copyright 2012, American Institute of Physics.)

be inserted either along the radial port or along the axis of the MC, where the magnetic field is almost zero. The surface of the probe is kept parallel to the wave vector to avoid any disturbance by microwaves [18,24]. Further, any possible disturbance from the microwaves in the dc measurements and any ac component is filtered out by introducing a pi-filter (Tusonix 4100-003) before extraction of the current from LP. The voltage drop across resistance R corresponding to the electron (or ion) current can either be fed to a digital oscilloscope (DSO; Tektronix TDS2014; path 2), or to a Boxcar's integrator (BCI; Stanford Research Systems SRS250; path 1), as indicated in Figure 17.1a. The trigger signal is applied from the microwave forward power, as shown in Figure 17.1a.

The temporal profile of the probe current is acquired in a DC-coupling mode using the DSO, with time averaging over 64 shots. The LP is operated in a fixed-voltage mode at $V_p = 10$ V for electron current measurement in the electron retardation region.

Figure 17.1c shows the electron current profile obtained using the BCI (open circles) and the DSO (continuous line). The microwave pulse is also shown in Figure 17.1c, with a pulse width of 20 μs. For this measurement, the gate width of the BCI is kept at 100 ns and the data is averaged over 1000 shots for reproducing the electron current temporal profile as observed in the oscilloscope.

Time-resolved LP characteristics and the spatial profile of the EPW are measured using the BCI. For time-resolved measurement of the probe characteristics, time delay of the BCI is adjusted with respect to the beginning of the microwave pulse, taken as $t = 0$. At each point in time, the probe bias is swept from −80 to +40 V. The plasma parameters are then calculated from these probe characteristics.

For spatiotemporal measurement of the EPW, as described toward the end of the section "Experimental Results," the BCI gate width is set at 5 ns. This resolution is smaller than the period of the plasma wave, and the spatial resolution is maintained at (Δz) 10 mm. At each axial location z, the electron current is measured by sweeping the time delay from 1 to 4 μs in steps of 5 ns, using the "external delay control" feature of the BCI and a personal computer (PC) employing a data-acquisition system (DAQ) and LABVIEW software. The probe is maintained at probe bias $V_p = 10$ V.

## Experimental Results

It is difficult to obtain reliable LP characteristics during the first few microseconds of the pulse ($t = 0$–4 μs), as the plasma is not properly formed and the densities are small. Hence, the plasma parameters were not possible to be estimated during this period. The experimental results of the excited EPW occurring within the pulse are discussed next. Figure 17.2a shows a typical temporal variation of the electron current as observed in the DSO. Three distinct stages of plasma evolution are depicted in Figure 17.2a. The expanded view of the EPW is shown in the inset. As soon as the microwave pulse is turned on (i.e., $t = 0$ μs), the EPWs are initiated and last for a few microseconds (~3 μs) of the

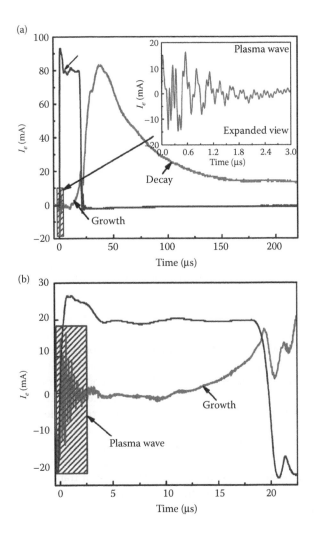

**FIGURE 17.2**
(a) Typical electron current profile indicating the three phases of the plasma evolution. Inset shows the expanded view from 0 to 3 µs showing the electron plasma wave and (b) magnified view of the electron current profile observed within the pulse (until 20 µs). (Reprinted with permission from S. Pandey, D. Sahu, and S. Bhattacharjee. 2012. Transition from interpulse to afterglow plasmas driven by repetitive short-pulse microwaves in a multicusp magnetic field, *Phys. Plasmas Lett.* **19**: 80703. Copyright 2012, American Institute of Physics.)

pulse-on time (Figure 17.2b, where an expanded view of the current profile inside the pulse is shown), and thereafter begins growth of the electron current initiated within the pulse and continues well beyond the end of the pulse (Figure 17.2a and b), as has been reported earlier [16–18]. The peak value of the electron current ($I_{max}$) is attained after the end of the pulse (around 40 µs from

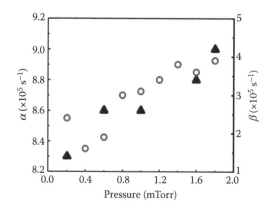

**FIGURE 17.3**

Variation of oscillation decay factor $\alpha(s^{-1})$ (▲) and electron current growth rate $\beta(s^{-1})$(○) with pressure. (Reprinted with permission from S. Pandey, D. Sahu, and S. Bhattacharjee. 2012. Transition from interpulse to afterglow plasmas driven by repetitive short-pulse microwaves in a multicusp magnetic field, *Phys. Plasmas Lett.* **19**: 80703. Copyright 2012, American Institute of Physics.)

start of the pulse, $t = 0$), depending upon the pressure, which is in conformity to earlier experimental results with shorter-duration pulses.

From the electron plasma oscillations, as shown in the inset of Figure 17.2a, an exponential decay factor $\alpha$ is obtained by considering the peak amplitude of the waves. Figure 17.3 shows that $\alpha$ increases with pressure, indicating that electron–neutral collisions play an important role in damping these waves. The growth rate $\beta$ of the electron current after the end of the plasma wave is estimated to be of the order of $10^5$ s$^{-1}$, as shown in Figure 17.3, and increases with pressure, possibly because of the increase in the ionization rate due to increase in neutral density. The present data extends the results toward lower pressure of an earlier work [17]. Table 17.1 summarizes some of the experimental results.

A representative set of probe characteristics after 4 μs is shown in Figure 17.4a. The temporal evolution of the plasma space potential $V_s(t)$, floating potential $V_f(t)$, plasma densities $n_e(t)$ and $n_i(t)$, and electron temperature $T_e(t)$

**TABLE 17.1**

Summary of Some of the Experimental Results

| Parameters | Unit | Value |
|---|---|---|
| Oscillation decay factor ($\alpha$) | s$^{-1}$ | $(8–9) \times 10^5$ |
| Oscillation decay time | μs | 2.7–4.8 |
| EPW frequencies | MHz | 3.8, 13.0 |
| Growth factor ($\beta$) | s$^{-1}$ | $(1.5–5) \times 10^5$ |
| Peak electron current ($I_{max}$) | mA | 4.5–130 |
| Time delay of $I_{max}$ from start of the pulse | μs | 23–69 |

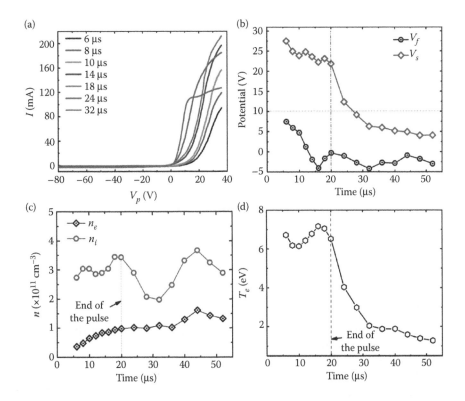

**FIGURE 17.4**

(a) A representative set of the Langmuir probe characteristic curves at different times of the microwave pulse from which the plasma parameters are obtained (at 1.0 mTorr). (b) Plasma space potential, $V_s(t)$, and floating potential, $V_f(t)$. (c) Plasma electron and ion densities, $n_e(t)$ and $n_i(t)$. (d) Electron temperature $T_e(t)$. (Reprinted with permission from S. Pandey, D. Sahu, and S. Bhattacharjee. 2012. Transition from interpulse to afterglow plasmas driven by repetitive short-pulse microwaves in a multicusp magnetic field, *Phys. Plasmas Lett.* **19**: 80703. Copyright 2012, American Institute of Physics.)

at 1.0 mTorr are shown in Figure 17.4b through d. It is observed that $V_f$ at $t \sim 6\,\mu s$ is around 7 V, and is close to the probe bias (10 V). It is considered that $V_f(t)$ during the time interval 1–3 μs oscillates around 10 V, which is the probe bias; therefore, one obtains positive and negative components of the current in EPW (Figure 17.2a). The plasma densities are estimated to be $n_i \sim (2.5\text{–}4.0) \times 10^{11} \text{cm}^{-3}$ and $n_e \sim (2\text{–}10) \times 10^{10} \text{cm}^{-3}$ during $t \sim 4\text{–}20\,\mu s$. The electron temperature $T_e$ acquires a high value ($\sim 7$ eV) within 4 μs of the pulse, but as the plasma density begins to rise, $T_e$ is seen to gradually decrease.

We report measurement of the electron current as shown in Figure 17.2 at a fixed probe bias ($V_P = 10$ V). Since the plasma potential $V_s(t)$ is varying with time (Figure 17.4b), the electron current profile $I_e(t)$ of Figure 17.2 is the electron saturation current only outside the pulse after $t \sim 25\,\mu s$, when $V_P \geq V_s$. However, below $t \sim 25\,\mu s$, the electron current in the retardation region is

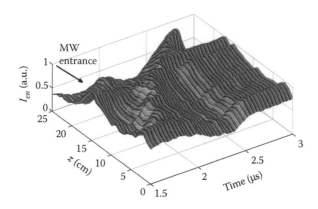

**FIGURE 17.5**

Spatiotemporal evolution of the normalized electron current ($I_{en}$) indicating electron plasma wave along the multicusp axis at different times from $t = 1.5–3\ \mu s$. (Reprinted with permission from S. Pandey, D. Sahu, and S. Bhattacharjee. 2012. Transition from interpulse to afterglow plasmas driven by repetitive short-pulse microwaves in a multicusp magnetic field, *Phys. Plasmas Lett.* **19**: 80703. Copyright 2012, American Institute of Physics.)

obtained. Our intention in this case is to obtain the nature of the electron current varying with time.

Attempts are made to measure the spatiotemporal profile of the EPW in the MC, and the results are presented in Figures 17.5 and 17.6. Time-resolved measurements are carried out using the BCI and a disk probe (5 mm

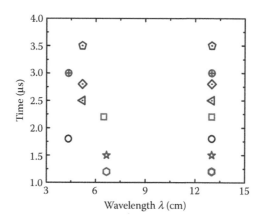

**FIGURE 17.6**

Time–wavelength distribution in the EPW as calculated from the spatiotemporal profile of the waves observed between $t = 1–4\ \mu s$. Identical symbols represent data corresponding to one particular time. (Reprinted with permission from S. Pandey, D. Sahu, and S. Bhattacharjee. 2012. Transition from interpulse to afterglow plasmas driven by repetitive short-pulse microwaves in a multicusp magnetic field, *Phys. Plasmas Lett.* **19**: 80703. Copyright 2012, American Institute of Physics.)

diameter) biased at 10 V. These results will be further improved upon by employing an electron energy analyzer in future experiments. Focus of the measurements is over the time duration of EPW.

Figure 17.5 shows the spatiotemporal profile of the electron current in the MC, which clearly indicates the presence of waves in space. During the time duration of 1.5–3 µs, large-amplitude waves propagating in space are observed, which is similar to the temporal profile of the EPW, as observed in the electron current shown in Figure 17.2. As time progresses until 3 µs, the peak of the current slowly gets enhanced, as well as gets shifted in position toward the microwave entrance side, indicating propagation of the waves. The wavelength of these waves consists mainly of two components, as shown in Figure 17.6, where time is plotted on the $y$ axis and the obtained wavelength is plotted on the $x$ axis. The mean lower wavelength is ~5.0 cm, while the higher component is ~13.0 cm. However, after 3 µs, the spatial current profile does not seem to change with time.

## Time–Frequency Analysis

A typical time series of the electron plasma oscillations is shown in Figure 17.7a. Conventional FFT results of the signal indicate the frequencies present in it; however, in order to understand the frequency variation with time, the EPWs are analyzed using the time–frequency analysis (TFA) [25]. Figure 17.7b shows the time–frequency distribution of the waves at 1.0 mTorr. The grey scale coding represents the power spectral density of the constituent frequencies. From Figure 17.7b, the simultaneous existence of two prominent frequency bands having frequencies centered around 3.8 and 13.0 MHz, respectively, and having a frequency spread $\Delta f \sim 4$ MHz can be seen. The two frequencies indicate that the plasma wave comprises primarily of two populations of electrons (identified by number densities $n_e1$ and $n_e2$, with mean energies of $T_e1$ and $T_e2$ in the plasma having thermal velocities ($v_{th} = (2k_BT_e/m_e)^{1/2}$) [26]. It is also seen that the power spectral density of these wave frequencies drops sharply within 0.4–1 µs, after which it has a slow decay until $t \sim 3$ µs.

## Model

### Temporal Variation of the Electron Current

When pulsed microwaves impinge on the gaseous medium, the free ambient electrons present in the MC are accelerated by the imposed electric field leading to the development of the EPW. At any instant, the applied wave

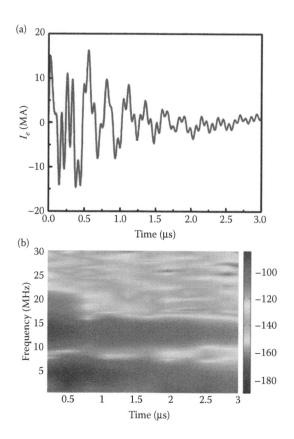

**FIGURE 17.7**

(a) Damped electron plasma oscillations until 3.0 μs. (b) Time frequency analysis (TFA) of the electron plasma waves, at 1.0 mTorr. (Reprinted with permission from S. Pandey, D. Sahu, and S. Bhattacharjee. 2012. Transition from interpulse to afterglow plasmas driven by repetitive short-pulse microwaves in a multicusp magnetic field, *Phys. Plasmas Lett.* **19**: 80703. Copyright 2012, American Institute of Physics.)

electric field $\vec{E}_a$ of frequency 2.45 GHz exerts force on the electrons and ions in opposite directions, and a region of space charge electric field $\vec{E}_{sc}$ is created. However, owing to the higher mass of ions, its motion may be neglected as in the current model. If $\vec{r}(t)$ is the instantaneous displacement of the electrons from its mean position, the net electric field experienced by the electrons in the plasma wave is $\vec{E}_a(t) - \vec{E}_{sc}(t)$. Furthermore, electrons also experience a frictional force $= -m_e v_{ce}(t)\vec{u}(t)$ due to the presence of background neutral gas particles. Here, $v_{ce}(t)$ is the electron–neutral elastic collision frequency, $m_e$ is the mass of the electron, and $\vec{u}(t) = d\vec{r}/dt$ is the instantaneous electron velocity. The frequency $v_{ce}(t) = N_g < \sigma(t)u(t) >$, where $<>$ symbolizes averaged value, $N_g$ is the neutral gas density, and $\sigma(t)$ is the elastic collision cross section for argon gas [27]. The force equation acting on an electron thus becomes

$$\vec{F} = -e\left\{\vec{E}_a(t) - \vec{E}_{sc}(t)\right\} - m_e v_{ce}(t)\vec{u}(t). \tag{17.1}$$

The space charge electric field $\vec{E}_{sc}$ is determined from Gauss's law as follows:

$$\vec{\nabla} \cdot \vec{E}_{sc}(t) = \frac{\rho(t)}{\varepsilon_0} = \frac{\{n_i(t) - n_e(t)\}e}{\varepsilon_0}. \tag{17.2}$$

Since the dominant mode of propagation of the microwaves (2.45 GHz) in the cylindrical chamber ($R = 6$ cm) is $TE_{11}$, the electric field $\vec{E}_a(t)$ in the central region of the plasma chamber and the space charge electric field $\vec{E}_{sc}$ are mainly along radial direction $\vec{r}$ while the propagation vector $\vec{k}$ is along the $z$ direction. For a probe kept at the center, the electron motion is primarily in the radial direction, as the force is along the radial direction; therefore, we may drop the vector notation. Considering the radial displacement to be $x$, the above equation simplifies to

$$E_{sc}(t) = \frac{-\{n_e(t) - n_i(t)\}e}{\varepsilon_0}x(t), \tag{17.3}$$

where $x(t)$ is the displacement of the electron from its mean position. Substituting Equation 17.3 into Equation 17.1 and replacing $\vec{u}$ by $dx(t)/dt$, the one-dimensional force equation takes the form of a forced-damped harmonic oscillator equation given by

$$m_e\frac{d^2x}{dt^2} + m_e v_{ce}(t)\frac{dx}{dt} + m_e \omega_r^2(t)x = qE_a(t), \tag{17.4}$$

where

$$\omega_r^2(t) = \frac{\{n_e(t) - n_i(t)\}e^2}{m_e \varepsilon_0}. \tag{17.5}$$

In the left-hand side of Equation 17.4, the second term acts as a damping force on the electrons due to the presence of electron–neutral elastic collisions, while the third term acts as the restoring force of the electrons.

In the experiment, the EM wave pulse has finite rise time ($t_1 = 370$ ns) and fall time ($\tau = 1.86$ μs); hence, $E_a(t)$ can be modeled as [18]

$$E_a(t) = \begin{cases} E_0 \exp\left(\dfrac{t - t_1}{t_1}\right), t < t_1 \\ E_0, \ t_1 \le t \le t_2 \\ E_0 \exp\left(-\dfrac{t - t_2}{\tau}\right), t > t_2 \end{cases}, \tag{17.6}$$

where $E_0$ is the amplitude of the electric field and $(t_2 - t_1) = 18.14$ µs is the period of constant electric field. However, the electric field $E_a(t)$ gets modified in the presence of plasma to $E_P(t)$ determined by the boundary condition, $\varepsilon_0 E_a(t) = \varepsilon_P(t) E_P(t)$. Here, $\varepsilon_0$ and $\varepsilon_P(t) = \varepsilon_0\{1 - \omega_P^2(t)/\omega^2\}$ are the permittivity of vacuum and plasma, respectively, and $\omega_P(t)$ is the electron plasma frequency. Because of elastic collisions, the electric field acting on the electrons will be further modified as $E_{eff}(t) = v_{ce}(\varepsilon_0/\varepsilon_P)E_a(t)/\sqrt{v_{ce}^2 + \omega^2}$ [17] and hence the effective microwave electric force on the electrons in the right-hand side of Equation 17.4 is $-eE_{eff}(t)\sin \omega t$.

The net charge density $[n_e(t) - n_i(t)]$ in Equation 17.5 is evaluated using the equation of continuity for each charged species [17]

$$\frac{\partial n_e(t)}{\partial t} = v_i(t)n_e(t) - D_a(t)\nabla^2 n_e(t), \tag{17.7a}$$

$$\frac{\partial n_i(t)}{\partial t} = v_i(t)n_e(t) - D_a(t)\nabla^2 n_i(t), \tag{17.7b}$$

where $v_i(t)$ is the ionization rate and $D_a = (e(T_e(t) + T_i(t))/m_e v_{ce}(t) + M v_{ci}(t)) \approx (eT_e(t)/m_e v_{ce}(t))$ is the ambipolar diffusion coefficient and is equivalent to the electron-free diffusion coefficient in the current experimental condition, since plasma is initially weakly ionized (i.e., $v_{ci}(t)$ can be neglected) and $T_e > T_i$. Taking the difference of the two equations in Equation 17.7, we obtain

$$\frac{\partial\{n_e(t) - n_i(t)\}}{\partial t} = -\frac{D_a(t)}{\Lambda^2}\{n_e(t) - n_i(t)\}, \tag{17.8}$$

where $\nabla^2$ is approximated as $1/\Lambda^2$, $\Lambda$ being the characteristic diffusion length, which is ~2.38 cm [18,21].

The temporal variation of the electron temperature $T_e(t)$ may be evaluated using the equation of power balance [18]

$$\frac{dT_e}{dt} = \frac{v_{ce}\{eE_a(t)\sin(\omega t)\}^2}{2m(\omega^2 + v_{ce}^2)} - \left(v_i E_i + v_{ex}E_{ex} + \frac{3m}{M}v_{ce}T_e\right), \tag{17.9}$$

where $E_i = 15.76$ eV and $E_{ex} = 11.6$ eV are ionization and average excitation threshold energies of the metastables ($^3P_2 = 11.5$ eV and $^3P_0 = 11.7$ eV), respectively, for argon atom; $M$ is the argon mass; and $v_{ex}(t)$ is the excitation frequency.

With the knowledge of $T_e(t)$ from Equation 17.9, Equations 17.4 and 17.8 are solved simultaneously in order to obtain electron velocity $u(t)$. Finally, the electron current density, $J_e(t) = en_e(t)u(t)$, is obtained by substituting $n_e(t)$ from Equation 17.7a. It is assumed that the electron distribution is a sum of two distributions, assumed to be Maxwellian. The initial electron densities and temperatures of these two populations are taken as $n_{e01} \sim 10^5 \, cm^{-3}$, $n_{e02} \sim 10^6 \, cm^{-3}$, $T_{e01} \sim 2 \, eV$, and $T_{e02} \sim 0.03 \, eV$. We further assume that these two populations do not interact (exchange energy) with each other in the time frame and that the mechanism of $T_e$ evolution for either of them is the same. The electron flux $J_{e1}$ and $J_{e2}$, corresponding to these populations, is obtained as discussed above. The final result $J_e(= J_{e1} + J_{e2})$ is the sum of these current densities. Figure 17.8a shows the electron current profile at 1.0 mTorr generated within the microwave pulse, while Figure 17.8b shows the expanded view of the current profile until 3 μs. We observe that there occur initial electron oscillations until ~3 μs of the incident pulse, after which the current begins to grow in agreement with the experimental results of Figure 17.2.

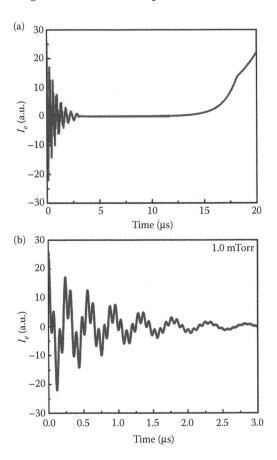

**FIGURE 17.8**
Modeling results on temporal profile of the electron current until: (a) 20 μs (within the microwave pulse) and (b) 3 μs of the microwave pulse. (Reprinted with permission from S. Pandey, D. Sahu, and S. Bhattacharjee. 2012. Transition from interpulse to afterglow plasmas driven by repetitive short-pulse microwaves in a multicusp magnetic field, *Phys. Plasmas Lett.* **19**: 80703. Copyright 2012, American Institute of Physics.)

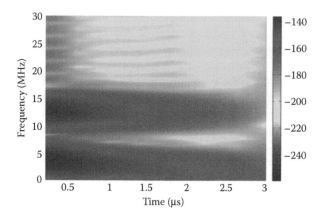

**FIGURE 17.9**
Time–frequency analysis of the EPW obtained from the model. (Reprinted with permission from S. Pandey, D. Sahu, and S. Bhattacharjee. 2012. Transition from interpulse to afterglow plasmas driven by repetitive short-pulse microwaves in a multicusp magnetic field, *Phys. Plasmas Lett.* **19**: 80703. Copyright 2012, American Institute of Physics.)

The EPW as generated from the model is analyzed with TFA in order to understand the variation of frequencies with time, and the result is shown in Figure 17.9. It clearly indicates the presence of two frequency bands centered around 3.5 and 12.8 MHz, the frequency spread being ~3 MHz for each band. The result also reflects the fact that the power spectral density of these frequency bands drops rapidly until $t \sim 1$ μs. These observations are in close agreement with the experimentally obtained results.

## Spatial Variation of the Electron Current

The spatial profile of the EPW cannot be explained by the above model, as it does not include the effect of the magnetic field $\vec{B}_0(r,\theta)$, within the MC confinement device. The presence of an MC magnetic field results in axial movement of the electrons due to the presence of $\vec{J} \times \vec{B}_0$ force. The $J_r$ and $B_{0\theta}$ components lead electrons to move in the axial ($z$) direction. In order to understand this effect, the electron dynamics are investigated in the MC field in the presence of bounded $TE_{11}$-mode microwave and electron–neutral collisions, using a Monte Carlo simulation code [28,29].

The vacuum $TE_{11}$ mode is modified due to the plasma dispersion; hence, the axial wave magnetic field amplitude can be expressed in terms of vacuum electric field amplitude $E_0$ for a cylindrical waveguide of radius $a$ as [30] $B_z(r,\theta) = (E_0/c)J_1(k_c r)\sin\theta$, where $k_c (=1.841/a)$ is the cutoff wave number for the $TE_{11}$ mode. The electric field in the presence of plasma gets modified as [11,31]

$$\vec{E}_P(\vec{r}) = E_{rP}(r,\theta)\hat{r} + E_{\theta P}(r,\theta)\hat{\theta}, \tag{17.10}$$

where

$$\vec{E}_{rP}(r,\theta) = \left(\frac{i\omega}{D_0}\right)\left[\frac{i}{r}\left(k_0^2 K_{22} - k_z^2\right)\right.$$

$$\left. + \frac{1}{r}\left(k_{20}K_{22} - k_z^2\right)\frac{\partial}{\partial\theta} + k_0^2 K_{12}\frac{\partial}{\partial r}\right]B_z(r,\theta) \tag{17.11a}$$

and

$$\vec{E}_{\theta P}(r,\theta) = \left(\frac{i\omega}{D_0}\right)\left[\frac{i}{r}k_0^2 K_{21} + \frac{1}{r}k_0^2 K_{21}\frac{\partial}{\partial\theta} + \left(k_0^2 K_{11} - k_z^2\right)\frac{\partial}{\partial r}\right]B_z(r,\theta) \tag{17.11b}$$

are the radial and polar components. Here, $K_{ij}$ are the components of the dielectric tensor derived for the $k \perp B_0$ mode of propagation [11,31] $D_0 = \left(k_0^2 K_{11} - k_z^2\right)\left(k_0^2 K_{22} - k_z^2\right) - k_0^4 K_{12}K_{21}$, $k_0$ is the wave number in free space, and $k_z$ is the axial wave number in the presence of plasma. The axial $k_z$ and radial $k_r$ wave numbers defined by $k_z(r,\theta)^2 = k_0^2 K_{22}(r,\theta) - k_c^2$ and $k_r(r,\theta)^2 = k_0^2 K_{11}(r,\theta) - k_z^2$ are implemented in the phase part $\exp i(\vec{k} \cdot \vec{r} - \omega t)$ of the wave.

The MC magnetic field, $\vec{B}_0(r,\theta)$, in cylindrical coordinates is implemented using the equations [32] $B_{or} = B_P(r/a)^{(n/2-1)}\sin(n\theta/2), B_{o\theta} = -B_P(r/a)^{(n/2-1)}\cos(n\theta/2)$, and $B_{oz} = 0$, where $n$ is the number of poles. The equation of motion for an electron subject to the above electric and magnetic fields, modified by the presence of the temporally growing plasma, is given by

$$\ddot{\vec{r}} = -(e/m_e)\left[\vec{E}_P(\vec{r})\left\{\exp i\left(\vec{k}\cdot\vec{r} - \omega t\right)\right\} + \dot{\vec{r}} \times \vec{B}_0(\vec{r}) + \vec{E}_{int}(\vec{r})\right], \tag{17.12}$$

where $\vec{r}$ is the position vector of the electron, $\vec{k}$ is the wave vector, and $\vec{E}_{int}(\vec{r})$ is the internal electric field generated due to the presence of electrons and ions in the volume. The Monte Carlo simulation is started with $n_{e0} = 10^4$ seed electrons that are present in the cylindrical volume defined by $a = 6.0$ cm and $L = 25.0$ cm. The electrons evolve together in time steps of $dt$ (~0.02 ns). The densities of electrons and ions are updated after each time step as they undergo collisions, gain energy, excite or ionize, or escape from the volume. The position and velocity vectors of the electrons are stored after 2 µs of time intervals, and phase space plot of the electron motion is obtained.

Figure 17.10a through d shows the phase space plots of the z-component of the electron velocity along the MC length with time. The distribution during the initial 2 µs indicates density striations and the development of the wave (Figure 17.5). At later times, the presence of a wave-like pattern in the axial electron distribution is demonstrated.

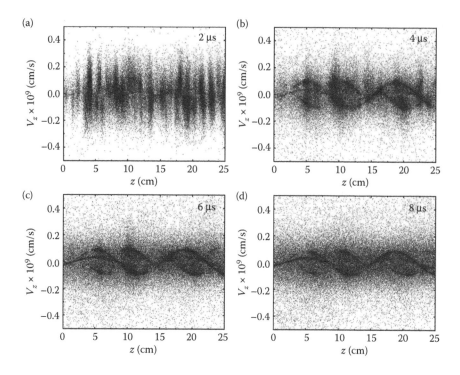

**FIGURE 17.10**
Time evolution of phase space plot of the electrons under the effect of pulsed microwave electric field (TE$_{11}$ mode) and MC magnetic field distribution at (a) $t = 2$ μs, (b) $t = 4$ μs, (c) $t = 6$ μs, and (d) $t = 8$ μs. (Reprinted with permission from S. Pandey, D. Sahu, and S. Bhattacharjee. 2012. Transition from interpulse to afterglow plasmas driven by repetitive short-pulse microwaves in a multicusp magnetic field, *Phys. Plasmas Lett.* **19**: 80703. Copyright 2012, American Institute of Physics.)

## Summary

In summary, when high-power, short-pulse microwaves impinge on a gaseous medium, EPWs are created that last for a few microseconds. TFA of the EPWs indicate simultaneous existence of two frequency bands centered around 3.8 and 13.0 MHz (with a frequency spread of 4 MHz), corresponding to two populations of electrons in the plasma wave. The temporal behavior of the electron current in the EPW is explained on the basis of a model obtained by considering the force on the electrons, the space charge field arising from charge separation between electrons and ions, electron–neutral collisions, and diffusion. It consequently leads to a forced damped oscillator equation, which can explain the electron kinetics reasonably well. The predictions from the model are in agreement with the experimental observations.

The spatial profile of the electron current suggests the longitudinality of the waves, which is explained by simulating electron dynamics in the MC magnetic field in the presence of bounded $TE_{11}$ mode microwaves, using a Monte Carlo simulation code.

## References

1. C. Rajyaguru, T. Fuji, H. Ito, N. Yugami, and Y. Nishida. 2001. Observation of ultrahigh-energy electrons by resonance absorption of high-power microwaves in a pulsed plasma, *Phys. Rev. E* **64**: 016403 (6 pages).
2. A.Y. Lee, Y. Nishida, N.C. Luhmann, Jr., S.P. Obenschain, B. Gu, M. Rhodes, J.R. Albritton, and E.A. Williams. 1982. Hot electrons produced by resonance absorption in a microwave-plasma interaction, *Phys. Rev. Lett.* **48**: 319–322.
3. N. Yugami, S. Kusawa, and Y. Nishida. 1994. Observation of temporal plasma-wave echoes in an ion-wave regime, *Phys. Rev. E* **49**: 2276–2281.
4. Md. Kamal-Al-Hassan, M. Starodubtsev, H. Ito, N. Yugami, and Y. Nishida. 2003. Excitation of ion-wave wakefield by the resonant absorption of a short pulsed microwave with plasma, *Phys. Rev. E* **68**: 036404.
5. J. Faure, Y. Glinec, A. Pukhov, S. Kiselev, S. Gordienko, E. Lefebvre, J.P. Rousseau, F. Burgy, and V. Malka. 2004. A laser–plasma accelerator producing monoenergetic electron beams, *Nature (London)* **431**: 541–544.
6. W.P. Leemans, B. Nagler, A.J. Gonasalves, C. Toth, K. Nakamura, C.G.R. Geedes, E. Esarey, C.B. Schroeder, and S.M. Hooker. 2006. GeV electron beams from a centimetre-scale accelerator, *Nat. Phys.* **2**: 696–699.
7. W.F. Divergilio, A.Y. Wong, H.C. Kim, and Y.C. Lee. 1977. Self-generated magnetic fields in the microwave plasma resonant interaction, *Phys. Rev. Lett.* **38**: 541–544.
8. D.K. Kalluri. 1988. On reflection from a suddenly created plasma half-space: Transient solution, *IEEE Trans. Plasma Sci.* **16**: 11.
9. M. Mirzaie, B. Shokri, and A.A. Rukhadze. 2010. The reflection of an electromagnetic wave from the self-produced plasma, *Phys. Plasmas* **17**: 012104.
10. M. Mirzaie, B. Shokri, and A.A. Rukhadze. 2008. The reflection index of an unsteady plasma, *J. Phys. D* **41**: 175210.
11. I. Dey and S. Bhattacharjee. 2011. Penetration and screening of perpendicularly launched electromagnetic waves through bounded supercritical plasma confined in multicusp magnetic field, *Phys. Plasmas* **18**: 22101.
12. L. Tonks and I. Langmuir. 1929. Oscillations in ionized gases, *Phys. Rev.* **33**: 195–210.
13. D. Bohm and E.P. Gross. 1949. Theory of plasma oscillations. B. excitation and damping of oscillations, *Phys. Rev.* **75**: 1864–1876.
14. D.H. Looney and S.C. Brown. 1954. The excitation of plasma oscillations, *Phys. Rev.* **93**: 965–969.
15. S. Bhattacharjee and H. Amemiya. 1998. Interpulse plasma of a high-power narrow-bandwidth pulsed microwave discharge, *J. Appl. Phys.* **84**: 115.
16. S. Bhattacharjee and H. Amemiya. 2000. Production of pulsed microwave plasma in a tube with a radius below the cutoff value, *J. Phys. D* **33**: 1104.

17. S. Bhattacharjee, H. Amemiya, and Y. Yano. 2001. Plasma build-up by short-pulse high-power microwaves, *J. Appl. Phys.* **89**: 3573.
18. S. Bhattacharjee, I. Dey, A. Sen, and H. Amemiya. 2007. Quasisteady state inter-pulse plasmas, *J. Appl. Phys.* **101**: 113311.
19. J.V. Mathew, I. Dey, and S. Bhattacharjee. 2007. Microwave guiding and intense plasma generation at sub-cutoff dimensions for focused ion beams, *Appl. Phys. Lett.* **91**: 41503.
20. M. Moissan and J. Pelletier (editors). 1992. *Microwave Excited Plasmas* (Elsevier, Amsterdam) p. 146.
21. S. Bhattacharjee and H. Amemiya. 1998. Microwave plasma in a multicusp cir-cular waveguide with a dimension below cutoff, *Jpn. J. Appl. Phys.* **37**: 5742.
22. M. Maeda and H. Amemiya. 1994. Electron cyclotron resonance plasma in mul-ticusp magnets with axial magnetic plugging, *Rev. Sci. Instrum.* **65**: 3751.
23. S. Bhattacharjee and H. Amemiya. 1999. Production of microwave plasma in narrow cross-sectional tubes; effect of the shape of cross section, *Rev. Sci. Instrum.* **70**: 3332.
24. J.V. Mathew, A. Chowdhury, and S. Bhattacharjee. 2008. Subcutoff microwave driven plasma ion sources for multi elemental focused ion beam systems, *Rev. Sci. Instrum.* **79**: 63504.
25. A.V. Oppenheim and R.W. Schafer. 1989. *Discrete-Time Signal Processing* (Prentice-Hall, Englewood Cliffs, NJ) p. 713.
26. F.F. Chen. 1984. *Introduction to Plasma Physics and Controlled Fusion* (Vol. 1, 2nd edition, Plenum, New York) p. 88.
27. M.A. Lieberman and A.J. Lichtenberg. 1994. *Principles of Plasma Discharges and Material Processing* (Wiley-Interscience, New York) p. 64.
28. I. Dey, J.V. Mathew, S. Bhattacharjee, and S. Jain. 2008. Sub-nanosecond electron transport in a gas in the presence of polarized electromagnetic waves, *J. Appl. Phys.* **103**: 83305.
29. S. Bhattacharjee and S. Paul. 2009. Random walk of electrons in a gas in the pres-ence of polarized electromagnetic waves: genesis of a wave induced discharge, *Phys. Plasmas* **16**: 104502.
30. D.M. Pozar. 2006. *Microwave Engineering* (3rd edition, John Wiley and Sons Inc., Singapore) p. 118.
31. I. Dey and S. Bhattacharjee. 2011. Anisotropy induced wave birefringence in bounded supercritical plasma confined in a multicusp magnetic field, *Appl. Phys. Lett.* **98**: 151501.
32. F.K. Azadboni, M. Sedaghatizade, and K. Sepanloo. 2010. Design studies of a multicusp ion source with femlab simulation, *J. Fusion Energ.* **29**(1): 5.
33. S. Pandey, D. Sahu, and S. Bhattacharjee. 2012. Transition from interpulse to afterglow plasmas driven by repetitive short-pulse microwaves in a multicusp magnetic field, *Phys. Plasmas* **19**: 80703.

# 18

## Transition from Interpulse to Afterglow Plasmas

## Introduction

In the power-off phase, plasmas generated by repetitive short-pulse microwaves in a multicusp (MC) magnetic field show a transitive nature from *interpulse* to *afterglow* as a function of pulse duration $t_w = 20$–$200$ µs. The ionized medium can be driven from a highly nonequilibrium to an equilibrium state inside the pulses, thereby dictating the behavior of the plasma in the power-off phase. Compared to afterglows, interpulse plasmas observed for $t_w < 50$ µs are characterized by a quasi-steady-state in electron density that persists for ~20–40 µs even after the end of the pulse, and has a relatively slower decay rate ($\sim 4.3 \times 10^4\,s^{-1}$) of the electron temperature, as corroborated by optical measurements. The associated electron energy probability function (EEPF) indicates depletion in low-energy electrons, which appear at higher energies just after the end of the pulse. The transition occurs at $t_w \sim 50$ µs, as confirmed by time evolution of integrated electron number densities obtained from the distribution function.

Pulse-modulated plasmas in the power-off phase exhibit interesting features that can be classified as an afterglow [1–6] or an interpulse plasma [7–9]. Afterglows are primarily characterized by a decaying plasma density ($n_o \sim 10^{10}$ cm$^{-3}$) and a small value of electron temperature $T_e \sim 1$–$2$ eV. This mode of plasmas in the power-off phase is employed in important applications such as electron–cyclotron–resonance ion sources (ECRIS) for extracting a higher charge-state ion current from the so-called afterglow peak [10–15]. Basic studies on afterglows have also been carried out extensively for improving plasma-assisted material processing [16–18], in view of improving growing film-surface quality. There have also been reports on understanding the underlying electron dynamics by generating time evolution of EEPF [19–23], mostly in rf-generated afterglows [20–23].

Interpulse plasmas, on the other hand, are an interesting plasma regime that has been earlier observed between pulses of high-power (60–100 kW), short-pulse (0.05–1.2 µs) microwaves [7–9], first reported by Bhattacharjee and Amemiya [7] in such plasmas using a radar microwave source. Contrary to afterglows, these plasmas are known to grow beyond the end of the pulse and are reported to have a quasi-steady-state period in the power-off phase between two successive pulses, with a relatively higher value of $T_e$ (~6–10 eV) undergoing a slow exponential decay [8]. Detailed measurements of the electron current and optical intensity, including modeling of the phenomena to predict time scales associated with interpulse plasma growth, quasi-steady-state, and decay, were carried out in Refs. [9] and [10]. There have been reports by other researchers on the determination of breakdown thresholds and for investigating microwave propagation through a gaseous medium (e.g., creation of artificially ionized layers in the atmosphere) [12–14].

Interpulse plasmas have been proposed to model ionospheric plasmas, for example, transition from a negative-ion-rich $D$-layer to a negative-ion-free $E$-layer [7]. Due to the presence of energetic electrons, it can be employed for generation of radicals [24] in plasma chemistry and production of high-current ion beams [25,26]. More recently, interpulse plasmas have been associated with excitation of electron plasma waves (EPWs) during initial plasma development within the pulse [27]. However, the fundamental question that arises is: under the same magnetic field configuration, what controls the plasma characteristics in the power-off phase of the microwave pulse leading to the two different natures of the plasma? One is interested to know the role of pulse duration ($t_w$), which governs the initial growth and subsequent evolution of electrons in the two cases and their subsequent effect on the plasma density and electron temperature in the power-off phase.

In this chapter, it is shown that: (i) the interpulse plasma state can be realized even with conventional microwave sources of 2.45 GHz of peak power 3.2 kW, having a repetition period of 2500 μs, and (ii) the transition from interpulse plasmas to afterglows as a function of pulse duration $t_w$ in the range 20–200 μs, which corresponds to an energy variation of 64–640 mJ.

## Experimental Setup

The experimental setup is shown in Figure 18.1a. It consists of a magnetron (MWG) that generates microwave pulses that are guided to a vacuum chamber (VC; having diameter $d = 20$ cm, length $l = 50$ cm) via a WR340 waveguide. A 12-pole MC ($d = 12$ cm, $l = 25$ cm) is used for plasma confinement in a minimum B-field configuration and is placed coaxially at a distance of 4.5 cm from the microwave entrance window (quartz) of the VC. Figure 18.1b shows a radial plot of the B-field, fitted with $B_o(r) = B_p(r/a)^{(n/2-1)}$, where ($n = 12$) and $a$ are the number of poles and radius of the MC, respectively, and $B_p$ is the magnetic field at the pole (surface) of the magnet (~0.45 T). The inset shows a Poisson simulation [28] of B-field lines in the MC cross section. It is seen

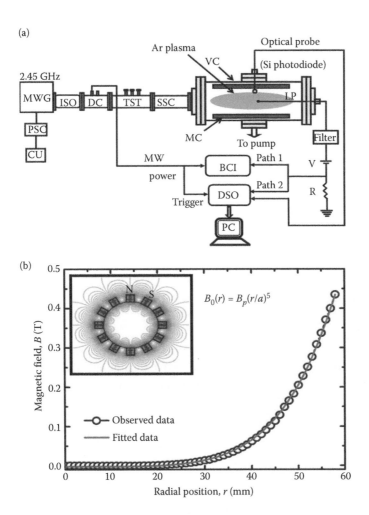

**FIGURE 18.1**
(a) Schematic of the experimental setup: CU: control unit, PSC: power supply and controller, MWG: microwave generator, ISO: isolator, DC: directional coupler, TST: triple stub tuner, SSC: straight section, VC: vacuum chamber, MC: multicusp, LP: Langmuir's probe, BCI: Boxcar's integrator, and DSO: digital oscilloscope. (b) Radial profile of a magnetic field in MC, shown in open circles along with the fitting, given as $B_0(r) = B_p(r/a)^5$, with continuous line. Inset shows cross-sectional view of the B-field lines in the MC using the Poisson simulation. (Reprinted with permission from S. Pandey, D. Sahu, and S. Bhattacharjee. 2012. Transition from interpulse to afterglow plasmas driven by repetitive short-pulse microwaves in a multicusp magnetic field, *Phys. Plasmas Lett.* **19**: 80703. Copyright 2012, American Institute of Physics.)

that the magnetic field is almost zero over a radius of 20 mm. The base pressure in the VC is maintained below $10^{-6}$ Torr before the experiment. Argon is used as the experimental gas at an operating pressure of 0.2–2.0 mTorr and is controlled with a mass flow controller (MKS Type 246). A Boxcar's integrator (BCI; Stanford Research Systems SRS250) is employed for time-resolved measurements (gate width 100 ns, averaging 1000). The plasma diagnostics are done using a planar Langmuir probe (LP) of $d = 4$ mm, an electron energy analyzer (EEA) probe [29], and a Si PIN photodiode (Hamamatsu Photonics S4349, spectral response range 190–1000 nm) with a fast response speed of 20 MHz. The plasma parameters are calculated from the current–voltage characteristics of the LP by sweeping its bias from −80 to +40 V. The EEA employed for generating EEPF consists of two grids and a collector, all having a diameter of 3 mm and spacing of 0.5 mm. The first grid is kept grounded so that a plasma sheath is formed at its boundary. The second grid is biased negatively ($V_d$) and is used as energy discriminator for the electrons. The electron current ($I_c$) is drawn by giving a positive bias to the collector. The EEPF is obtained by taking the first derivative of the collector current $I_c$ with respect to the discriminator bias $V_d$ [30]. All the measurements are taken at the center of the MC, where the magnetic field is almost zero.

## Results

Figure 18.2a and b illustrates the temporal profile of plasma density $n_e$ ($t$) and electron temperature $T_e$ ($t$) for $t_w = 20$ and 100 µs, respectively, at a pressure of 1.4 mTorr. An expanded view of $n_e$ ($t$) for $t_w = 20$ µs is indicated in the inset of Figure 18.2a. The $n_e$ ($t$) at the end of pulse is higher in the case of 100 µs owing to higher average power available for electron heating and ionization. Figure 18.2a shows that $n_e$ increases within the pulse, reaching a peak value ($7.0 \times 10^{10}$ cm$^{-3}$) at the end of the pulse, which is maintained for about 60 µs (*three times the pulse duration*) in the power-off phase, as shown in the inset, followed by a slow decay. The $T_e$ is ~4.8 eV at the pulse end and then decays exponentially, attaining 1 eV in ~90 µs. It may be noted that in earlier experiments [7–9] where higher powers were available, $T_e$ was relatively higher (~6–10 eV) at the end of the pulse and took a longer time to decay. Furthermore, slow decay of $T_e$ ($t$) suggests that the high-energy electrons are still present and continue the ionization process in the power-off phase. Earlier observation of a quasi-steady-state in the power-off phase supports this viewpoint [7–9].

At a larger $t_w = 100$ µs (Figure 18.2b), $n_e$ rises exponentially as the pulse is turned on, attaining a steady-state value ~$1.2 \times 10^{11}$ cm$^{-3}$ within the pulse in ~75 µs and then decays off exponentially after the pulse ends—typical of afterglow plasmas. The attainment of a steady state within the pulse and then an afterglow has been reported earlier in dc [1,2] and rf plasmas [13–15]. In this case, a steady state is attained within the pulse and the plasma electrons (ions) reach a state of equilibrium; therefore, after the end of the pulse, $T_e$ and $n_e$ both decrease. The presence of an afterglow peak in $n_e$, as described by many authors [13–15], is also evident in Figure 18.2b at the end of the pulse. A

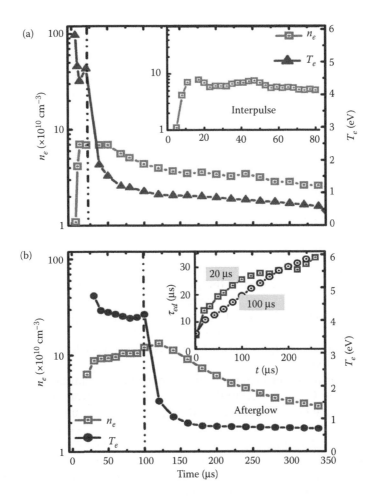

**FIGURE 18.2**
The $n_e$ (t) and $T_e$ (t) profile at $t_w$ (a) 20 μs, where inset shows an expanded view of $n_e$ (t) indicating interpulse plasma, and (b) 100 μs indicating afterglow plasma, where the inset shows the temporal profile of $\tau_{ed}$ after the end of the pulse (t = 0) in the two cases. (Reprinted with permission from S. Pandey, D. Sahu, and S. Bhattacharjee. 2012. Transition from interpulse to afterglow plasmas driven by repetitive short-pulse microwaves in a multicusp magnetic field, *Phys. Plasmas Lett.* **19**: 80703. Copyright 2012, American Institute of Physics.)

similar trend in $n_e$ (t) is obtained for $t_w > 100$ μs as well. A comparison of the decay of $T_e$ (t) in the two cases (cf. Figure 18.2) indicates that $T_e$ (t) ~ 0.75 eV is attained much faster in the case of afterglows.

The MC magnetic field plays an important role in plasma confinement in the two cases. A comparison of the variation of the electron diffusion time $\tau_{ed} = \Lambda^2/D_\perp$ (where $\Lambda$ and $D_\perp$ are the characteristic diffusion length and diffusion coefficient perpendicular to the magnetic field) in the above two cases is shown in the inset of Figure 18.2b for a magnetic field, $B \sim 10$ G near the center of the MC. The trend is the same at other values of the magnetic field away from the center of the MC. Here, $t = 0$ implies the end of the pulse and $D_\perp = eT_e/(mv_{en} + Mv_{in})(1 + \gamma B^2)$, where $v_{en}$ and $v_{in}$ are electron– and ion–neutral collision frequency, $m$ and $M$ are electron and ion mass, and $\gamma = e^2/(mMv_{en}v_{in})$. It is observed that $\tau_{ed} \sim 10$–$40$ μs, calculated by taking $T_e(t)$ and $n_e(t)$ profiles for $t_w = 20$ and $100$ μs. However, while it increases almost linearly for $t_w = 100$ μs, the increase is more like an exponential for $t_w = 20$ μs until $t \sim 200$ μs, after which both get converged. Higher value of $\tau_{ed}$ for $t_w = 20$ μs just after $t = 0$ indicates longer confinement of electrons in interpulse plasmas as compared to afterglows, and it is also confirmed from EEA measurements that it is the hot electrons that stay for longer periods in interpulse plasmas, as discussed later.

Figure 18.3a and b shows the EEPF profile at different time intervals after the end of the pulse, for $t_w = 20$ and $100$ μs, respectively. It may be noted that high-energy tail electrons having energies greater than $150$ eV are realized in the discharge. In interpulse plasma ($t_w = 20$ μs case), as $t$ increases from $20$ to $40$ μs just after the end of the pulse, there is a drop in the population of low-energy electrons and a corresponding increase in the population of high-energy electrons (see inset of Figure 18.3a). After $40$ μs, electrons of all energies begin to decrease. However, for $t_w = 100$ μs (Figure 18.3b), the high- as well as the low-energy electrons decrease continuously after the end of the pulse ($t > 100$ μs).

The number of hot electrons $N_h$ in an energy range can be evaluated by integrating the EEDF ($= \sqrt{\varepsilon} \times$ EEPF, where $\varepsilon$ is the electron energy) from a certain $E_i$ to $E_f$, where subscripts $i$ and $f$ represent initial and final values of the energy. $N_h$ is calculated in the energy range $15$–$150$ eV, and is normalized with the value obtained at the end of the pulse when $t_w = 20$ μs. Figure 18.4a shows the temporal profile of normalized $N_h$ values ($N_{hn}$) with the pulse duration $t_w$ as a parameter. The end of the pulse is denoted by $t = 0$. For afterglows ($t_w = 50$–$100$ μs), $N_{hn}$ is a maximum at $t = 0$ and then undergoes an exponential decrease, and it is observed that although $N_{hn}$ at $t = 0$ is higher for $t_w = 100$ μs, its subsequent decay is much faster as compared to lower $t_w$ values. At $t_w = 40$ μs, the maximum value of $N_{hn}$ is maintained until $t \sim 10$ μs and then $N_{hn}$ starts to decrease. As $t_w$ is further lowered, one attains the interpulse regime where the maximum value of $N_{hn}$ is realized later, after the end of the pulse (e.g., $t_w = 20$ μs). The transition between the two regimes is seen to occur at $t_w = 50$ μs. The high-energy electrons at $t_w = 20$ μs (interpulse), $50$ μs (transition state), and $100$ μs (afterglow) are further compared by plotting the temporal profile of collector current $I_{cn}$ (normalized to unity) of EEA at discriminator bias $V_d = 100$ V, from the end of the pulse ($t = 0$), as shown in Figure 18.4b. It clearly shows that the decay rate of $I_{cn}$ is faster at $t_w = 100$ μs ($6.1 \times 10^4$ s$^{-1}$) than that at $t_w = 20$ μs ($4.1 \times 10^4$ s$^{-1}$), implying that hot electrons persist for longer periods in interpulse plasmas after the end of the pulse.

Figure 18.5 shows the variation of the total optical intensity $I_{op}(t)$ (normalized to unity) versus time after the end of the pulse ($t = 0$), with $t_w$ as a

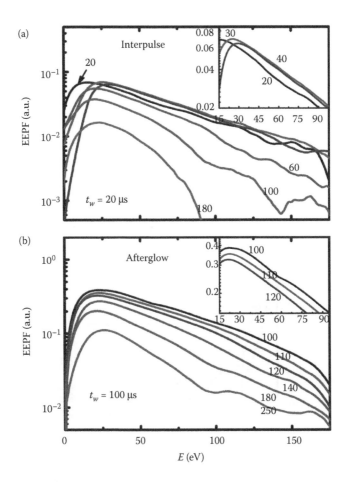

**FIGURE 18.3**

Time evolution of EEPF after the end of the pulse ($t = 0$) for $t_w$ (a) 20 μs and (b) 100 μs. Inset shows magnified view of EEPF within 90 eV of electron energy. (Reprinted with permission from S. Pandey, D. Sahu, and S. Bhattacharjee. 2012. Transition from interpulse to afterglow plasmas driven by repetitive short-pulse microwaves in a multicusp magnetic field, *Phys. Plasmas Lett.* **19**: 80703. Copyright 2012, American Institute of Physics.)

parameter indicated as highlighted numbers in the plots. The decrease in the intensity can be fitted with exponentials of the form $I_o \exp(-t/\tau)$, with the decay time constant $\tau$ showing a sharp decrease from 640 to 486 μs, as $t_w$ increases from 20 to 50 μs and then it decreases slowly with further increase in $t_w$. Furthermore, $I_{op}(t)$ depends on the excitation rate $\Gamma(t) = n_e(t) \langle \sigma_{ex}(t) \cdot v(t) \rangle$, where $\sigma_{ex}(t)$ is the excitation cross section, and $v(t)$ is the electron velocity. In its simplest form, $\Gamma \propto n_e (T_e/E_{ex})^2$, with $E_{ex}$ being the threshold excitation energy [9]. The ratio of $T_e(t)$ for $t_w = 20$ and 100 μs can then be written in terms of its $n_e(t)$ (Figure 18.2) and $I_{op}(t)$ profiles of Figure 18.5, as

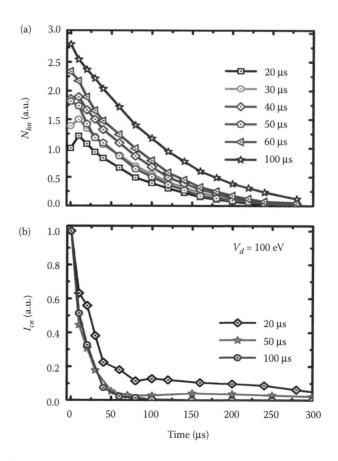

**FIGURE 18.4**

Temporal variation of (a) normalized number of hot electrons $N_{hn}$ having energy lying between 15 and 150 eV and (b) EEA collector current (normalized to unity) by keeping $V_d$ at 100 V for $t_w = 20$ µs (interpulse), 50 µs (transition state), and 100 µs (afterglow). Here, $t = 0$ implies the end of the pulse. (Reprinted with permission from S. Pandey, D. Sahu, and S. Bhattacharjee. 2012. Transition from interpulse to afterglow plasmas driven by repetitive short-pulse microwaves in a multicusp magnetic field, *Phys. Plasmas Lett.* **19**: 80703. Copyright 2012, American Institute of Physics.)

$$T_{20,100}(t) = \frac{T_{e,20}(t)}{T_{e,100}(t)} = \sqrt{\frac{n_{e,100}(t)\, I_{op,20}(t)}{n_{e,20}(t)\, I_{op,100}(t)}},$$ where subscripts 20 and 100 stand for

$t_w = 20$ and 100 µs, respectively. The relation is used to compare $T_e(t)$ profile in interpulse plasma ($t_w = 20$ µs) and afterglow ($t_w = 100$ µs). The resulting ratio $T_{20,100}(t)$ is seen to lie in the range ~1.0–1.4, after the end of the pulse ($t = 0$), and is shown in the inset of Figure 18.5. It first increases slowly after $t = 0$, attains a maximum, and then begins to drop after $t = 50$ µs. The drop in $T_{20,100}(t)$ indicates that the decay rate of $T_e(t)$ for $t_w = 100$ µs is faster in comparison to that for $t_w = 20$ µs in this period. Finally, after $t = 200$ µs, $T_{20,100}(t)$

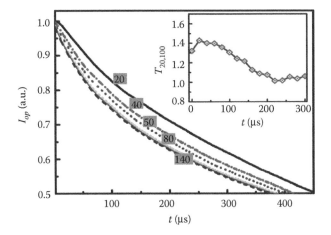

**FIGURE 18.5**

Temporal decay of total (normalized) optical intensity $I_{op}$ (t) after the end of the pulse ($t = 0$), as $t_w$ is varied from 20 to 140 μs, shown as numbers in the plots. A plot of temporal variation in ratio $T_{20,100}$ (t) calculated from $n_e$ (t) and $I_{op}$ (t) is shown in the inset. (Reprinted with permission from S. Pandey, D. Sahu, and S. Bhattacharjee. 2012. Transition from interpulse to afterglow plasmas driven by repetitive short-pulse microwaves in a multicusp magnetic field, *Phys. Plasmas Lett.* **19**: 80703. Copyright 2012, American Institute of Physics.)

tends to saturate to its minimum, indicating that $T_e$ (t) does not change with time anymore, in both the cases, as is also evident from Figure 18.2. The ratio of $T_e$ (t) calculated from LP experiments for $t_w = 20$ and 100 μs (Figure 18.2) shows a similar trend, with $T_{20,100}$ (t) lying in the range ~1.1–1.4.

<hr/>

## Summary

In summary, pulsed plasmas generated by repetitive short-pulse microwaves in an MC magnetic field are investigated for $t_w$ ~ 20–200 μs. Such plasmas in the power-off phase are known to be governed by the presence of energetic electrons, which are found to be a function of $t_w$, and determine whether the plasma will exhibit interpulse or an afterglow state. At $t_w < 50$ μs, $n_e$ (t) shows a quasi-steady-state condition after the end of the pulse for a period greater than $t_w$ itself, unlike the afterglows observed for $t_w > 50$ μs. The evolution of EEPF with time suggests an increase in $N_{hn}$ after the end of the pulse, in case of interpulse plasmas. Furthermore, the EEA probe results indicate that the decay rate of hot electrons (>100 eV) are faster in afterglows as compared to interpulse plasmas. The results of the LP measurements of electron temperature agree reasonably with those obtained from optical intensity ratios. To conclude, an interesting evolution of the state of a plasma from an interpulse obtained at shorter pulse duration ($t_w < 50$ μs) to an afterglow obtained at longer pulse duration ($t_w > 50$ μs) is observed experimentally for the first time. By tuning $t_w$, the population of energetic electrons and hence the ionization can be controlled in the power-off phase.

# References

1. R.J. Freiberg and L.A. Weaver. 1968. Microwave investigation of the transition from ambipolar to free diffusion in afterglow plasmas, *Phys. Rev.* **170**: 336–341.
2. M.M. Pejovic, J.P. Karamarkovic, G.S. Ristic, and M.M. Pejovic. 2008. Analysis of neutral active particle loss in afterglow in krypton at 2.6-mbar pressure, *Phys. Plasmas* **15**: 013502.
3. P. Diomede, S. Longo, and M. Capitelli. 2006. Charged particle dynamics and molecular kinetics in the hydrogen postdischarge plasma, *Phys. Plasmas* **13**: 113505.
4. V.A. Godyak, R.B. Piejak, and B.M. Alexandrovich. 2002. Electron energy distribution function measurements and plasma parameters in inductively coupled argon plasma, *Plasma Sources Sci. Technol.* **11**: 525.
5. H.-C. Lee, M.-H. Lee, and C.-W. Chung. 2010. Experimental observation of the transition from nonlocal to local electron kinetics in inductively coupled plasmas, *Appl. Phys. Lett.* **96**: 041503.
6. A. Agarwal, P.J. Stout, S. Banna, S. Rauf, and K. Collins. 2012. Decreasing high ion energy during transition in pulsed inductively coupled plasmas, *Appl. Phys. Lett.* **100**: 044105.
7. S. Bhattacharjee and H. Amemiya. 1998. Interpulse plasma of a high-power narrow-bandwidth pulsed microwave discharge, *J. Appl. Phys.* **84**: 115.
8. S. Bhattacharjee and H. Amemiya. 2000. Production of pulsed microwave plasma in a tube with a radius below the cutoff value, *J. Phys. D: Appl. Phys.* **33**: 1104.
9. S. Bhattacharjee, I. Dey, A. Sen, and H. Amemiya. 2007. Quasi steady state interpulse plasmas, *J. Appl. Phys.* **10**: 113311.
10. K. Langbein. 1996. Experimental investigation of the afterglow of the pulsed electron cyclotron resonance discharge, *Rev. Sci. Instrum.* **67**: 1334.
11. G.A. Askaryan, G.M. Batanov, I.A. Kossyi, and A.Yu. Kostinskii. 1991. *Sov. J. Plasma Phys.* **17**: 48.
12. C.A. Sullivan, W.W. Destler, J. Rodgers, and Z. Segalov. 1988. Short-high-microwave propagation in the atmosphere, *J. Appl. Phys.* **63**: 5228.
13. W.M. Bollen, C.L. Yee, A.W. Ali, M.J. Nagurney, and M.E. Read. 1983. High-power microwave energy coupling to nitrogen during breakdown, *J. Appl. Phys.* **54**: 101.
14. S.P. Kuo, Y.S. Zhang, and P. Kossey. 1990. Propagation of high power microwave pulses in air breakdown environment, *J. Appl. Phys.* **67**(6): 2762–2766.
15. C.E. Hill, D. Küchler, F. Wenander, and B.H. Wolf. 2000. Effect of a biased probe on the afterglow operation of an ECR4 ion source, *Rev. Sci. Instrum.* **71**: 863.
16. L. Maunoury, L. Adoui, J.P. Grandin, F. Noury, B.A. Hube, E. Lamour, C. Prigent et al. 2008. Afterglow mode and the new micropulsed beam mode applied to an electron cyclotron resonance ion source, *Rev. Sci. Instrum.* **79**: 02A313.

17. A. Georg, J. Engemann, and A. Brockhaus. 2002. Investigation of a pulsed oxygen microwave plasma by time-resolved two-photon allowed laser-induced fluorescence, *J. Phys. D: Appl. Phys.* **35**: 875.

18. A. Brockhaus, G.F. Leu, V. Selenin, Kh. Tarnev, and J. Engemann. 2006. Electron release in the afterglow of a pulsed inductively-coupled radiofrequency oxygen plasma, *Plasma Sources Sci. Technol.* **15**: 171.

19. M. Meško, Z. Bonaventura, P. Vašina, V. Kudrle, A. Tálský, D. Trunec, Z. Frgala, and J. Janča. 2006. An experimental study of high power microwave pulsed discharge in nitrogen, *Plasma Sources Sci. Technol.* **15**: 574.

20. S. Samukawa. 1996. Pulse-time-modulated electron cyclotron resonance plasma etching with low radio-frequency substrate bias, *Appl. Phys. Lett.* **68**: 316.

21. P. Subramonium and M.J. Kushner. 2004. Pulsed inductively coupled chlorine plasmas in the presence of a substrate bias, *Appl. Phys. Lett.* **79**: 2145 2001; P. Subramonium and M.J. Kushner. 2004. Pulsed plasmas as a method to improve uniformity during materials processing, *J. Appl. Phys.* **96**: 82.

22. O.V. Vozniy and G.Y. Yeom. 2009. High-energy negative ion beam obtained from pulsed inductively coupled plasma for charge-free etching process, *Appl. Phys. Lett.* **94**: 231502.

23. R. Hugon, G. Henrion, and M. Fabry. 1996. Time resolved determination of the electron energy distribution function in a DC pulsed plasma, *Plasma Sources Sci. Technol.* **5**: 553.

24. C.A. DeJoseph, V.I. Demidov, and A.A. Kudryavtsev. 2007. Nonlocal effects in a bounded low-temperature plasma with fast electrons, *Phys. Plasmas* **14**: 057101.

25. R.R. Arslanbekov, A.A. Kudryavtsev, and L.D. Tsendin. 2001. Electron-distribution-function cutoff mechanism in a low-pressure afterglow plasma, *Phys. Rev. E* **64**: 016401.

26. A. Maresca, K. Orlov, and U. Kortshagen. 2002. Experimental study of diffusive cooling of electrons in a pulsed inductively coupled plasma, *Phys. Rev. E* **65**: 056405.

27. G. Wenig, M. Schulze, P. Awakowicz, and A.V. Keudell. 2006. Modelling of pulsed low-pressure plasmas and electron re-heating in the late afterglow, *Plasma Sources Sci. Technol.* **15**: S35.

28. M.J. Kushner. Pulsed plasma-pulsed injection sources for remote plasma activated chemical vapor deposition, 1993. *J. Appl. Phys.* **73**: 4098.

29. S. Bhattacharjee, T. Nakagawa, Y. Nomiya, Y. Ikegami, M. Kase, A. Goto, and Y. Yano. 2002. Power absorption and intense collimated beam production in the pulsed high-power microwave ion source at RIKEN, *Rev. Sci. Instrum.* **73**: 620.

30. J.V. Mathew, I. Dey, and S. Bhattacharjee. 2007. Microwave guiding and intense plasma generation at subcutoff dimensions for focused ion beams, *Appl. Phys. Lett.* **91**: 041503.

31. S. Pandey, D. Sahu, and S. Bhattacharjee. 2012. Transition from interpulse to afterglow plasmas driven by repetitive short-pulse microwaves in a multicusp magnetic field, *Phys. Plasmas Lett.* **19**: 80703.

# 19

## Introduction to Focused Ion Beams

### Focused Ion Beams

Ion beam tools are becoming increasingly important with the rapid development of nanotechnology. Focused ion beam (FIB) systems are widely used for nanostructuring, nanoimplantations, deposition, material research, and more [1–3]. However, commercially available liquid metal ion source (LMIS)-based FIB systems are capable of providing only gallium (Ga) ions as output. This limits its functionality and applicability in areas where FIBs of other elements are required. Moreover, contamination issues associated with the use of Ga ions can be significant in certain applications [4].

The heavier mass of the Ga ions is found to cause impact-induced damage while carrying out circuit modifications on prototype chips [1]. The nondestructive secondary ion imaging of a surface is possible only with light ions [5]. In FIB systems, contamination is inevitable due to the use of liquid–metal ions, so an inert gas ion source is very much desired. For example, when LMIS is used for sputtering of copper, a $Cu_3 Ga$ phase alloy can be formed, which is particularly resistant to milling and contributes to the uneven profile [6]. Ga ions not only change the electrical properties but also can affect the magnetic properties of the devices. Ga staining due to deposition of Ga ions in the quartz substrate during FIB repair of a photomask is another important issue that limits the use of LMIS-based FIB systems. It has been demonstrated that when a Ga ion beam is used for photomask repair, implanted Ga ions can absorb 73% of incident 248 and 193 nm ultraviolet light, but only 0.7% of the incident light is absorbed when krypton ion beam is used for mask repair [7]. In direct deposition of insulator or metallic material using an FIB system, with the aid of a certain precursor gas, organic contaminants as well as Ga atoms are found to be included in the deposited film [8].

It would be of great significance if reliable, stable, and alternative sources for FIBs of other elements could be developed, which will open flood gates for research and development of nanodevices. Some of the many interesting possibilities include efficient sputtering or patterning of surface materials by krypton ions [9], single-cell irradiation studies of biological samples with inert gas ions such as $Ar^+$ [10], generation of high-resistivity silicon layers by hydrogen ion implantation in silicon substrate [11], surface modification of aluminum by oxygen ion implantation, and direct nanoscale nitridation by $N^+$ implantation [8,12].

There have been efforts worldwide to develop miniature gaseous plasma-based ion sources for producing multi-element FIBs [9]. Notable among them are the Penning ion sources [10], filament-driven plasma ion sources [11], and recently, mini *rf*-driven ion sources [8]. The Penning sources are known to have a smaller brightness and higher energy spread, and filament-based ion sources have a lower current density and limited lifetime. *rf* ion sources have indicated that close-to-desired beam brightness is possible [12]; however, the main concerns are instabilities due to rapid switching of the plasma between capacitive and inductive modes, which influence the plasma density [13].

We have therefore considered the development of a microwave plasma ion source for FIB systems, which can provide a high-density plasma in a compact cross-section by overcoming the waveguide (plasma chamber) cut-off limitation. A small object size of the plasma source is crucial for achieving good beam focusing (demagnification factor), and a stable high-density plasma is important for beam current.

It was demonstrated earlier that microwave plasmas can be realized in a narrow circular waveguide with a sub-cutoff radius for the fundamental waveguide mode [5–8,14]. In the case of continuous mode microwaves, plasmas with densities above the ordinary mode (O-mode) cutoff density ($10^{-11}$ cm$^{-3}$) could be produced and maintained at a low-pressure regime ($10^{-3}$–$10^{-5}$ Torr) until an axial length of more than 30 cm [5,7]. The pulsed mode operation enabled plasma sustenance at relatively higher pressures ($10^{-3}$–1 Torr) [1,8].

### Microwave Plasma Sources

In this research, we have investigated the possibility of employing the compact sub-cutoff microwave-driven multicups plasmas developed in Chapters 5, 6, and 7 for continuous mode microwaves and Chapters 9 and 10 for pulsed mode, for FIB applications for the first time [15,16]. Efforts have been made to develop such a system for applications of multi-element FIBs.

---

## References

1. A.L. Giannuzzi and A.F. Stevie (editors). 2005. *Introduction to Focused Ion Beams: Instrumentation, Theory, Techniques and Practice* (Springer, New York).
2. V.A. Stanishevsky (editor). 2004. *Encyclopedia of Nanoscience and Nanotechnology*, Vol. 3 (American Scientific Publishers, California).

3. T. Ishitani, T. Ohnishi, and T. Yaguchi. 2008. A.A. Tseng (editor), *Nanofabrication: Fundamentals and Applications* (World Scientific, Singapore).

4. J. Orloff (editor). 2009. *Handbook of Charged Particle Optics* (2nd edition, CRC Press, New York).

5. S.K. Guharay, E. Sokolovsky, and J. Orloff. 1999. Characteristics of ion beams from a Penning source for focused ion beam applications. *J. Vac. Sci. Technol. B* **17**: 2779.

6. X. Jiang, Q. Ji, A. Chang, and K.N. Leung. 2003. Mini rf-driven ion sources for focused ion beam systems. *Rev. Sci. Instrum.* **74**: 2288.

7. L. Scipioni, D. Stewart, D. Ferranti, and A. Saxonis. 2000. Performance of multicusp plasma ion source for focused ion beam applications. *J. Vac. Sci. Technol. B* **18**: 3194.

8. K. Edinger, J. Melngailis, and Orloff. 1998. Study of precursor gases for focused ion beam insulator deposition. *J. Vac. Sci. Technol. B* **16**: 3311.

9. S. Reyntjens and R. Puers. 2004. A review of focused ion beam applications in microsystem technology. *J. Micromech. Microeng.* **11**: 287.

10. J. Orloff. 1993. High-resolution focused ion beams. High-resolution focused ion beams. *Rev. Sci. Instrum.* **64**: 1105.

11. J. Melngailis. 1987. Focused ion beam technology and applications. *J. Vac. Sci. Technol. B* **5**: 469.

12. V.N. Tondare. 2005. Quest for high brightness, monochromatic noble gas ion sources. *J. Vac. Sci. Technol. A* **23**: 1498.

13. Q. Ji, K.-N. Leung, T.-J. King, X. Jiang, and B.R. Appleton. Development of focused ion beam systems with various ion species. 2005. *Nucl. Instr. Meth. Phys. Res. B* **241**: 335.

14. Y. Lee, W.A. Barletta, K.N. Leung, V.V. Ngo, P. Scott, M. Wilcos, and N. Zahir. 2001. Multi-aperture extraction system with micro-beamlet switching capability. *Nucl. Instr. Meth. Phys. Res. A* **474**: 86.

15. J.V. Mathew, I. Dey, and S. Bhattacharjee. 2007. Microwave guiding and intense plasma generation at subcutoff dimensions for focused ion beams. *Appl. Phys. Lett.* **91**: 041503.

16. J.V. Mathew, A. Chowdhury, and S. Bhattacharjee. 2008. Subcutoff microwave driven plasma ion sources for multielemental focused ion beam systems. *Rev. Sci. Instrum.* **79**: 063504.

# 20

## Experimental Setup for Basic Plasma Studies

### Introduction

In this chapter, the experimental setup used for basic plasma experiments for ion beam extraction is described. The plasma (ion) source, including the microwave (MW) system, vacuum chamber, pumping system, multicusp plasma confinement device, probes for plasma diagnostics, and other experimental accessories, is described in detail. The plasma source from which ions are extracted and focused forms one of the most important components of the multi element FIB system.

Figure 20.1 shows a schematic of the experimental apparatus comprising the plasma (ion) source. The apparatus consists of a vacuum chamber (VC), inside which the near-circular multicusp (MC) is placed. VC is made of stainless steel with 20 cm diameter and 40 cm length. The chamber is evacuated to pressures below $10^{-7}$ Torr by a turbomolecular pump (TMP) backed by a diaphragm pump (DP). Argon is generally used as a test gas, and other gases such as Ne, Kr, $H_2$, and so on are also used for the plasma experiments. Ar is introduced into the chamber through the gas inlet (GI) and a stable flow is maintained by using a mass flow controller (MKS 1179 A). The plasma (ion) source is normally operated in the pressure range of 0.1–0.6 mTorr. Continuous-mode MWs are launched into the VC through a quartz window (W). The standard rectangular waveguide (WG) has been terminated at W mounted on a flange. The probes for plasma diagnostics (Langmuir's probe, ion energy analyzer probe, etc.) have been inserted through the radial ports P1, P2, and P3, and through the axial port P4. View ports can be attached to any of these radial or axial ports for plasma observation.

### Basic Plasma Experiments

In this chapter, the basic plasma experiments for optimizing the source performance of the FIB system are described. The basic studies of the plasma

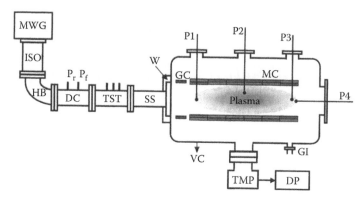

**FIGURE 20.1**
Schematic of the experimental apparatus for basic plasma studies: MWG: microwave generator, ISO: isolator, HB: H-bend, DC: directional coupler, TST: triple stub tuner, SS: straight-section waveguide, W: quartz window, GC: guiding cylinder, MC; multicusp waveguide, VC: vacuum chamber, TMP: turbomolecular pump, DP: diaphragm pump, GI: gas inlet, P1, P2, P3, and P4: probe insertion ports.

(ion) source include (i) measurement of plasma parameters using the Langmuir probe, (ii) study of the effect of different MC configurations, (iii) effect of the size of the MC for making the source compact, and (iv) finding the optimum pressure and MW power levels of operation. The radial and axial plasma uniformity in the source has been looked at, which is important for deciding the geometry and location of the ion extraction electrodes. Physical mechanisms of generation and sustenance of high-density plasmas in sub-cutoff dimensional MC waveguides are investigated. The propagation of waves inside the sub-cutoff MC waveguide is analyzed using electric field probe measurements.

## Measurement of Plasma Parameters

The different plasma parameters of the plasma (ion) source have been measured along the radial and axial directions of the MC using a Langmuir probe to obtain the spatial distribution of plasma density, electron temperature, and so on.

Figure 20.2 shows the radial variation of plasma (ion) density, $N_+$ and MC magnetic field, and $B$ from the center ($r = 0$) to the magnet surface ($r = 4$ cm), which is indicated as the "wall." The probe is inserted along port P2 (Figure 20.1). The radial variation of $B$ exhibits a magnetic bottle where the region $r < 1$ cm is almost magnetic field free. The ECR magnetic field $B_{ECR}$ (=875 G) lies around 2.2 cm. $N_+$ is rather uniform for $r < 2$ cm and then decreases toward the wall. In the central region, $N_+ \sim 10^{11}$ cm$^{-3} > N_c$, where $N_c$ is the cutoff density

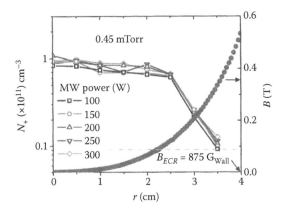

**FIGURE 20.2**
Radial variations of plasma (ion) density $N_+$ and magnetic field $B$ at 0.45 mTorr and different MW powers.

determined from the electron plasma frequency ($\omega_{pe}$). For 2.45 GHz MWs, $N_c = 0.745 \times 10^{11}$ cm$^{-3}$. In Figure 20.2, $N_+$ is plotted at a fixed neutral pressure of 0.45 mTorr for different MW powers, varying in the range 100–300 W. A small increase in $N_+$ with MW power in this power range can be noted.

Figure 20.3 shows the radial profile of electron temperature, $T_e$ for the same plasma conditions in Figure 20.2. $T_e$ increases from the center of the MC toward the ECR zone and then decreases toward the wall of the MC. The peaks near the ECR point show that resonance absorption of the incident MWs and electron heating is occurring in this region. The electron temperature in the central region is ~7 eV and at the ECR region it increases to almost double, ~15 eV.

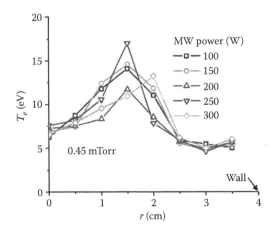

**FIGURE 20.3**
Radial variation of electron temperature, $T_e$ at 0.45 mTorr and different MW powers.

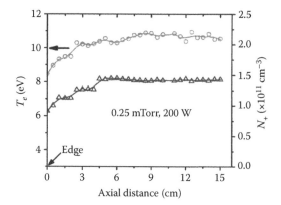

**FIGURE 20.4**
Axial variation of plasma (ion) density, $N_+$ and electron temperature, $T_e$ at 0.25 mTorr and 200 W, MW power.

Figure 20.4 shows the axial variation of $N_+$ and $T_e$ from the edge ($z = 0$) of the MC up to the center ($z = 15$ cm). Port P4 is used for the axial measurements (Figure 20.1). $N_+$ and $T_e$ are nearly constant in the bulk plasma and start to decrease at a distance of ~5 cm from the edge of the MC. $N_+$ varies from $1.3 \times 10^{11}$ cm$^{-3}$ in the bulk plasma to $0.9 \times 10^{-11}$ cm$^{-3}$ at the edge. $T_e$ varies from 11 to 9 eV. The results indicate that the plasma is axially uniform over a major portion of the MC, except for the edges.

The effect of end plugging of MC is shown in Figure 20.5, where the axial variation of $N_+$ is compared with and without end plugging. In an end-plugged condition, the polarity of the end magnets in the MC is reversed. With end plugging, a better plasma density ($N_+ \sim 0.9 \times 10^{11}$ cm$^{-3}$) is obtained

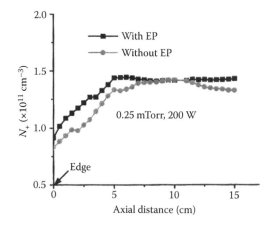

**FIGURE 20.5**
Axial variation of plasma (ion) density, $N_+$ at 0.25 mTorr and 200 W, MW power with end plugging (EP) and without EP of multicusp.

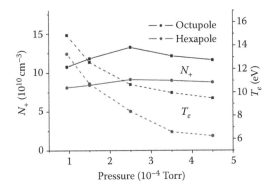

**FIGURE 20.6**
Variation of the plasma (ion) density, $N_+$ and electron temperature, $T_e$ with discharge pressure for the two multicusp configurations. (Reprinted with permission from J.V. Mathew, A. Chowdhury, and S. Bhattacharjee. 2008. Subcutoff microwave driven plasma ion sources for multi elemental focused ion beam systems, *Rev. Sci. Instrum.* **79**: 063504. Copyright 2008, American Institute of Physics.)

at the plasma emission surface from where the ion beam is extracted. This shows that an end-plugging mechanism is effective in axial plasma confinement and maintains a good plasma density in the beam extraction region.

Two MC geometries, namely, octupole and hexapole, have been compared to study the effect of magnetic confinement on plasma parameters. $N_+$ and $T_e$ plots in Figure 20.6 show that the octupole confinement is more efficient because of better plasma density and electron heating. The more poles in the MC, the better the plasma confinement, and ECR action will be more efficient due to the increase in the number of resonance zones. The central minimum-B volume will be larger in the case of octupole than hexapole and hence advantageous for ion beam extraction. Although decapole or dodecapole configurations could provide a better plasma confinement scheme than octupole, there are size constraints in designing the MC at near sub-cutoff dimensions with an ample gap between the magnets to insert the radial probes.

Figure 20.6 also shows the effect of neutral pressure on plasma parameters. It is seen that there is an optimum pressure, ~0.25 mTorr, where the plasma density $N_+$ is maximum and then $N_+$ decreases on either sides. The electron temperature is found to decrease with increase in discharge pressure as the electron–neutral collision frequency increases with increase in neutral pressure.

## Sub-Cutoff Dimensional Plasma Sources

In Chapters 5, 6, and 7 for the continuous mode and Chapters 9 and 10 for the pulsed mode microwaves, it was demonstrated that MW plasmas can be realized and sustained in MC waveguides with the cross sections below cutoff values for the fundamental waveguide mode [1–6]. Both circular and square cross-sectional MCs were studied [2,3]. The reduction in size of the MC is favorable because it increases the power density deposited in the plasma, thereby enhancing the plasma density. Also, a small source size is crucial for beam focusing (demagnification factor) and hence important for FIB systems. In sub-cutoff dimensional waveguides, plasmas with densities above the ordinary mode (O-mode) cutoff density ($10^{11}$ cm$^{-3}$) could be produced and maintained with continuous-mode MWs (cw), at a low-pressure regime ($10^{-3}$–$10^{-5}$ Torr), until an axial length of more than 30 cm [1–3,5]. The pulsed-mode operation enabled plasma sustenance at relatively higher pressures ($10^{-3}$–1 Torr) [4,7].

The production of plasma was explained on the basis of MW field penetration at the entrance of the MC waveguide [2,4]. The plasma sustenance was based on the fact that, once the plasma is created, a change in refractive index near the magnetized plasma region close to MC helps waves to propagate with a reduced wavelength. In the case of pulsed waves, the waves are refracted by the dense plasma at the entrance, which diverges from the axis and is reflected by the MC inner wall to be maximum at a location along the axis, due to phase mixing [4]. In the following, we carry out further investigations of some of the mechanisms in the case of cw MWs described in Chapters 2, 5, 6, and 7.

## Plasma Generation

When MWs of angular frequency $\omega$ in free space are sent into a circular MC waveguide of a radius below the cutoff value for the dominant empty waveguide mode, $TE_{11}$, the waves decay exponentially (evanescent) inside the waveguide. The MW field varies axially, as $E(z) = E_0 \exp(-\gamma z)$, where $\gamma$ is the propagation constant given by

$$\gamma = \sqrt{k_c^2 - k^2} = k_c a \sqrt{\frac{1}{a^2} - \frac{1}{a_c^2}}, \tag{20.1}$$

where $k$ is the wave number, $k_c$ is the cutoff wave number, and $a$ is the below-cutoff radius of the waveguide. For MWs of 3 GHz, $a_c$ is 2.93 cm. For $a < a_c$, $\gamma$ is positive and the waves are damped. We consider a sub-cutoff dimension of $a \sim 2.87$ cm for the calculations. $E_0$ is the maximum field at the waveguide entrance given by $E_0 = P/(c\varepsilon_0 A)$, where $P$ is the peak power of the launched waves, $c$ is the light velocity, and $A$ is the area of cross section of the MC waveguide.

If the electric field amplitude of the evanescent wave satisfies the threshold electric field ($E_{th}$) condition required for breakdown, breakdown occurs until an axial distance $z_{th}$ inside the waveguide, and the plasma is formed up to that axial distance [4,7]. An expression for $E_{th}$ can be obtained as follows [8–13]:

$$E_{th}^2 = \frac{8\lambda U_i \left(\omega^2 + v_c^2\right)}{3\Lambda \left(e/m_e\right)}, \tag{20.2}$$

where $U_i$ is the ionization potential (15.76 V for argon), $\lambda$ is the mean-free path of electron, and $\Lambda$ is the characteristic diffusion length for breakdown given by [8,12,14]

$$\frac{1}{\Lambda^2} = \left(\frac{2.4}{a}\right)^2 + \left(\frac{\pi}{L}\right)^2, \tag{20.3}$$

where $L$ is the initial breakdown length until the plasma is formed.

The axial distance $z_{th}$ until the breakdown threshold field $E_{th}$ is satisfied is given by

$$z_{th} = \frac{1}{k_c}\frac{\xi}{\sqrt{\xi^2 - 1}}\ln\left(\frac{E_0}{E_{th}}\right), \tag{20.4}$$

where $\xi = a_c/a$.

In Figure 20.7, $z = 0$ indicates the entrance of the waveguide. It may be noted that the field penetrates to the furthest extent in the waveguide in the case of $TE_{11}$ mode, since $TE_{11}$ mode has a lower cutoff frequency than $TM_{01}$ and $TE_{01}$ modes. The threshold field, $E_{th}$ at 1 Torr, is shown with a dotted line. Without a magnetic field, high-power pulsed waves can be used for breakdown. For argon at 1 Torr and pulsed waves of 100 kW peak power, $E_{th}$ is satisfied up to a distance of ~5.5 cm, until which the plasma can form initially.

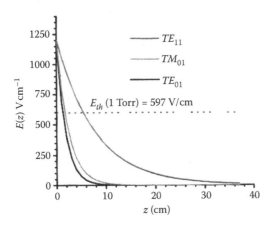

**FIGURE 20.7**
Axial decay of electric field $E(z)$ inside a sub-cutoff multicusp waveguide for three main waveguide modes $TE_{11}$, $TM_{01}$, and $TE_{01}$, respectively.

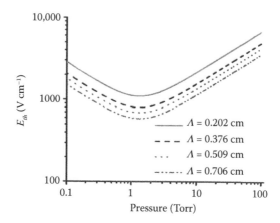

**FIGURE 20.8**

Variation of threshold electric field, $E_{th}$, with pressure for some characteristic diffusion lengths, $\Lambda$.

Figure 20.8 shows the variation of a threshold electric field $E_{th}$ with pressure and $\Lambda$ as a parameter. The symmetry of the curve about the minimum is in contrast with the asymmetric Paschen curve for dc field breakdown [11,14]. The minimum of the threshold field occurs at ~1 Torr. The characteristic diffusion length, $\Lambda = 0.706$, corresponds to $L \sim 5.5$ cm (Figure 20.7).

The extent of breakdown $z_{th}$ at different pressures for the waveguide modes $TE_{11}$, $TM_{01}$, and $TE_{01}$ is shown in Figure 20.9. It can be seen that the extent of breakdown $z_{th}$ exhibits a maximum at ~1 Torr and decreases on either sides with pressure. The fundamental waveguide mode, $TE_{11}$, dominates over other modes.

At low-pressure regimes, plasma can be generated with medium-power continuous mode MWs, like the ones used in the experiments, by the help of resonant power absorption. The MC waveguide is always kept at a distance from the quartz window, through which the MWs enter the VC. The ECR points in front of the MC waveguide resonantly absorb the MWs and initiate plasma at the entrance, even though the MC dimension is below cutoff. The physics of wave penetration and sustenance of the plasma are discussed below.

### Plasma Sustenance

Once the plasma is initiated, a typical radial plasma (ion) density profile in the MC waveguide is shown in Figure 20.10 [2], which is the zeroth-order diffusion mode.

The radial plasma (ion) density $N_+(X)$ can be fitted with a Bessel-type distribution, and the fitting equation is given by [3,4]

$$N_+(X) = N_0 J_0\left(2.4X^5\right) \tag{20.5}$$

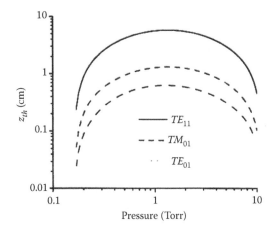

**FIGURE 20.9**
Variation of the breakdown extent $z_{th}$ with pressure for the waveguide modes $TE_{11}$, $TM_{01}$, and $TE_{01}$, respectively.

and

$$B(X) = B_0 X^4,$$
(20.6)

where the normalized radius $X = r/a$, $N_0$ is the central plasma density, $J_0$ is the Bessel function of the first kind of order zero, and $B_0$ is the surface magnetic field. The magnetic field profile conforms to a decapole multicusp configuration. The analytical expression for radial plasma density profile is obtained by

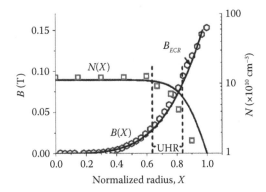

**FIGURE 20.10**
Radial variations of magnetic field $B(X)$ and plasma (ion) density $N_+(X)$ at 0.01 Torr. The open symbols show the experimental data and the solid lines show the fitting relations. (Reprinted with permission from J.V. Mathew, I. Dey, and S. Bhattacharjee. 2007. Microwave guiding and intense plasma generation at sub-cutoff dimensions for focused ion beams, *Appl. Phys. Lett.* **91**: 041503. Copyright 2007, American Institute of Physics.)

introducing the perpendicular diffusion constant, $D_\perp = D_0/(1 + \omega_{ce}^2\tau^2)$, in the density distribution [14–16]. $D_0$ is the diffusion coefficient in the case when $B = 0$.

The plasma thereby forms an inhomogeneous and anisotropic medium inside the waveguide, with a radially varying refractive index $n$. Under infinite plasma approximation, from Equation 20.3 [2–4]

$$n(X)^2 = 1 - \frac{\alpha(X)^2\left[(1 - j\delta) - \alpha(X)^2\right]}{(1 - j\delta)^2 - \beta(X)^2 - \alpha(X)^2(1 - j\delta)}, \tag{20.7}$$

the radial variation of plasma (ion) density and magnetic field are incorporated in $\alpha(X)$ and $\beta(X)$, where $X$ is the normalized radius of the chamber. Correspondingly, the wavelength of the waves in the waveguide can be obtained from Equation 20.8 [17]

$$\left(\frac{k_g(X)}{k_0}\right)^2 = n(X)^2 - \left(\frac{k_c}{k_0}\right)^2, \tag{20.8}$$

where $k_g$ and $k_0$ are the wave numbers in the magnetized plasma in a waveguide and free space, respectively. For 3 GHz MWs, $k_0 = 0.6283$ cm$^{-1}$ and $k_c = 0.642$ cm$^{-1}$ (circular waveguide) [18].

Figure 20.11 shows the radial distribution of a refractive index at different pressures. The refractive index in the central region (where the plasma density remains almost constant) is always less than unity, the free space value. However, at the periphery, the refractive index increases to values greater than unity, indicating the presence of propagation windows near the wall of

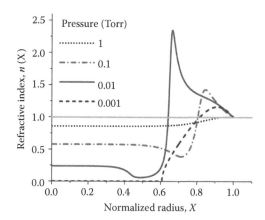

**FIGURE 20.11**
Radial distribution of refractive index inside the multicusp at different pressures. (Reprinted with permission from J.V. Mathew, I. Dey, and S. Bhattacharjee. 2007. Microwave guiding and intense plasma generation at sub-cutoff dimensions for focused ion beams, *Appl. Phys. Lett.* **91**: 041503. Copyright 2007, American Institute of Physics.)

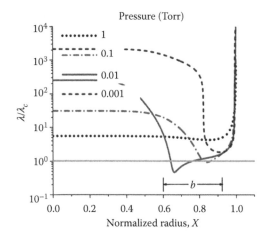

**FIGURE 20.12**
Radial variation of normalized wavelength $\lambda_g/\lambda_c$ at different pressures. (Reprinted with permission from J.V. Mathew, I. Dey, and S. Bhattacharjee. 2007. Microwave guiding and intense plasma generation at sub-cutoff dimensions for focused ion beams, *Appl. Phys. Lett.* **91**: 041503. Copyright 2007, American Institute of Physics.)

the MC. From the radial variation of normalized wavelength in Figure 20.12, it can be seen that the wavelengths decrease to very small values in these annular windows, even below the cutoff wavelength, and thereby waves can pass through the periphery of the central, overdense plasmas.

These narrow propagation windows in the MC periphery can be correlated to the resonance phenomena in MW plasmas. A plot of UHR ($\alpha^2 + \beta^2 = 1$) and ECR ($\beta^2 = 1$) resonances, as a function of the MC cross section, is shown in Figure 20.13. The null line corresponds to resonance. The ECR ($B_{ECR} = 1072$ G)

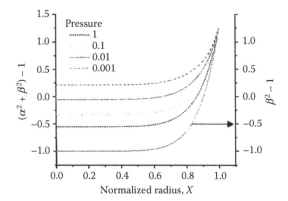

**FIGURE 20.13**
Plot of UHR ($\alpha^2 + \beta^2 = 1$) and ECR ($\beta^2 = 1$) resonances as a function of normalized radial distance in the multicusp. The null line corresponds to resonance.

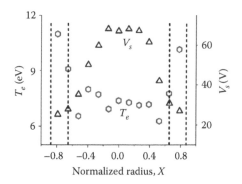

**FIGURE 20.14**
Radial variation of electron temperature, $T_e$, and plasma space potential, $V_s$, inside the multicusp. (Reprinted with permission from J.V. Mathew, I. Dey, and S. Bhattacharjee. 2007. Microwave guiding and intense plasma generation at sub-cutoff dimensions for focused ion beams, *Appl. Phys. Lett.* **91**: 041503. Copyright 2007, American Institute of Physics.)

is located at $\sim X = 0.9$. It may be noted that at higher pressures the wave propagation regions corroborate with UHR resonance region. However, at lower pressures ($<10^{-3}$ Torr), for $v < \omega_c$, the propagation window shifts toward the ECR resonance, as shown in Figure 20.12.

Figure 20.14 shows the radial variation of electron temperature, $T_e$, and plasma space potential, $V_s$, inside the MC. The electron temperature is almost constant ($\sim 7$ eV) in the central region, while it rises to above 10 eV near the wall of the MC. This clearly indicates the presence of resonance phenomenon at the periphery, by which plasma heating occurs. The plasma space potential is $\sim 70$ V in the central region, where the plasma density is high and it decays to low values ($\sim 25$ V) toward the periphery.

Figure 20.15 shows the radial variation of a wave electric field inside the plasma medium, measured using an electric field probe described in Chapter 3. The electric field profile has a lower value in the central region, and it peaks near the peripheral region. At the central region, since the plasma density is close to cutoff, MWs get attenuated and therefore the field intensity is small. The higher field intensity at the periphery shows a propagation region. By comparing Figure 20.15 to Figure 20.11, it can be noted that the maximum of the electric field occurs at the UHR region [19].

## Conclusion

The Langmuir probe measurements provide a good picture of the plasma profile inside the MC waveguide. The radial plasma density distribution

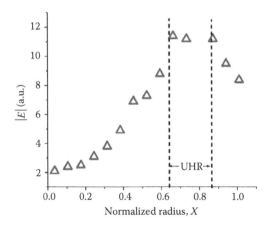

**FIGURE 20.15**

Radial variation of wave electric field intensity inside the multicusp. (Reprinted with permission from J.V. Mathew, I. Dey, and S. Bhattacharjee. 2007. Microwave guiding and intense plasma generation at sub-cutoff dimensions for focused ion beams, *Appl. Phys. Lett.* **91**: 041503. Copyright 2007, American Institute of Physics.)

shows uniform distribution over a radius of ~2 cm, with $N_+ \geq 10^{11}$ cm$^{-3}$ at the center of MC. Electron temperature $T_e$ is ~6–7 eV, and it peaks to ~15 eV near the periphery of the MC, where electron heating due to ECR and UHR happens. The axial profile of plasma density and electron temperature shows that both $N_+$ and $T_e$ remain constant in the bulk plasma and decrease toward the edge of the MC. The end-plugging of MC is effective in axial plasma confinement and maintains a high plasma density in the beam extraction region. The plasma density does not vary much with increase in MW power, while for the present experimental geometry, an optimum pressure of ~0.25 mTorr is found where $N_+$ is maximum. The study of the plasma parameters indicates that the octupole MC configuration provides better plasma confinement and plasma heating, which makes it more efficient than the hexapole geometry.

The mechanisms of MW guiding, plasma generation, and sustenance by MWs launched into a narrow cross-sectional chamber with sub-cutoff dimension for the dominant waveguide mode $TE_{11}$ have been studied. After the initial breakdown, plasma is formed until a finite distance in the waveguide. The refractive index of the medium is modified and has a radial variation. The waves can propagate through the peripheral plasma with a reduced wavelength due to the presence of an upper hybrid resonance region. The UHR region is where the waves can transfer the maximum energy to the plasma. The possibility of sustaining such narrow cross-sectional, high-density plasmas is favorable for compact MW-based multi-element FIB systems.

# References

1. H. Amemiya and M. Maeda. 1996. Multicusp type machine for electron cyclotron resonance plasma with reduced dimensions, *Rev. Sci. Instrum.* **67**: 769.

2. S. Bhattacharjee and H. Amemiya. 1998. Microwave plasma in a multicusp circular waveguide with a dimension below cutoff, *Jpn. J. Appl. Phys.* **37**: 5742.

3. S. Bhattacharjee and H. Amemiya. 1999. Production of microwave plasma in narrow cross sectional tubes: Effect of the shape of cross section, *Rev. Sci. Instrum.* **70**: 3332.

4. S. Bhattacharjee and H. Amemiya. 2000. Production of pulsed microwave plasma in a tube with a radius below the cut-off value, *J. Phys. D: Appl. Phys.* **33**: 1104.

5. R. Geller. 1996. *Electron Cyclotron Resonance Ion Sources and ECR Plasmas* (Institute of Physics, Bristol).

6. S. Bhattacharjee. 1999. Production of microwave plasma in a waveguide with a dimension below cutoff, Doctoral dissertation, Graduate School of Science and Engineering, Saitama University, Japan.

7. S. Bhattacharjee, H. Amemiya, and Y. Yano. 2001. Plasma buildup by short-pulse high-power microwaves, *J. Appl. Phys.* **89**: 3575.

8. A. Kraszewski. 1967. *Microwave Gas Discharge Devices* (Lliffe, London).

9. A.D. MacDonald and S.C. Brown. 1949. High frequency gas discharge breakdown in hydrogen, *Phys. Rev.* **76**: 1634.

10. A.D. MacDonald and J.H. Matthews. 1956. Electrical breakdown in argon at ultrahigh frequencies, *Can. J. Phys.* **34**: 395.

11. A.D. MacDonald, D.U. Gaskell, and H.N. Gitterman. 1963. Microwave breakdown in air, oxygen and nitrogen, *Phys. Rev.* **130**: 1841.

12. S. Krasik, D. Alpert, and A.O. McCoubrey. 1949. Breakdown and maintenance of microwave discharges in Argon, *Phys. Rev.* **76**: 722.

13. Y.P. Raizer. 1991. *Gas Discharge Physics* (Springer, Berlin).

14. M. Moissan and J. Pelletier (editors). 1992. *Microwave Excited Plasmas* (Elsevier, Amsterdam).

15. U. Jordan, D. Anderson, L. Lapiere, M. Lisak, T. Olsson, J. Puech, V.E. Semenov, J. Sombrin, and R. Tomala. 2006. On the effective diffusion length for microwave breakdown, *IEEE Trans. Plasma Sci.* **34**: 421.

16. S.A. Self and H.N. Ewald. 1966. Static theory of a discharge column at intermediate pressures, *Phys. Fluids* **9**: 2486.

17. S. Takeda. 1994. Propagation of waves through magnetized plasmas in waveguides, *Jpn. J. Appl. Phys.* **33**: 757.

18. D.K. Cheng. 1989. *Field and Wave Electromagnetics* (Pearson Education, Singapore).

19. J.V. Mathew, I. Dey, and S. Bhattacharjee. 2007. Microwave guiding and intense plasma generation at sub-cutoff dimensions for focused ion beams, *Appl. Phys. Lett.* **91**: 041503.

20. J.V. Mathew, A. Chowdhury, and S. Bhattacharjee. 2008. Subcutoff microwave driven plasma ion sources for multi elemental focused ion beam systems, *Rev. Sci. Instrum.* **79**: 063504.

# 21

## Basic Beam Studies: Extraction and Ion Energy Distribution

### Introduction

In this chapter, the ion energy distribution in the plasma source both in the bulk plasma and near the plasma meniscus from where the beam is extracted is studied. Measurements are taken along both the axial and radial directions in the plasma source using ion energy analyzer (IEA) probes. The effect of end-plugging of the multicusp (MC) on ion energy distribution has been investigated. A small quadrupole magnetic filter (QF) is introduced at the extraction end of the MC, and its influence on ion energy spread is investigated. The local dependence of ion energy spread on mass of the ionic species, wave, and neutral pressure has also been looked at. The effect of the geometrical acceptance angle on the IEA probe measurements is discussed.

Thereafter, some basic beam experiments, including the total beam current measurements and the study of the effect of beam extraction on ion energy spread at the plasma meniscus, are described. The beam-current measurements are made for three different gas species, Ar, Kr, and $H_2$, with different plasma electrode apertures and extraction potentials up to $-5$ kV. The variation in ion current density and ion energy spread with gas pressure and microwave power is investigated. The effect of secondary electron contribution to the total beam current has been looked at by using three different target materials.

### Experiment

The design and construction of an IEA probe is described in Chapter 3. Ion energy distribution is measured both along the axial and radial directions in the plasma source. For axial measurements, port P2 is used, while for radial

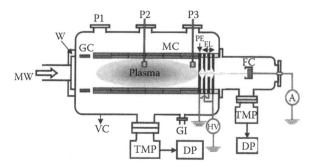

**FIGURE 21.1**

Schematic of the experimental apparatus. MW, microwaves; W, quartz window; GC, guiding cylinder; MC, multicusp; PE, plasma electrode; FC, Faraday's cup; VC, vacuum chamber; GI, gas inlet; A, ammeter; HV, high-voltage power supply; P1, P2, P3, ports; TMP, turbo molecular pump; DP, diaphragm pump.

measurements, the ports P2 and P3 are used (cf. Figures 21.1). In radial measurements, the size of the probe limits the movement of the probe to ~2 cm from the center of the MC. The data are taken at different neutral pressures ranging from 0.1 to 0.5 mTorr, and wave powers from 100 to 250 W.

## Total Beam Current Measurements

Figure 21.1 shows the schematic of the experimental setup for total beam current measurements. The plasma source part is the same as the one used for the basic plasma experiments. For beam extraction, the MC is closed at one end using the plasma electrode (PE), which has a central aperture. To measure the total extractable ion current, a Faraday cup (FC) of ~5 cm diameter is kept at a distance of ~1 cm from the PE. The FC consists of an Al target plate, which can be floated to high potentials. SS and Cu plates are also used as target materials. Three PE aperture sizes, 1, 5, and 8 mm have been experimented. The extraction potential is provided by a high-current (150 mA), DC-regulated power supply and is varied in the range from 0 to –5 kV. The beam current is measured using the inbuilt ammeter of the power supply.

## Ion Energy Spread Measurements

Figure 21.2 shows the schematic of the experimental setup for ion energy spread measurements. The ion energy distribution function (IEDF) is measured in the bulk plasma and in plasma meniscus region using IEA probes inserted along the radial ports P2 and P3, respectively. *In situ* measurements of IEDF and beam current are made. The extraction system consists of the PE along with a three-electrode, single-Einzel lens (EL) system for beam focusing. The first and third electrodes of the EL are held at high potential and varied up to –5 kV, while the middle electrode is grounded with the PE. The

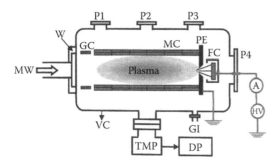

**FIGURE 21.2**
Schematic of the experimental apparatus. MW, microwaves; W, quartz window; GC, guiding cylinder; MC, multicusp; PE, plasma electrode; EL, Einzel's lens; FC, Faraday's cup; VC, vacuum chamber; GI, gas inlet; A, ammeter; HV, high-voltage power supply; P1, P2, P3, P4, ports; TMP, turbo molecular pump; DP, diaphragm pump.

triplet serves the purpose of both beam acceleration and focusing. Differential pumping is employed for the extraction system to minimize the extent of beam ion neutral collisions, by using a turbo molecular pump backed with a diaphragm pump. The extraction system is isolated from the plasma chamber except for the 1 mm aperture in the PE, through which ions are extracted. The beam current is measured using a Keithley 2001 multimeter.

## Results and Discussion

The dependence of ion energy spread on discharge pressure and microwave powers has been investigated in the bulk plasma. Figure 21.3 shows the variation of ion energy spread (solid symbols) and mean ion energy, $E_{mean}$ (open symbols) with pressure (solid lines) and wave power (dashed lines) for Ar ions. $\Delta E$ tends to increase from 8.6 to 12.3 eV, with increase in wave power from 100 to 250 W. $E_{mean}$ remains almost constant with varying microwave power, implying there is no significant variation in the sheath thickness, $\eta \propto \lambda_D$, through which the ions are accelerated toward the grid in this power regime [1]. $\Delta E$ decreases with increasing pressure from 13 eV at 0.15 mTorr to 8.41 eV at 0.45 mTorr gas pressure, and may be attributed to the decrease in ion temperature, $T_i$, with increase in number density of neutrals. As the pressure is increased from 0.15 to 0.45 mTorr, the mean free path of ions ($\lambda_i \approx$ 1/330 p for Ar, where $p$ is the pressure in Torr [2]) reduces from 35 to 5 cm due to collisions. Consequently, the ions are accelerated to relatively low kinetic energies; thereby, the higher-pressure regime is dominated by a group of low-energy ions, which is reflected in the shift in $E_{mean}$ to lower values. A similar trend has been observed in the case of Kr ions also.

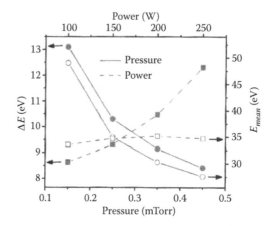

**FIGURE 21.3**

Variation of ion energy spread (solid symbols) and mean ion energy (open symbols) with neutral pressure (solid lines) and wave power (dashed lines) for Ar ions. For power scan, the neutral pressure is held at 0.25 mTorr, and for pressure scan microwave power is kept at 150 W.

The plasma is axially confined in the MC by magnetic-end plugging (EP). The effect of EP has been studied using IEA probe and compared with the Langmuir probe data. A 2-D Poisson [3] simulation of the magnetic field profile of EP is shown in Figure 21.4, where the polarity of the magnets is represented by alphabetical letters N and S.

Figure 21.5 shows the axial variation of plasma (ion) density, $N_+$ (circle), ion energy spread, $\Delta E$ (square) and mean ion energy, $E_{mean}$ (triangle) with EP (solid symbols) and without EP (open symbols) for Ar ions at neutral pressure of 0.25 mTorr and 200 W MW power. The microwave entry side of the MC is end-plugged in both cases. The axial confinement of plasma by EP yields a higher plasma density (~$10^{11}$ cm$^{-3}$) at the plasma meniscus. The variation of $\Delta E$ is similar to the density profile. With EP, $\Delta E$ decreases from ~10 to 4.84 eV for Ar ions, and from ~8 to 3.7 eV for Kr ions from the bulk plasma region toward the plasma meniscus. The ion energy spread in the beam extraction

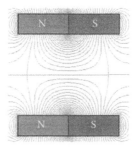

**FIGURE 21.4**

2D Poisson simulation of magnetic end-plugging.

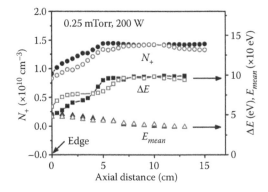

**FIGURE 21.5**
Axial variation of plasma (ion) density, $N_+$ (circle), ion energy spread, $\Delta E$ (square), and mean ion energy, $E_{mean}$ (triangle) with end-plugging (EP; solid symbols) and without EP (open symbols) for Ar ions at a pressure of 0.25 mTorr and a power of 200 W. (Reprinted with permission from J.V. Mathew and S. Bhattacharjee. 2009. Ion energy distribution near a plasma meniscus for multielement focused ion beams, *J. Appl. Phys.* **105**: 096101. Copyright 2009, American Institute of Physics.)

region clearly turns out to be comparable to that of LMIS. The mean energy increases from the bulk plasma toward the meniscus by ~30% for end-plugged configuration and ~25% for an un-end plugged case. The increase in $E_{mean}$ toward the meniscus can be attributed to the gradient in $N_+$ near the meniscus, which facilitates the flow of higher-energy ions toward the edge.

Figure 21.6 shows a typical ion (Ar) energy distribution function (solid symbols) at the extraction region of the MC for 0.25 mTorr pressure and 200 W power with a mean energy of 53 eV. The experimentally obtained IED

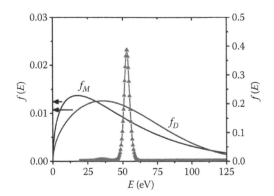

**FIGURE 21.6**
Comparison of ion energy distribution at the extraction region of the multicusp for 0.25 mTorr pressure and 200 W power (solid symbols) with the Maxwellian ($f_M$) and the Druyesteyn ($f_D$) energy distributions having the same mean energy. (Reprinted with permission from J.V. Mathew and S. Bhattacharjee. 2009. Ion energy distribution near a plasma meniscus for multielement focused ion beams, *J. Appl. Phys.* **105**: 096101. Copyright 2009, American Institute of Physics.)

has been compared with standard-model energy distribution functions of the same mean energy given by [4]

$$f(E) = \frac{m\Gamma(5/2m)^{3/2}\sqrt{E}}{\Gamma(3/2m)^{5/2}E_{mean}^{3/2}}\exp\left\{-\frac{m\Gamma(5/2m)^m E^m}{\Gamma(3/2m)^m E_{mean}^m}\right\}, \qquad (21.1)$$

where $\Gamma$ is the complete gamma function and $m$ is an integer 1 and 2. $m = 1$ corresponds to Maxwellian, $f_M$ and $m = 2$ corresponds to Druyvesteyn, $f_D$. The comparison shows that the population of low- and high-energy tail ions are deficient in the IED as compared to $f_M$ and $f_D$. The ions posses medium energy, and the distribution is largely monoenergetic with a smaller energy spread of ~5 eV.

Unlike RF plasmas, microwave plasmas exhibit a monoenergetic behavior for IEDF, which is quite important from the point of view of FIB applications. The variation of ion energy spread as a function of wave frequency may be expressed as [5,6]

$$\Delta E = \left(4eV_s\tau_w\right)/\left(\pi\tau_{ion}\right), \qquad (21.2)$$

where $V_s$ is the voltage drop across the plasma sheath, $\tau_w$ is the wave period, and $\tau_{ion}$ is the time taken by the ions to transverse the sheath, which is a function of sheath thickness, $s$, and the mean sheath voltage, $V_s'$.

$$\tau_{ion} = 3s\left(m_i/2eV_s'\right)^{1/2}. \qquad (21.3)$$

For $N_+ \sim 10^{11}$ cm$^{-3}$, $T_e \sim 10$ eV, $s \sim 0.25$ mm, and for $V_s \approx V_s' \sim 35$ V, we obtain for waves of 2.45 GHz, $\tau_{ion}/\tau_w$ is ~100, and hence the ions will take many wave cycles to cross the sheath, and therefore the effect of the applied electromagnetic field is negligible.

Figure 21.7 shows the variation of ion energy spread with microwave frequency as given by Equation 21.3. It is evident that at microwave frequencies (~2000 MHz) the energy spread decreases drastically to ~1 eV. The monoenergetic nature of ions in microwave plasmas is evident from the fact that ions in the microwave plasma do not respond to the incident high-frequency microwaves ($f_0 = 2.45$ GHz), since the plasma ion frequency $f_{pi}$ lies in the MHz range ($f_{pi} \sim 10$ MHz $\ll f_0$, for $N_+ \sim 10^{11}$ cm$^{-3}$). On the other hand, the plasma electron frequency, $f_{pe} \sim 2.85$ GHz, is comparable to that of the microwave frequency and helps in the generation and sustenance of plasma by upper hybrid resonance (UHR) phenomenon [7].

The effect of the magnetic filter on ion energy distributions has been studied using a QF inserted inside the MC at the extraction end. A 2-D Poisson simulation of the magnetic field profile of the QF inside the larger MC is shown in Figure 21.8. The effect of QF on the energy spread of Ar ions is shown in Figure 21.9, where the IED is plotted with QF (solid triangle) and

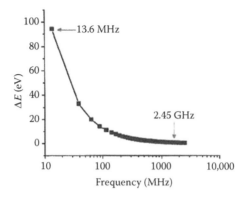

**FIGURE 21.7**
Variation of ion energy spread, Δ*E* with wave frequency.

**FIGURE 21.8**
2D Poisson simulation of quadrupole magnetic filter (QF) inside the multicusp (MC).

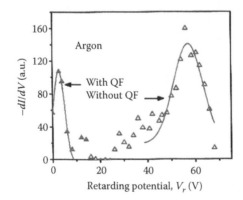

**FIGURE 21.9**
Comparison of ion energy distribution near the plasma meniscus, with magnetic quadrupole filter (QF) (solid triangle) and without QF (open triangle) for Ar ions at 0.15 mTorr and 200 W.

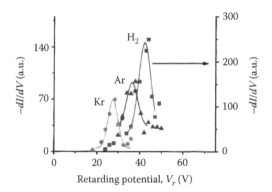

**FIGURE 21.10**
Comparison of ion energy distribution for three plasmas: Ar, Kr, and $H_2$ without QF at 0.35 mTorr and 150 W. Right $y$ axis shows $H_2$.

without QF (open triangle). The measurements were taken at a distance of ~4 cm from the edge of the MC.

With the introduction of the filter, $\Delta E$ decreases from 12 to 5.7 eV (58%). The considerable shift in $E_{mean}$ from 57 to 3 eV suggests that the QF can effectively prevent the high-energy ions from entering into the extraction region. A calculation of the diffusion constant with and without the magnetic filter indicated that the radial ion diffusion $D_\perp$ is reduced by a factor of ~100 in the presence of the filter [2].

The effect of the mass of ionic species on the energy spread is investigated without the QF in place, as shown in Figure 21.10 for Kr (5.22 eV), Ar (6.08 eV), and $H_2$ (6.54 eV) ions at 0.15 mTorr and 15 W.

In accordance with Equation 21.2, the results show that $\Delta E$ decreases with increase in ion mass. The peak of the distribution shifts toward lower-energy regime with increase in ion mass, possibly because the ions with higher mass attain lower velocities for the same potential drop in the sheath.

## Ion Energy Distribution along the Radial Direction

Figure 21.11 shows the radial variation ion energy spread, $\Delta E$, in the beam extraction region. The radial magnetic field profile and radial variation of plasma (ion) density, $N_+$, and electron temperature, $T_e$, are also shown alongside for completeness. $r = 0$ corresponds to the axis of the MC and at $r = 0$ the radial ion energy spread is ~5 eV, which agrees with the axial ion energy spread data. $\Delta E$ increases in the radial direction and appears to peak near the electron–cyclotron resonance (ECR) zone. The increase in ion energy spread toward the ECR zone indicates that the ions are also affected by the plasma-heating mechanism happening in that region. The size of the IEA probe restricts the radial motion of the probe beyond $r = 2$ cm.

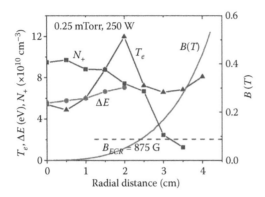

**FIGURE 21.11**

Radial variation of ion energy spread, $\Delta E$ (circle), electron temperature, $T_e$ (triangle), and plasma (ion) density, $N_+$ (square) in the beam extraction region. The solid line shows the radial variation of magnetic field. (Reprinted with permission from J.V. Mathew, S. Paul, and S. Bhattacharjee, 2010. Ion energy distribution near a plasma meniscus with beam extraction for multi-element focused ion beams, *J. Appl. Phys.* **107**: 093306. Copyright 2010, American Institute of Physics.)

## Total Beam Current Measurements

Measurements of the total extracted ion current density, $J_i$, are made for gas species such as Ar, Kr, and $H_2$ for different discharge pressures, microwave powers, and PE apertures. The ion current density at the source is taken as the extracted ion current divided by the PE aperture area.

Figure 21.12 shows the dependence of ion current density, $J_i$, on extraction potential, $V$, at different pressures for Ar gas at 270 W microwave power

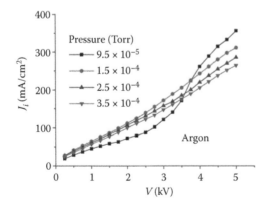

**FIGURE 21.12**

Extracted ion current density, $J_i$, versus extraction potential, $V$, at different pressures for Ar gas at 270 W MW power and 5 mm PE aperture. (Reprinted with permission from J.V. Mathew, A. Chowdhury, and S. Bhattacharjee. 2008. Subcutoff microwave driven plasma ion sources for multi elemental focused ion beam systems, *Rev. Sci. Instrum.* **79**: 063504. Copyright 2008, American Institute of Physics.)

and 5 mm PE aperture. The optimum pressure of operation is found to be ~0.15 mTorr since $J_i$ exhibits a maximum value at this pressure. The behavior at 0.095 mTorr may possibly be due to a mode change leading to a change in plasma density.

In Figure 21.13, $J_i$ for Kr, Ar, and $H_2$ are compared at 0.25 mTorr, the microwave power is 270 W, and 5 mm PE aperture size. As predicted by Child–Langmuir's law, $J_i$ shows an inverse relationship with ion mass. In both cases, we find that the extracted ion current density is far from saturation at −5 kV.

The variation of $J_i$ with microwave power is shown in Figure 21.14 with extraction potential kept at −4 kV and the PE aperture of 5 mm. The operating pressure is 0.25 mTorr. $J_i$ tends to increase with microwave power and the increase is more prominent for lighter gas atoms.

Figure 21.15 shows the dependence of $J_i$ on PE aperture size at 270 W MW power. At −5 kV, $J_i$ is a maximum for 1 mm aperture. The high current density is promising for a focused ion beam system. Although the increase in aperture size should increase the beam angular divergence, the angular intensity $dI/d\Omega$ remains almost constant because of the high current. This is advantageous for large-volume milling in FIB applications as compared to commercial LMIS, which fails at higher beam currents (>1 nA) [8].

The change in the total ion current density, $J_t$, with extraction potential, $V$, for different target materials in the FC such as aluminum, copper, and stainless steel is next carried out to study the contribution of secondary electrons. Figure 21.16 shows the variation of $J_i$ with $V$ for three different targets of the same area. The contribution of secondary electrons seems to be small—up to −5 kV extraction potential—which is in accordance with earlier studies [9,10]. The secondary electron yield, $\gamma$, is given by $\gamma = (I_T/I_i - 1)$, where $I_T$ is the total current and $I_i$ is the true ion current [11]. For a 5 kV ion beam, $\gamma$ for

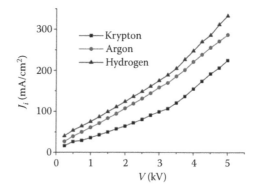

**FIGURE 21.13**
Ion current density, $J_i$, as a function of extraction potential, $V$, for three different gases at 270 W MW power and 5 mm PE aperture. (Reprinted with permission from J.V. Mathew, A. Chowdhury, and S. Bhattacharjee. 2008. Subcutoff microwave driven plasma ion sources for multi elemental focused ion beam systems, *Rev. Sci. Instrum.* **79**: 063504. Copyright 2008, American Institute of Physics.)

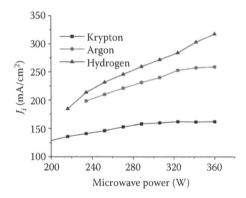

**FIGURE 21.14**
Ion current density, $J_i$, as a function of microwave power for three gas species, namely, Kr, Ar, and $H_2$. (Reprinted with permission from J.V. Mathew, A. Chowdhury, and S. Bhattacharjee. 2008. Subcutoff microwave driven plasma ion sources for multi elemental focused ion beam systems, *Rev. Sci. Instrum.* **79**: 063504. Copyright 2008, American Institute of Physics.)

Al, Cu, and SS are approximately 3.8, 1.9, and 1.7, respectively [10]. From the secondary electron yield data, the true ion current density, $J_{ion}$, is deduced to be ~2 A/cm² at –5 kV.

From $J_{ion}$, we can estimate the reduced-source side brightness, $B_{rs}$ of the plasma ion source, which is related to the ion temperature, $T_i$ (in eV), by [12,13]

$$B_{rs} = eJ_{ion}/T_i Am^{-2}sr^{-1}V^{-1}. \tag{21.4}$$

From the above results, considering a $J_{ion}$ ~2 A/cm² for 1 mm PE aperture and a $T_i$ of ~0.2 eV, a reduced source side brightness $B_{rs}$ of ~$10^5$ Am$^{-2}$ sr$^{-1}$ V$^{-1}$ may be predicted, which is close to that of LMIS [13].

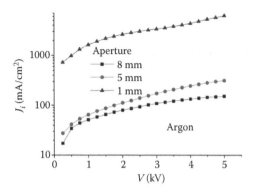

**FIGURE 21.15**
Variation of extracted ion current density, $J_i$, with extraction potential, $V$, for three aperture sizes of 1, 5, and 8 mm. (Reprinted with permission from J.V. Mathew, A. Chowdhury, and S. Bhattacharjee. 2008. Subcutoff microwave driven plasma ion sources for multi elemental focused ion beam systems, *Rev. Sci. Instrum.* **79**: 063504. Copyright 2008, American Institute of Physics.)

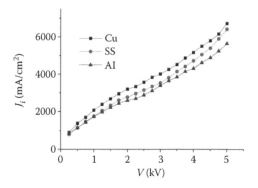

**FIGURE 21.16**

Extracted ion current density, $J_i$, as a function of extraction potential, $V$, for three different target materials. SS (stainless steel), Al, and Cu at 270 W microwave power and 0.25 mTorr. (Reprinted with permission from J.V. Mathew, A. Chowdhury, and S. Bhattacharjee. 2008. Subcutoff microwave driven plasma ion sources for multi elemental focused ion beam systems, *Rev. Sci. Instrum.* **79**: 063504. Copyright 2008, American Institute of Physics.)

## Ion Energy Spread Measurements

During beam extraction, it is difficult to measure the axial ion energy distribution because of the pressure of the extraction system. We have therefore measured the IEDF from the radial direction while the beam extraction is in progress, to study the effect of beam extraction on an ion energy spread. The IEDF has been studied with two identical IEA probes employed in the bulk plasma and in the beam extraction region, as shown in Figure 21.2. As we move away from the center of MC along the radial direction, the increase in radial magnetic field can introduce some anisotropy in the ion energy distribution. In the plasma meniscus region ($z \sim 2.5$ cm, from the edge of the MC), the IEA probe is kept almost near the axis of the MC, at a radial distance of $r \sim 0.5$ cm.

Figure 21.17 shows the variation of beam current (open symbols) with extraction potential at different neutral pressures. There is a threshold potential, $\sim$–2 kV, beyond which the beam current increases sharply. The Child–Langmuir's law has been fitted with the experimental data as shown by the dashed line in Figure 21.17. The beam current at –5 kV extraction potential for 1 mm PE aperture is ~7 μA. This current is at least three orders of magnitude larger at the substrate than commercially available FIB systems.

Also shown in Figure 21.17 is the variation of ion energy spread (solid symbols) with extraction potentials up to –5 kV at different gas pressures. Figure 21.17 shows that $\Delta E$ remains almost constant as the ions are being extracted. This is significant since the chromatic aberrations of the electrostatic focusing lens arising from the inherent energy spread of the plasma ion source do not increase with the increase in extraction potentials, which will be very helpful to retain the spot size [14]. The presence of an extraction electrode

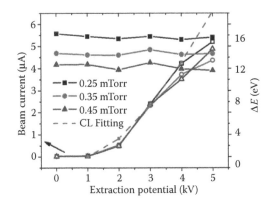

**FIGURE 21.17**
Variation of beam current (open symbols) and ion energy spread (solid symbols) with extraction potential at different pressures. (Reprinted with permission from J.V. Mathew, S. Paul, and S. Bhattacharjee. 2010. Ion energy distribution near a plasma meniscus with beam extraction for multi-element-focused ion beams, *J. Appl. Phys.* **107**: 093306. Copyright 2010, American Institute of Physics.)

system is found to increase the ion energy spread near the plasma meniscus, as compared to the case when the extraction system is not placed while IEA probe measurements are made. The increase in spread can be attributed to the formation of electromagnetic standing waves induced by reflections from the PE [15]. However, this can be overcome by operating at higher gas pressures. The increase in operating gas pressure has a twofold advantage. The electromagnetic standing waves are found to be damped by the increase in collisionality due to increase in pressure [15]. Also, the increase in pressure leads to increase in ion-neutral collisions, thereby decreasing the ion energy spread. Figure 21.17 shows that $\Delta E$ decreases from ~17 to 11 eV as the pressure is varied from 0.25 to 0.45 mTorr.

## Conclusion

Ion energy distributions have been studied along the axial and radial directions near the plasma meniscus of the microwave plasma source. The magnetic field in the ion beam extraction region is tailored by (i) EP the MC magnets and (ii) by employing a quadrupole filter. The axial ion energy spread measurements show that EP is effective in reducing the ion energy spread, and QF is found to reduce the ion energy spread by 58% and the mean energy by 95% (~3 eV). The ion energy distribution at the plasma meniscus is found to be nearly monoenergetic with a spread of ~4 eV for Kr ions and ~5 eV for Ar ions, which is comparable to that of Ga-based LMIS.

The radial ion energy spread in the plasma meniscus is comparable to the axial spread at the center of the MC, $r = 0$, while it increases toward the ECR zone, indicating that the ions are also affected by the ECR plasma-heating mechanism.

The minimum-B field at the center of the MC and the least disturbance to the ions in the plasma by the incoming microwaves ($f_0 \gg f_{pi}$) help in obtaining monoenergetic and low-energy spread ions at the meniscus. The application of microwaves instead of radio waves is found to be advantageous in this respect. The comparable or even less-energy spread with respect to LMIS is promising for reduced chromatic aberration in multi-element-focused ion beams with an adequate current density and uniform spot size.

The total beam current measurements and the study of the effect of beam extraction on ion energy spread at the plasma meniscus are carried out. The variation of ion current density with extraction potentials up to –5 kV is measured for three different gas species, Ar, Kr, and $H_2$, and three PE apertures. Also, the dependence of an extracted ion current density on gas pressure and microwave power has been looked at. The ion-current density is found to be far from saturation at –5 kV. Excluding the secondary electron contributions, for 1 mm PE aperture, a source side ion current density of ~2 $A/cm^2$ has been obtained.

The radial ion energy distribution has been measured in the extraction region (plasma meniscus) of the MC, while the ions are being extracted. The energy spread is found to have a weak dependence on beam extraction, which is of importance from a point of view of focused beam spot size. With the extraction system in place, there is an increase in radial spread, possibly due to the formation of standing waves in the MC and can be overcome by operating at higher neutral pressures. The beam current at –5 kV extraction potential for 1 mm PE aperture is ~7 μA, considering (a) losses during focusing, this current is at least three orders of magnitude larger at the substrate than commercially available FIB systems, and (b) the contribution of secondary electrons is rather small at these voltages for substrate material of stainless steel.

## References

1. S. Bhattacharjee and H. Amemiya. 1998. Microwave plasma in a multicusp circular waveguide with a dimension below cutoff, *Jpn. J. Appl. Phys.* **37**: 5742.
2. M.A. Lieberman and A.J. Lichtenberg. 1994. *Principles of Plasma Discharges and Material Processing* (Wiley–Interscience: New York).
3. J.H. Billen and L.M. Young. 2002. *Poisson Superfish, LA-UR-96-1834* (Los Almos Nat. Lab. Rep., NM).
4. M. Maeda and H. Amemiya. 1994. Electron cyclotron resonance plasma in multicusp magnets with a checkered pattern, *Jpn. J. Appl. Phys.* **33**: 5032.

5. E. Kawamura, V. Vahedi, M.A. Lieberman, and C.K. Birdsall. 1999. Ion energy distributions in rf sheaths; Review, analysis and simulation, *Plasma Sources Sci. Technol.* **8**: R45.
6. T. Panagopoulos and D.J. Economou. 1999. Plasma sheath model and ion energy distribution for all radio frequencies, *J. Appl. Phys.* **85**: 3435.
7. J.V. Mathew, I. Dey, and S. Bhattacharjee. 2007. Microwave guiding and intense plasma generation at sub-cutoff dimensions for focused ion beams, *Appl. Phys. Lett.* **91**: 041503.
8. N.S. Smith, D.E. Kinion, P.P. Tesch, and R.W. Boswell. 2007. A high brightness plasma source for focused ion beam applications, *Microsc. Microanal.* **13** (**Suppl 2**): 180.
9. B. Szapiro, J.J. Rocca, and T. Prabhuram. 1988. Electron yield of glow discharge cathode materials under helium ion bombardment, *Appl. Phys. Lett.* **53**: 358.
10. W. En and N.W. Cheung. 1996. A new method for determining the secondary electron yield dependence on ion energy for plasma exposed surfaces, *IEEE. Trans. Plasma Sci.* **24**: 1184.
11. Y. Chutopa, B. Yotsombat, and I.G. Brown. 2003. Measurement of secondary electron emission yields, *IEEE Trans. Plasma Sci.* **31**: 1095.
12. V.N. Tondare. 2005. Quest for high brightness, monochromatic noble gas ion sources, *J. Vac. Sci. Technol.* A **23**(6): 1498.
13. L. Scipioni, D. Stewart, D. Ferranti, and A. Saxonis. 2000. Performance of multi-cusp plasma ion source for focused ion beam applications, *J. Vac. Sci. Technol. B* **18**: 3194.
14. J.V. Mathew, S. Paul, and S. Bhattacharjee. 2010. Ion energy distribution near a plasma meniscus with beam extraction for multi element focused ion beams, *J. Appl. Phys.* **107**: 093306.
15. I. Dey and S. Bhattacharjee. 2008. Experimental investigation of standing wave interactions with a magnetized plasma in a minimum-B field, *Phys. Plasmas* **15**: 123502.
16. J.V. Mathew and S. Bhattacharjee. 2009. Ion energy distribution near a plasma meniscus for multielement focused ion beams, *J. Appl. Phys.* **105**: 096101.
17. J.V. Mathew, A. Chowdhury, and S. Bhattacharjee. 2008. Subcutoff microwave driven plasma ion sources for multi elemental focused ion beam systems, *Rev. Sci. Instrum.* **79**: 063504.

# 22

## Multielement Focused Ion Beam

### Introduction

In this chapter, simulations for designing the beam column of the ME-FIB system are discussed. Widely known and commercially available beam simulation codes, namely, AXCEL-INP and SIMION, are employed for obtaining beam trajectory and emittance. Einzel lenses (ELs) are designed for electrostatic focusing of the ion beam. The effect of various parameters such as electrode geometry, applied extraction voltage, space charge in the beam, shape of the plasma boundary, and so on, on beam optics is looked at. The plasma parameters obtained from previous experiments are used in the simulations. This study has been of immense importance in the design and further fabrication of ion beam optics.

### Beam Column of MEFIB

The ion beam column of an FIB system consists of the extraction and focusing electrodes. The extraction system basically involves the plasma electrode and extraction electrode as described in the "Results and Discussion" section. Once the beam is extracted, focusing is carried out using the electrostatic EL system. Usually, a double-EL system is incorporated, along with beam-limiting apertures to control the beam size and current [1]. The electrode assembly in the beam column has to be designed to obtain an ion beam of adequate current density that can be eventually focused to a submicron to nanometer size spot. Space charge forces try to dominate when the charged particle beam is extracted from the plasma source. Computational methods are therefore required for the design and optimization of a suitable beam extraction and focusing system.

## Beam Simulation Codes

To simulate the beam extraction and focusing, AXCEL-INP [2] and SIMION [3] codes are used. The code AXCEL-INP, which uses the finite difference method (FDM), is a Vlasov solver that includes the solution of the Poisson equation and particle distribution function. From initial input conditions, a 2D potential map is created; particle trajectories are then calculated and a space charge map is generated. By iterative approach, the map is then used to recalculate the potentials and new particle trajectories until self-consistent particle and potential distributions are found. For the beam simulation, input parameters such as ion mass and charge, ion current density, electron and ion temperatures, plasma potential, and more are used [2].

SIMION uses a highly modified fourth-order Runge–Kutta technique to solve Laplace's equation for computing electrostatic or magnetostatic potential fields. Although SIMION does not support the Poisson solutions to field equations, it employs charge repulsion methods that can estimate certain types of space charge and particle repulsion effects. Ion trajectories are a result of electrostatic and space charge repulsion forces on the basis of the current position and velocity of the ions. Input parameters used for the beam simulation are number of ions, ion mass and charge, ion source position, and initial kinetic energy of the ions, etc. [3].

## Results and Discussion

### Single-EL System

Figure 22.1 shows a typical ion beam trajectory simulation and emittance plot for beam extraction and focusing of a single-EL using AXCEL-INP code.

The equipotential lines in Figure 22.1a show the formation of a double convex-type electrostatic lens with the three-electrode EL system. This is a decelerating–accelerating type of lens with the first $(E_1)$ and last $(E_3)$ electrodes of EL kept at high potential (–15 kV) and the middle electrode $(E_2)$ kept at ground potential. The plasma electrode (PE) is kept at ground potential. The beam at the PE acts as the object for the EL. The focal point of the beam is ~1 cm from the last electrode, $E_3$. The important beam simulation parameters are described in Table 22.1, including the plasma parameters, which are obtained from our experiments.

For the linear transformations of elliptical emittance diagrams in beam optics (as shown in Figure 22.1b), the ellipse is specified by the Twiss parameters $\alpha$, $\beta$, $\gamma$, and $\varepsilon$ that give the equation for the ellipse, centered on the $x$–$x'$ coordinate axis in the form, $\gamma x^2 + 2\alpha x x' + \beta x'^2 = \varepsilon$ [4]. The parameters $\beta$ and

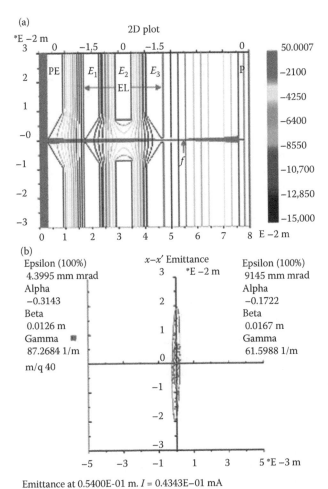

**FIGURE 22.1**
AXCEL-INP simulation of (a) ion trajectory and (b) beam emittance for ion extraction and focusing using a single-EL. PE, plasma electrode; EL, Einzel's lens; $E_1$, $E_2$, $E_3$, three electrodes of Einzel's lens; f, beam focal point; P, target plate. "alpha," "beta," and "gamma" are the Twiss parameters. (Reprinted with permission from J.V. Mathew and S. Bhattacharjee, 2011. Compact electrostatic beam optics for multi-element focused ion beams: Simulation and experiments, *Rev. Sci. Instrum.* **82**: 013501. Copyright 2011, American Institute of Physics.)

$\gamma$ determine the eccentricity of the ellipse, $\alpha$ being associated with the inclination of the axis of the ellipse and $\varepsilon$ is the product of the semi-axes of $1/\pi$ times the area of the ellipse.

The two ellipses in the x–x' emittance plot shown in Figure 23.1b correspond to rms emittance and the effective or 100% emittance. The effective emittance encompasses all the particles in the beam. The rms emittance is defined by the relation.

**TABLE 22.1**

Beam Simulation Parameters for the Single-EL System
Shown in Figure 22.1

| Parameters of a Single-EL Ion Beam Column | |
|---|---|
| Plasma electrode (PE) aperture | 1 mm |
| Thickness of electrodes | 6 mm |
| Distance between PE and $E_1$ | 8 mm |
| Aperture of $E_1$ and $E_3$ | 3 mm |
| Interelectrode distance of EL | 6 mm |
| Potential of $E_1$ and $E_3$ | −15 kV |
| Potential of $E_2$ | 0 V |
| Gas | Argon |
| Ion current density | 60 A/m² |
| Electron temperature | 10 eV |
| Ion temperature | 2 eV |

$$\varepsilon_{rms,x} = \sqrt{\langle x^2 \rangle \langle x'^2 \rangle - \langle xx' \rangle^2} \qquad (22.1)$$

From the Twiss parameters in Figure 22.1b, the beam divergence at the focal point is calculated to be ~1°, and the beam spot size is ~300 µm. The beam has a slope of ~85°, which shows a nearly focused beam, with rms beam emittance of ~0.9 mm mrad.

The single EL in Figure 22.1 serves the purpose of both beam extraction and focusing, with the first electrode of EL acting as the extractor electrode. Figure 22.2 shows the variation of ion beam current with extraction potential (potential applied on electrodes $E_1$ and $E_3$) for the single-lens system shown in Figure 22.1 using AXCEL-INP. The beam current is measured at a distance of ~1 cm from the last electrode ($E_3$). The extracted beam current starts to saturate at ~−13 kV, and hence −15 kV potential is used for the simulations in Figure 22.1. A Child–Langmuir fitting ($I \propto V^{3/2}$) is given to the simulation data. The beam current of ~7 µA obtained at −5 kV, using a single EL shown in Figure 23.6, is in agreement with the simulation results.

For Ar plasma of density, $N_+ \sim 10^{11}$ cm⁻³ [5,6] and electron temperature, $T_e \sim 10$ eV, the effective plasma (ion) current density $J_p$ is found to be ~ 50 Am⁻². The space–charge, limited-vacuum current density, $J_{cl}$ governed by Child–Langmuir's law, is limited by the effective plasma ion current density, $J_p$, which leads to saturation in extracted current density as shown in Figure 22.2.

## Double-EL System

To focus the ion beam to a finer spot, a second set of EL is employed in the beam line. Simulation results for a double-EL system with $EL_1 = -15$ kV and $EL_2 = -30$ kV separated by a gap of 18 mm show that the beam divergence

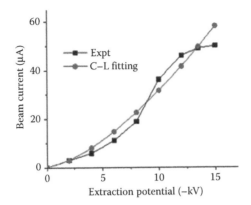

**FIGURE 22.2**
Variation of beam current with extraction potential for the single-Einzel lens system shown in Figure 22.1 using AXCEL-INP simulation. (Reprinted with permission from J.V. Mathew and S. Bhattacharjee, 2011. Compact electrostatic beam optics for multi-element focused ion beams: Simulation and experiments, *Rev. Sci. Instrum.* **82**: 013501. Copyright 2011, American Institute of Physics.)

reduces from ~0.9 mm mrad (for the single-lens system) to ~0.7 mm mrad with the introduction of the second EL set. The beam spot size is ~200 μm and the focal point is ~1 cm from $L_3$. The beam slope is ~88°, which indicates a nearly focused beam with beam current ~0.045 mA. A beam-limiting aperture, S, is normally used between the two ELs to control the final beam spot size and current. The beam at the beam-limiting aperture acts as the object for the second lens.

Figure 22.3 shows the simulation of a double-Einzel system with a beam-limiting aperture, S, of 0.5 mm diameter kept between the two ELs, $EL_1$ and $EL_2$. S is floated to –20 kV in accordance with the equipotential lines surrounding it to avoid any divergence of the beam. It is seen that the beam emittance reduces from ~0.7 to ~0.5 mm mrad by the introduction of S. The beam spot size remains almost the same, ~200 μm. The beam simulation parameters are listed in Table 22.2, including the plasma parameters that are obtained from our experiments.

An important parameter that affects the beam spot size is the ion temperature, $T_i$. Figure 22.4 shows the variation of beam spot size with ion temperature. The potential on the beam-limiting aperture is kept at –20 kV. The beam spot size is found to increase from ~100 to 300 μm as $T_i$ increases from 0.2 to 3.5 eV. The beam current is not found to change with the variation in $T_i$.

A comparison of focusing action of the double-EL system shown in Figure 22.3 for four different ion species, $H_2$, Ne, Ar, and Kr, is given in Table 22.3. The plasma density is taken to be ~$10^{11}$ cm$^{-3}$. The comparison is based on the Twiss parameters $\alpha$, $\beta$, $\gamma$, and $\varepsilon$ obtained from the emittance plots. Focal point (FP) indicates the distance along the beam axis from the last electrode ($L_3$), where the beam emittance is lowest. For the lighter gas, $H_2$, the beam

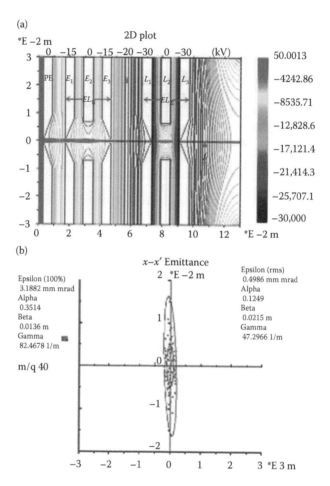

**FIGURE 22.3**
AXCEL-INP simulation of (a) ion trajectory and (b) beam emittance, for ion extraction and focusing using a double-EL system, with a beam-limiting aperture (S) introduced between the Einzel lenses, $EL_1$ and $EL_2$ ($EL_1$, $EL_2$, $EL_3$ and $L_1$, $L_2$, $L_3$), are electrodes of $EL_1$ and $EL_2$, respectively. ($f$) Denotes the beam focal point. (Reprinted with permission from J.V. Mathew and S. Bhattacharjee, 2011. Compact electrostatic beam optics for multi-element focused ion beams: Simulation and experiments, *Rev. Sci. Instrum.* **82**: 013501. Copyright 2011, American Institute of Physics.)

divergence (obtained from $\gamma$) is largest, while the beams spot size (obtained from $\beta$) is minimum. The heavier ions focus at nearly the same point along the beam axis (~1.7 cm), while the focal point of $H_2$ is closer to the last electrode.

It is evident that beam-limiting apertures with aperture sizes <100 μm will only be effective in reducing the beam spot size further, which is a mechanically challenging job. By varying the potential applied to the beam-limiting aperture, the beam spot size can be controlled. This is investigated next.

**TABLE 22.2**

Beam Simulation Parameters for the Double-EL
Shown in Figure 22.3

| Parameters of a Double-EL Ion Beam Column | |
|---|---|
| Gas | Argon |
| Ion current density | 60 A/m$^2$ |
| Plasma electrode (PE) aperture | 1 mm |
| Thickness of electrodes | 6 mm |
| Distance between PE and $E_1$ | 8 mm |
| Aperture of $E_1$ and $E_3$ | 1.5 mm |
| Interelectrode distance of $EL_1$ | 6 mm |
| Potential of $E_1$ and $E_3$ | –15 kV |
| Potential of $E_2$ | 0 V |
| Potential of $S$ | –20 kV |
| Aperture of $S$ | 0.5 mm |
| Distance between $EL_1$–$S$ and $S$–$EL_2$ | 9 mm |
| Aperture of $L_1$ and $L_3$ | 1.5 mm |
| Interelectrode distance of $EL_2$ | 6 mm |
| Potential of $L_1$ and $L_3$ | –30 kV |
| Potential of $L_2$ | 0 V |
| Electron temperature | 10 eV |
| Ion temperature | 2 eV |

Figure 22.5 shows the effect of grounding the beam-limiting aperture. Only energetic ions are able to penetrate the potential barrier created by the grounded $S$. This is reflected in one order decrease in a focused-beam current. The beam emittance decreases from 0.5 mm mrad, as in the earlier case, to 0.13 mm mrad with a beam spot size of ~10 μm. The focal point is ~1 cm

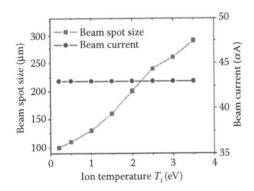

**FIGURE 22.4**
Variation of beam spot size and beam current with ion temperature, $T_i$. (Reprinted with permission from J.V. Mathew and S. Bhattacharjee, 2011. Compact electrostatic beam optics for multi-element focused ion beams: Simulation and experiments, *Rev. Sci. Instrum.* **82**: 013501. Copyright 2011, American Institute of Physics.)

**TABLE 22.3**

Comparison of Twiss Parameters for Four Different Gases Using the Double-EL System Shown in Figure 22.3

| Ion (Z) | $\varepsilon$ | $\alpha$ | $\beta$ | $\gamma$ | $J_i$(A/m²) | FP (cm) |
|---|---|---|---|---|---|---|
| H₂ (I) | 0.517 | −0.3166 | 0.0086 | 128.592 | 300 | 0.8 |
| Ne (20) | 0.5623 | −0.8467 | 0.0347 | 49.4219 | 80 | 1.7 |
| Ar (40) | 0.4986 | 0.129 | 0.0215 | 47.29 | 60 | 1.7 |
| Kr (85) | 0.5324 | −0.5965 | 0.0356 | 38.0364 | 40 | 1.6 |

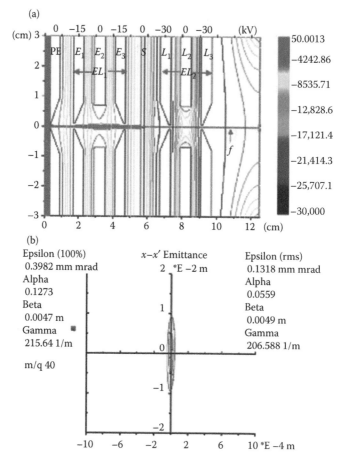

Emittance at 0.1128 m. $I = 0.5176\text{E}{-}02$ mA

**FIGURE 22.5**

AXCEL-INP simulation of (a) ion trajectory and (b) beam emittance, for ion extraction and focusing using a double-EL system, with beam-limiting aperture (S) kept at ground potential. (*f*) Denotes the beam focal point. (Reprinted with permission from J.V. Mathew and S. Bhattacharjee, 2011. Compact electrostatic beam optics for multi-element focused ion beams: Simulation and experiments, *Rev. Sci. Instrum.* **82**: 013501. Copyright 2011, American Institute of Physics.)

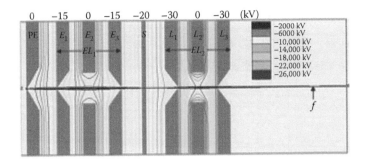

**FIGURE 22.6**
(a) SIMION simulation of beam trajectory for ion extraction and focusing using a double-Einzel lens system using a beam-limiting aperture (S) with electrode parameters exactly similar to that in Figure 22.3. (*f*) Denotes the beam focal point. (Reprinted with permission from J.V. Mathew and S. Bhattacharjee, 2011. Compact electrostatic beam optics for multi-element focused ion beams: Simulation and experiments, *Rev. Sci. Instrum.* **82**: 013501. Copyright 2011, American Institute of Physics.)

from $L_3$. With 1 mm PE aperture and 0.5 mm beam-limiting aperture, this is the smallest spot size we could obtain in the present set-up. The ion temperature used in this simulation is 2 eV.

Figure 22.6 shows the beam trajectory using SIMION simulation code for the double-lens system with electrode parameters exactly similar to that in Figure 22.3. The space charge effect is approximately taken care of in the simulation by a beam current (~1 μA), since in the presence of space charge the radial electric field acting on the beam can be written as

$$E_r = \frac{I_0}{2\pi\varepsilon_0 v_0 r} = \frac{I_0}{2\pi\varepsilon_0 r \left(2e/m_i\right)^{1/2} V^{1/2}}, \tag{22.2}$$

where $I_0$ is the total beam current, and $v_0 = (2eV/m_i)^{1/2}$ is the axial ion velocity with $V$, the extraction voltage.

The plasma source is taken to be a spherical distribution with its center coinciding with the center of the PE aperture. The beam is found to focus at a farther distance (~3.75 cm) from the last electrode, $L_3$. The beam spot size is ~200 μm. When the slit potential is set to 0 V, similar to the configuration in Figure 22.5, the spot size gets reduced to ~20 μm. This shows that both AXCEL-INP and SIMION give almost comparable results.

## Tolerance Studies

The double-EL design in Figure 22.3 with potential distribution, $EL_1 = -15$ kV, $S = -20$ kV, and $EL_2 = -30$ kV, has been optimized for the electrode geometrical

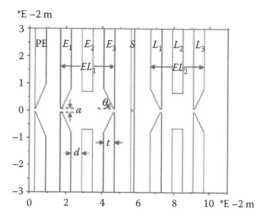

**FIGURE 22.7**

Electrode geometry parameters for the beam tolerance study. $a$, electrode aperture; $d$, inter-electrode distance; $t$, electrode thickness; ($\theta$), electrode angle. (Reprinted with permission from J.V. Mathew and S. Bhattacharjee, 2011. Compact electrostatic beam optics for multi-element focused ion beams: Simulation and experiments, *Rev. Sci. Instrum.* **82**: 013501. Copyright 2011, American Institute of Physics.)

parameters, and a detailed tolerance study has been carried out. The PE aperture and beam-limiting aperture are fixed at 1 and 0.5 mm, respectively. The parameters used for the tolerance study are shown in Figure 22.7. The change in Twiss parameters $\alpha$, $\beta$, $\gamma$, and $\varepsilon$ with the variation in electrode geometrical parameters is looked at. The effect of ±1 mm variation in electrode thickness, electrode aperture, interelectrode distance, and ±1° variation in electrode angle are studied. The beam emittance is found to be more sensitive on the precision of grounded electrode apertures of the ELs ($E_2$ and $L_2$), interelectrode distances of EL 2 ($EL_2$) electrode apertures of EL 2 ($EL_2$), and electrode angle. However, it is not much influenced by small variation in electrode thickness, electrode apertures of EL 1 ($EL_1$), and the distance between the ELs and beam-limiting aperture ($S$). The beam current is ~45 µA for all cases except for the case where $S$ is grounded, where it reduces to ~5 µA. As described earlier, the grounded beam-limiting aperture gives a minimum emittance. The results of the tolerance study show that the design of the double ELS is quite robust in terms of its performance with regard to small changes in the design parameters.

## Conclusion

The conceptual design of the MEFIB system has been discussed in this chapter. An electrostatic EL system is designed using AXCEL-INP and SIMION codes for micron size-focused beams. The parameters used in simulations such as ion current density, electron and ion temperatures, initial kinetic energy of the ions, and so on, are taken from the experimental data. The results for single-EL system, double-EL system, and the advantage of a beam-limiting aperture are discussed in detail. The dependence of beam spot size on ion temperature is demonstrated. The focusing action of the lens system for different gas species is studied. Simulation results show that a 1 mm beam at the PE can be focused to 10 μm using a double EL with a 0.5 mm grounded beam-limiting aperture. The simulations of AXCEL-INP and SIMION codes are found to give similar results. A detailed tolerance study for the double EL system with the beam-limiting aperture is presented, where the effect of ±1 mm variation in electrode thickness, electrode aperture, interelectrode distance, and ±1° variation in electrode angle are studied.

# References

1. J. Orloff (editor). 2009. *Handbook of Charged Particle Optics* (2nd edition, CRC Press, New York).
2. P. Spadtke, AXCEL-INP, *Junkerstr.* **99**: 65205 (Wiesbaden, Germany).
3. D.J. Manura and D.A. Dahl. 2008. *SIMION Version 8.0 User's Manual* (Scientific Instrument Services, Inc., New York).
4. R. Talman. 2006. *Accelerator X-Ray Sources* (Wiley-VCH Verlag GmbH Co. KGaA, Weinheim).
5. J. V. Mathew, A. Chowdhury, and S. Bhattacharjee. 2008. Subcutoff microwave driven plasma ion sources for multi elemental focused ion beam systems, *Rev. Sci. Instrum.* **79**: 63504.
6. J.V. Mathew and S. Bhattacharjee. 2009. Ion energy distribution near a plasma meniscus for multielement focused ion beams, *J. Appl. Phys.* **105**: 96101.
7. J.V. Mathew and S. Bhattacharjee. 2011. Compact electrostatic beam optics for multi-element focused ion beams: Simulation and experiments, *Rev. Scientific Instrum.* **82**: 13501.

# 23

## Focused Ion Beam Experiments

## Introduction

In this chapter, the FIB experiments, including the measurements of beam current, beam profile, and beam spot size, are presented and discussed. The focused beam current is measured using the Faraday cup (NEC-FC 50) and the variation of beam current with electrode potentials and different ion species are studied. In beam profile measurements, two different techniques are employed to intercept the beam: (i) knife edge and (ii) wire, as discussed in Chapter 22. Both the intercepted current by the knife edge and wire and the transmitted current recorded in the Faraday cup are used for the diagnostics. The beam spot size is measured by the micrography of craters formed by the FIB impinging on copper and aluminum substrates.

## Results and Discussion

### Focused Beam Current Measurements

The focused beam current measurement using the Faraday cup is described in the section "Beam Profile Measurements." The beam blanker is kept in the retracted position.

The measurements are taken in two ways: (a) potentials on the Einzel lens 1, $EL_1$, beam-limiting aperture, $S$, and Einzel lens 2, $EL_2$, are varied simultaneously in the radio 1:2:3, using a voltage divider (Figure 23.1) and a single HV power supply. (b) Potential on $S$ and $EL_2$ are kept fixed in the ratio 2:3 at –12 kV and –18 kV, respectively, while the potential on $EL_1$ is varied independently.

The voltage divider in the ratio 1:2:3 is initially designed for the proposed beam column with potentials $EL_1 = –15$ kV, $S = –20$ kV, and $EL_2 = –30$ kV. For the present experiments, the maximum operating potential is kept at –18 kV. For these potentials also, focusing action of the beam column has been demonstrated. Case (b) is studied to identify the potential of $EL_1$, for fixed

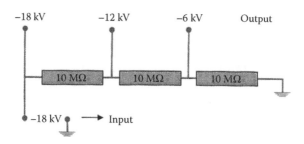

**FIGURE 23.1**
Schematic of voltage divider used in focused beam current measurements.

potentials of $S$ and $EL_2$, with which maximum current yield is obtained. These two cases of focused beam current measurements are discussed in detail in the sections below.

## Case A

Figure 23.2 shows the variation of a focused beam current with extraction potential, which corresponds to the potential on $EL_2$, while potentials on $EL_1$ and $S$ are varied simultaneously with $EL_2$ in the ration 1:2:3. In Figure 23.2a, two different gases, Ar and Kr, at 0.2 mTorr discharge pressure and 250 W power are compared. The beam current increases with extraction potential and reaches saturation at ~1.6 µA for Ar ions at ~−8 kV extractions potential and at ~0.8 µA for Kr ions at ~−6 kV extraction potential. The beam current yield of Kr is approximately half that of Ar ions, as expected. Figure 23.2b shows the variation of a focused beam current with extraction potential for Kr ions at two microwave powers, 250 and 350 W, while the pressure is maintained at 0.35 mTorr. The beam current increases with increase in microwave

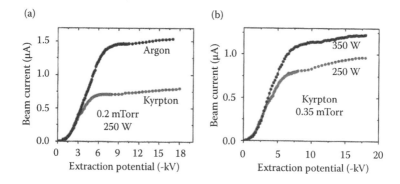

**FIGURE 23.2**
Variation of focused beam current with extraction potential for (a) argon and krypton gases and (b) two microwave powers for krypton gas. The extraction potential corresponds to the potential applied on Einzel's lens 2, $EL_2$. The potentials on $EL_1$, $S$, and $EL_2$ are varied simultaneously in the ratio 1:2:3.

power, indicating that if the power is increased further there is a possibility of obtaining higher beam current. The variation of focused beam current with extraction potential for Kr ions at two gas pressures, 0.2 and 0.35 mTorr, at 250 power can be compared from Figure 23.2a and b. The beam current remains almost constant except for a small increase in saturation current at 0.35 mTorr.

Figure 23.3 shows the variation of focused beam current with extraction potential, which corresponds to the potential on $EL_1$, while the potentials on S and $EL_2$ are kept fixed at –12 and –18 kV, respectively. A different trend in beam current profile is observed in this case. The beam current tends to increase with extraction potential, reaches a maximum, and then decreases. SIMION simulation of a beam column shown in Figure 23.4 (with the beam-limiting aperture [S] region zoomed in) explains this phenomenon. For potentials of $EL_1 > |-2.7\ kV|$, with $S = -12\ kV$ and $EL_2 = -18\ kV$, the beam-limiting aperture S starts blocking the beam (Figure 23.4b), which causes the decrease in beam current, as observed in experiments.

Hence, for fixed potentials, $S = -12\ kV$ and $EL_2 = -18\ kV$, the maximum current yield is obtained for $EL_1 \sim -2.7\ kV$. Therefore, this combination of potentials has been used for further experiments on beam profile and beam spot size measurements. Figure 23.3a shows that at 350 W power, $H_2$ plasma gives the maximum ion current of ~10.2 μA, while Ar ion current peaks at ~ 2.5 μA and Kr ion current at ~1.5 μA. The beam current of ~2.5 μA obtained at –2.7 kV extraction potential is in agreement with the I–V characteristics obtained from simulation for a single Einzel lens system, shown in Figure 23.3. The current yield can be further enhanced by increasing the microwave power as

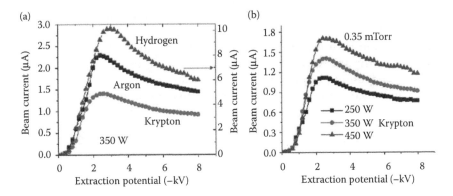

**FIGURE 23.3**
Variation of focused beam current with extraction potential for (a) argon and krypton gases at 0.25 mTorr pressure and 350 W power and hydrogen gas at 0.8 mTorr pressure and 350 W power and (b) three different microwave powers at 0.35 mTorr pressure for krypton gas. The extraction potential corresponds to the potential applied on Einzel's lens 1, $EL_1$, while S and $EL_2$ are held at –12 and –18 kV, respectively. (Reprinted with permission from J.V. Mathew and S. Bhattacharjee, 2011. Compact electrostatic beam optics for multi-element focused ion beams: Simulation and experiments, *Rev. Sci. Instrum.* **82**: 13501. Copyright 2011, American Institute of Physics.)

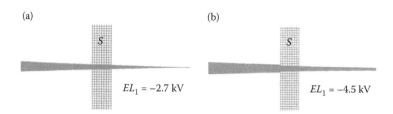

**FIGURE 23.4**
SIMION simulation of the beam column (the beam-limiting aperture, $S$ region is zoomed in) for two potentials on $EL_1$ (a) $EL_1 = -2.7$ kV and (b) $EL_1 = -4.5$ kV, with $S = -12$ kV and $EL_2 = -18$ kV. (Reprinted with permission from J.V. Mathew and S. Bhattacharjee, 2011. Compact electrostatic beam optics for multi-element focused ion beams: Simulation and experiments, *Rev. Sci. Instrum.* **82**: 013501. Copyright 2011, American Institute of Physics.)

indicated in Figure 23.3b. Since we have not employed cooling for the multicusp system, long-time operation (>30 min) at microwave power >450 W can lead to demagnetization of the multicusp magnets, and hence we restricted to ≤250 W for normal operations. A comparison of the change in focused beam current with extraction potential for Kr ions at two different gas pressures, 0.25 mTorr and 0.35 mTorr, and 350 W power from Figure 23.3a and b show an almost identical trend, similar to case (a) for 250 W power.

## Beam Profile Measurements

For the beam profile and beam spot size measurements, the potential distribution: $EL_1 = -2.7$ kV, $S = -12$ kV, and $EL_2 = -18$ kV are used, as discussed earlier. Figure 23.5 shows the SIMION beam column simulation for the above-mentioned potential configuration. Figure 23.5a shows the beam simulation and Figure 23.5b shows the emittance plot for the beam at three points along the axis, given by "a," "b," and "c" in Figure 23.5a and b. "a" represents a converging beam, "b" a focused beam, and "c" a diverging beam. Beam emittance is plotted with the beam divergence ($\tan^{-1} [v_x/v_z]$), where beam axis is ($z$) against beam ($x$) position. Ideally, at the focal point, the beam emittance should be a straight line parallel to the $y$ axis, centered at $x = 0$, but the finite spread in Figure 23.5b can be attributed to the aberrations in beam optics. Also, the emittance plot shows a beam divergence of ~1° at the focal point. Figure 23.5c shows the beam profile at the focal point "b." The focal point is obtained at ~1.8 cm from $L_3$ electrode.

The beam profile measurements are carried out using (i) knife-edge and (ii) wire methods [1,2]. A schematic of the beam profile measurement employing wire is shown in Figure 23.6a. Photographs of knife edge and wire along with the Teflon holders are shown in Figure 23.6b. The precise movement of the knife edge and wire is controlled by 1 μm precision linear feed-through. The knife edge, made of the thin SS sheet of ~0.2 mm thickness and 2 cm length, is attached to the linear feed-through and cuts the beam at different

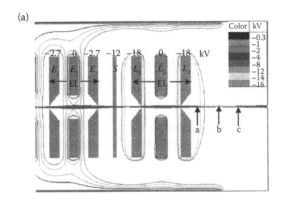

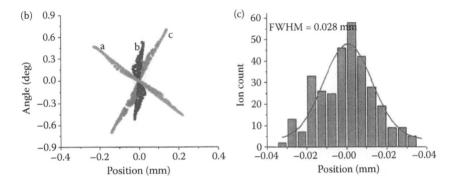

**FIGURE 23.5**
(a) SIMION simulation of focused beam, (b) beam emittance at three points along the axis of the beam "a," "b," and "c," and (c) the beam profile at the focal point. (Reprinted with permission from J.V. Mathew and S. Bhattacharjee, 2011. Compact electrostatic beam optics for multi-element focused ion beams: Simulation and experiments, *Rev. Sci. Instrum.* **82**: 013501. Copyright 2011, American Institute of Physics.)

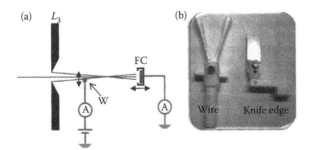

**FIGURE 23.6**
(a) Schematic of beam profile measurement using wire method. W: 20 μm wire used for scanning the beam, $L_3$: last electrode of Einzel's lens 2, $EL_2$, A: ammeter, FC: Faraday's cup. (b) Photographs of knife edge and wire along with the Teflon holders. These holders are attached to the linear feed-through.

positions along the beam axis after the last electrode ($L_3$). The intercepted beam current and transmitted beam current collected by the Faraday cup are measured using the current meters. Wire methods use a 20 μm Tungsten wire to intercept the beam with the same 1 μm precision linear motion, and the beam current through the wire is measured. The speed of motion of the motorized linear feed-through can be varied in the range of 10–100 μm/s, and the beam current from the knife edge and Faraday cup are acquired in the computer using LabVIEW with a precision of 10 data points per second.

Since the linear feed-through is constrained to move along the radial direction, "L"-shaped holders of different lengths, shown in Figure 23.6b, have been used to measure the beam profile at different points along the beam axis. So it is difficult to determine the exact focal point using this technique. Also, there is some error in positioning the holders while replacing the knife edges and wires. The emission of secondary electrons causes fluctuations in detected current during the experiments, with the kind of potentials used. The wire method seems to be more accurate, since the fluctuations due to secondary emission are at minimum due to the small dimension of the wire.

Figure 23.7 shows the beam profile obtained using a knife edge at four points along the beam axis, 0.7, 2.2, and 3 cm from the last electrode $L_3$, for argon plasma at 0.35 mTorr pressure and 250 W power. The measurements are taken in steps of ~0.7 cm from $L_3$. The smallest FWHM of 70 μm is obtained at ~1.5 cm from $L_3$.

Figure 23.8 shows the beam profile obtained using the wire method at four points along the beam axis, 0.8, 1.6, 2.2, and 3 cm from the last electrode, $L_3$, for argon plasma at 0.35 mTorr pressure and 250 W power at 1.6 cm; the

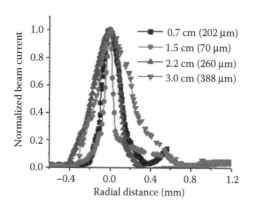

**FIGURE 23.7**
Beam profiles at four points along the beam axis using knife-edge method, for an argon beam operated at 0.35 mTorr pressure and 250 W microwave power. The FWHM gives an indication of the beam size and is given in brackets. (Reprinted with permission from J.V. Mathew and S. Bhattacharjee, 2011. Compact electrostatic beam optics for multi-element focused ion beams: Simulation and experiments, *Rev. Sci. Instrum.* **82**: 013501. Copyright 2011, American Institute of Physics.)

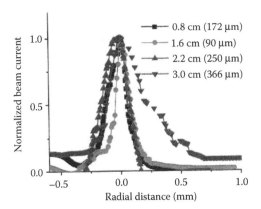

**FIGURE 23.8**
Beam profiles at four points along the beam axis using wire method, for an argon beam operated at 0.35 mTorr pressure and 250 W microwave power. The FWHM gives an indication of the beam size and is given in brackets.

FWHM of beam profile gives ~90 μm. The beam profiles obtained using both knife-edge and wire methods are almost consistent.

A comparison of the beam profiles obtained by both the methods is shown in Figure 23.9. An approximate fitting shows that the beam focal point falls close to 1.4 cm from the last electrode, $L_3$ (which closely matches with the SIMION simulations shown in Figure 23.5) with a minimum beam size of ~40 μm, as denoted by point "f" in Figure 23.9.

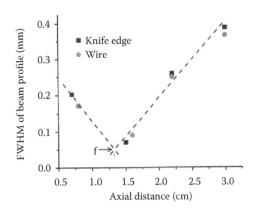

**FIGURE 23.9**
Comparison of beam profiles obtained at different points along the beam axis by knife-edge and wire methods. An approximate fitting is given by the dashed line, and "f" denotes the approximate focal point of the beam.

## Beam Spot Size Measurements

In beam spot size measurements, metallic substrates like copper and aluminum are exposed to the beam and the craters formed by beam irradiation are imaged using an optical microscope (Carl Zeiss, Axio Imager. A1m, 2 μm resolution). The substrate holder is mounted on a linear motion feed-through of 200 mm stroke length and 0.1 mm precision as described in the section "Beam Profile Measurements" (see Figure 23.8). The substrate holder can be rotated ±70° with ~5° precision. The beam spots are measured at different positions along the axis, by keeping the substrate at different angular positions. The potential distribution used for the experiments is the same as that in Figure 23.5.

Figure 23.10a and b shows, microscope images of Cu and Al substrates exposed to argon beam, at 0.35 mTorr and 250 W for ~30 s and ~60 s, respectively. The substrate is positioned at ~1.5 cm from the last electrode, $L_3$. A halo region of ~50 μm is observed in both cases. Al image shows a core region of ~25 μm diameter due to longer irradiation time and Al being a soft metal. This is in agreement with the simulation shown in Figure 23.5. The Cu substrate in Figure 23.10c has been exposed to the beam for ~2 min and the spot formed by the beam is much deeper in that case. Although the beam is spread on a larger region, the core region is ~25 μm in size. The precision of linear motion of the substrate holder and backlash error prevented the exact determination of the focal point of the FIB.

With ~25 μA Ar ions, focused to 25 μm spot, an image side ion current density of ~1 A/cm² is obtained. This can be greatly enhanced mainly in two ways: (i) Bringing down the focused beam size to submicron dimensions by (a) reducing the aperture of a plasma electrode and beam-limiting slit, (b) controlling the potential of the Einzel lens system, (c) capillary guiding of the beam through nanocapillaries [3], and (d) using magnetic beam confinement elements such as an octupole. (ii) Improving the effective plasma (ion) current density ($J_p$) by (a) operating at higher microwave powers, neutral pressures, and so on, and (b) using techniques like

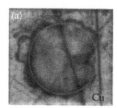 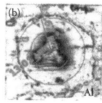 

**FIGURE 23.10**

Beam spots obtained on copper and aluminum substrates using argon beam, at 0.35 mTorr and 250 W. Microscope images of beam spots obtained on copper and aluminum substrates using argon beam, at 0.35 mTorr and 250 W where (a) on Cu exposed for ~30 s. The spot size is ~24.01 mm (b) on Al exposed for ~60 s where the inner spot size is ~13.01 mm and the outer one is ~ 21.67 mm and (c) on Cu exposed for 120 s and the spot size is ~11.95 mm.

slow-wave dielectric liners inside the multicusp to increase the wave–plasma interaction.

## Conclusion

The focused beam has been characterized using focused beam current, beam profile, and beam spot size measurements. Focused beam currents of ~10 µA have been obtained using multi-element ions. Beam profile measurements using knife-edge and wire methods are used to identify the beam diameter at the focal point of the FIB. The focal point is obtained at a distance of ~1.5 cm from the last electrode. Beam spot size measurements by the micrography of craters formed by FIB on metallic substrates show a core region diameter of ~25 µm, which is in good agreement with the simulation results. Submicron dimension beams can be obtained by reducing the aperture of the plasma electrode and beam-limiting slit, by using magnetic beam confinement elements, and by controlling the potential of the Einzel lens system.

## References

1. M. Reiser. 1994. *Theory and Design of Charged Particle Beams* (Wiley, New York).
2. Y. Ishii, A. Isoya, T. Kojima, and K. Arakawa. 2003. Estimation of keV submicron ion beam width using a knife-edge method. *Nucl. Instr. Meth. Phys. Res. B* **211**: 415.
3. N. Stolterfoht, J.H. Bremer, V. Hoffmann, R. Hellhammer, D. Fink, A. Petrov, and B. Sulik. 2002. Transmission of 3 keV $N_e^{7+}$ ions through nanocapillaries etched in polymer foils: Evidence for capillary guiding. *Phys. Rev. Lett.* **88**: 133201.
4. J.V. Mathew and S. Bhattacharjee. 2011. Compact electrostatic beam optics for multi-element focused ion beams: Simulation and experiments, *Rev. Sci. Instrum.* **82**: 13501.

# 24

## Summary and Future Prospects

A summary of the important results of MEFIB (multi-element-focused ion beam) is presented in this chapter. An MEFIB system using a microwave multicusp plasma source with a dimension at or below cutoff has been developed. The development is an outcome of basic research carried out in the generation of compact microwave plasmas. Further investigations have been carried out on the utilization of the compact plasmas as an ion source, beam extraction measurements, design and optimization of beam optics using computational tools, fabrication and installation of the complete MEFIB system, and finally the focused beam diagnostics.

The mechanism of generation and sustenance of plasmas in sub-cutoff dimensional multicusp waveguides has been studied. The propagation of waves in plasma waveguides is constrained by the plasma density and waveguide geometry cutoffs. In sub-cutoff dimensional waveguides, where the waves are evanescent inside the multicusp, plasma can be initiated if the threshold electric field for microwave breakdown is satisfied up to a certain distance inside the waveguide. High-power pulsed microwaves can be used in high-pressure regimes for plasma initiation. In low-pressure regimes, by using a guiding cylinder in front of the multicusp, an electron–cyclotron resonance (ECR) mechanism helps in plasma initiation in front of the multicusp. Once the plasma is generated at the multicusp entrance, due to the presence of the radially varying magnetic field, the refractive index of the medium shows a drastic variation near the periphery of the multicusp, where the upper hybrid resonance (UHR) zone occurs. Microwaves can then propagate through this annular window with a reduced wavelength, and thereby uniform plasma is sustained. This phenomenon has been experimentally verified by the electric field measurements inside the multicusp, which shows a peak in the UHR zone.

The Langmuir probe measurements provide a good picture of the plasma profile inside the waveguide. The radial plasma density distribution shows a high-density, uniform central plasma of radius ~2 cm with $N_+ \geq 10^{11}$ cm$^{-3}$. At the center of the multicusp, electron temperature $T_e$ is ~6–7 eV and its peaks ($T_e \sim 15$ eV) near the periphery of the multicusp, where electron heating due to ECR and UHR happens. The plasma remains axially uniform over a major portion of the multicusp, except at the edges (~3 cm). The end-plugging mechanism of multicusp is effective in axial plasma confinement and maintains a good plasma density in the beam extraction region. A comparison of multicusp geometries (octupole and hexapole) based on plasma parameters shows that octupole multicusp configuration provides a better plasma confinement and higher plasma density than hexapole geometry due to the increase in the number of poles, and thereby more number of resonance zones for plasma heating.

Another important plasma parameter that decides the performance of the extracted beam is ion energy distribution. The ion energy distribution at the plasma meniscus is found to be nearly monoenergetic with a spread of ~4 eV for Kr ions and ~5 eV for Ar ions, which is comparable to that of Ga-based LMIS. The axial ion energy spread measurements show that end-plugging is effective in reducing the ion energy spread in the plasma meniscus, and the introduction of a quadrupole magnetic filter in the beam extraction region is found to reduce the ion energy spread by 58% and the mean energy by 95%. It is found that by reducing the geometrical acceptance angle (from 45° to 10°) of the ion energy analyzer (IEA) probe by introducing a capillary tube at the probe entrance, close to unidirectional distribution can be obtained with a spread that is smaller by at least 1 eV. The minimum-$B$ field at the center of the multicusp and the least disturbance to the ions in the plasma by the incoming microwaves ($f_0 \gg f_{pi}$) help in obtaining monoenergetic and low-energy-spread ions at the meniscus. The application of microwaves instead of radio waves is found to be advantageous in this respect. The comparable or even less energy spread with respect to LMIS is promising for reduced chromatic aberration in a multi-element-focused ion beam with adequate current density and uniform spot size.

Once the plasma source is optimized based on plasma parameter studies, the total beam current measurements are carried out. The extracted ion current density has been measured for three different gas species, Ar, Kr, and $H_2$, and different plasma electrode (PE) apertures with extraction potentials up to –5 kV. The variations in ion current density with gas pressure and microwave power have been looked at, and the obtained ion current density is found to be far from saturation, at –5 kV. Excluding the secondary electron contribution for 1 mm PE aperture, a source side current density of ~2 A/cm² has been obtained, from which a reduced source side brightness, $B_{rs}$ of ~$10^5$ A m⁻² sr⁻¹ V⁻¹, can be predicted for the plasma source.

The radial ion energy distribution has been measured in the extraction region (plasma meniscus) of the plasma (ion) source to study the effect of beam extraction on ion energy spread at the plasma meniscus. With the extraction system in place, the radial spread is found to increase, possibly due to the formation of standing waves in the multicusp. This can be overcome by operating at higher neutral pressures. It is found that ion energy distribution has a weak dependence on the extraction potential, which is favorable for obtaining a stable focused beam current.

The next step in the research was the design of ion beam optics for focusing the ion beam using electrostatic Einzel lens systems. Widely known and commercially available simulation codes, namely, AXCEL-INP and SIMION, are used for the design. The parameters used in simulations such as plasma source current density, electron and ion temperatures, and initial kinetic energy of the ions, and so on, are taken from the results of the basic plasma research. Different combinations of electrostatic lenses have been studied, including single- and double-Einzel lens systems with and without beam-limiting aperture, to focus the ion beam. For a 1 mm beam at the plasma electrode aperture, the rms beam emittance of the focused beam is found to reduce from ~0.9 mm mrad for a single-Einzel lens system $(EL_1)$ to ~0.5 mm mrad for double-Einzel lens assembly $(EL_1$ and $EL_2)$ with a beam-limiting aperture (S), using a potential distribution of $EL_1 = -15$ kV, S $= -20$ kV, and $EL_2 = -30$ kV. It is found that the beam spot size can be further reduced by controlling the potential applied on S. Thus, the beam emittance has been brought down to ~0.1 mm mrad with a corresponding beam spot size of ~10 µm at the focal spot, with ~5 µA beam current and 30 keV energy by keeping S at ground potential, by restricting the passage of ions based on their energy. Dependence of beam spot size on ion temperature and the focusing action of the lens system for different gas species is looked at. The double-Einzel lens design has been optimized for different electrode geometrical parameters and a detailed tolerance study has been conducted. The effect of ±1 mm variation in electrode thickness, electrode aperture, and interelectrode distance and ±1° variation in electrode angle are studied. The beam emittance is found to be more sensitive on the precision of grounded electrode apertures of the Einzel lenses ($E_2$ and $L_2$), interelectrode distances of $EL_2$, electrode apertures of $EL_2$, and electrode angles. However, it is not much influenced by small variations in electrode thickness, electrode apertures of $EL_1$, and the distance between the Einzel lenses and beam-limiting aperture (S).

The complete multi-element FIB system was fabricated and installed based on the output from the basic research on plasma source and beam optics designs. Different tests have been carried out to study the performance of the integrated system, including testing the vacuum integrity, microwave coupling tests, plasma generation, optical alignment of the whole system, high-voltage tests, and beam extraction and focusing onto substrates.

The focused ion beam has been characterized using focused beam current, beam profile, and beam spot-size measurements. The focused beam current is measured using the Faraday cup and the variation of beam current with extraction potentials and different ion species in studies. Focused beam currents of ~10 µA have been obtained for multi-element ions with a beam energy of 18 keV. Knife edge and wire methods using a 1 µm precision linear feed-through are used to obtain the beam profile at different points along the beam axis. The focal point is ~1.5 cm from the last electrode and the focused beam spot size is ~40 µm with ~2.5 µA of $Ar^+$ ion beam current, as determined from the beam profile data. Beam spot size measurements are carried out by impinging focused ion beams on metallic substrates like Cu and Al for ~60 s and by the micrography of craters formed on the substrates. The spots show a core region of diameter, ~25 µm on the substrates, which is in agreement with the simulation results.

There is much scope of further research in optimizing the ME FIB system for experiments. The employment of a 5-axis stage is required for the precise measurement of a beam focal point and positioning the substrates. Also, beam deflectors will be required for scanning the beam over the substrate. The beam column is not yet fully operational with higher Einzel lens potentials as per the proposed simulation designs. Once it can be operated at the full capability, large beam currents of >50 µA are expected for heavier ions like Ar. Even in the present state with ~2.5 µA for Ar ions, focused to ~25 µm spot, beam current density of ~1 A/cm² is obtained, which can be further enhanced mainly by improving effective plasma ion current density ($J_p$) and submicron focusing of the beam. $J_p$ can be enhanced by (a) operating at higher microwave powers, neutral pressures, etc., and (b) using techniques like slow-wave dielectric liners inside the multicusp to increase the wave plasma interaction. Submicron beams can be obtained by (a) reducing the aperture of plasma electrode and beam-limiting slit, (b) controlling the potential of the Einzel lens system, (c) capillary guiding of the beam through nanocapillaries, and (d) using magnetic beam confinement elements such as an octupole. Other focusing techniques such as guiding the beam through low-density plasma can also be looked at. Only gaseous ions have been tested in this research, which can be extended to solid and volatile ions.

# Index

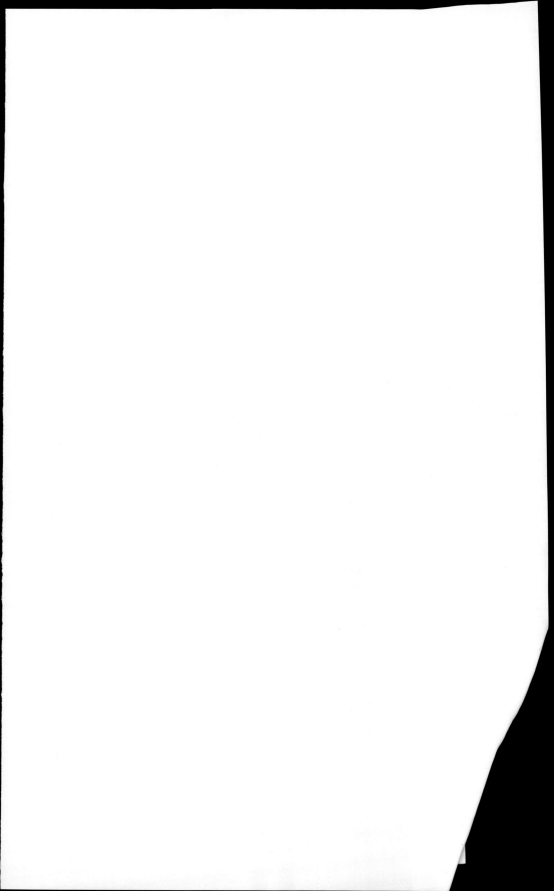